Introduction to Modern Portfolio Optimization With NUOPT and S-PLUS

Bernd Scherer   R. Douglas Martin

Introduction to Modern Portfolio Optimization With NUOPT and S-PLUS

Springer

Bernd Scherer
Deutsche Asset Management
Frankfurt 60325
Germany

R. Douglas Martin
Department of Statistics
University of Washington
Seattle, WA 98195-4322
USA

Library of Congress Cataloging-in-Publication Data
Scherer, Bernd Michael.
Introduction to modern portfolio optimization with NUOPT and S-PLUS / Bernd Scherer, R. Douglas Martin.
p. cm.
Includes bibliographical references and index.

1. Portfolio management—Data processing. I. Martin, Douglas R. II. Title.
HG4529.5.S325 2005
332.6′0285′53—dc22 2004058911

ISBN-13: 978-1-4419-1934-2 Printed on acid-free paper.
e-ISBN-13: 978-0-387-27586-4

Printed in the United States of America.

9 8 7 6 5 4 3 2

springer.com

To Katja, Jean, and Julia

and

In deep appreciation and fond memory of John W. Tukey

Preface

Purpose of Book

This book was written to expose its readers to a broad range of modern portfolio construction methods. It provides not only mathematical expositions of these methods, but also supporting software that gives its readers valuable hands-on experience with them. It is our intention that readers of the book will be able to readily make use of the methods in academic instruction and research, and to quickly build useful portfolio solutions for finance industry applications. The book is "modern" in that it goes well beyond the classical constrained mean-variance (Markowitz) portfolio optimization and benchmark tracking methods, and treats such topics as general utility function optimization, conditional-value-at-risk (CVaR) optimization, multiple benchmark tracking, mixed-integer programming for portfolio optimization, transaction costs, resampling methods, scenario-based optimization, robust statistical methods (such as robust betas and robust correlations), and Bayesian methods (including Bayes-Stein estimates, Black-Litterman, and Bayes factor models via Markov Chain Monte Carlo (MCMC)).

The computing environment used throughout the book consists of special limited-use S-PLUS® software that is downloadable from Insightful Corporation as described later in this Preface, specifically: S-PLUS, the S-PLUS Robust Library, the S+NUOPT™ optimization module, and the S+Bayes™ Library. In addition, we have provided approximately 100 S-PLUS scripts, as well as relevant CRSP sample data sets of stock returns, with which the user can recreate many of the examples in the book. The scripts represent, in effect, a large set of recipes for carrying out basic and advanced portfolio construction methods. The authors believe these recipes, along with real as well as artificial data sets, will greatly enhance the learning experience for readers, particularly those who are encountering the portfolio construction methods in the book for the first time. At the same time, the script examples can provide a useful springboard for individuals in the finance industry who wish to implement advanced portfolio solutions.

Stimulation for writing the present book was provided by Scherer's *Portfolio Construction and Risk Budgeting* (2000), which discusses many of the advanced portfolio optimization methods treated here. One of us (Martin) had given a number of talks and seminars to quant groups on the use robust statistical methods in finance, and based on the enthusiastic response, we felt the time was ripe for inclusion of robust methods in a book on portfolio construction. It also seemed apparent, based on the recent increase in academic research and publications on Bayes methods in finance, the intuitive appeal of Bayes methods in finance, and the hint of a groundswell of interest among practitioners, that the time was ripe to include a thorough introduction to modern Bayes methods in a book on portfolio construction. Finally, we wanted to augment the current user documentation for S+NuOPT to demonstrate the many ways S+NuOPT can be effectively used in the portfolio game.

Intended Audience

This book is intended for practicing quantitative finance professionals and professors and students who work in quantitative areas of finance. In particular, the book is intended for quantitative finance professionals who want to go beyond vanilla portfolio mean-variance portfolio construction, professionals who want to build portfolios that yield better performance by taking advantage of powerful optimization methods such as those embodied in S+NuOPT and powerful modern statistical methods such as those provided by the S-Plus Robust Library and S+Bayes Library. The book is also intended for any graduate level course that deals with portfolio optimization and risk management. As such, the academic audience for the book will be professors and students in traditional Finance and Economics departments, and in any of the many new Masters Degree programs in Financial Engineering and Computational Finance.

Organization of the Book

Chapter 1. This introductory chapter makes use of the special NuOPT functions `solveQP` and `portfolioFrontier` for basic Markowitz portfolio optimization. It also shows how to compute Markowitz mean-variance optimal portfolios with linear equality and inequality constraints (e.g., fully-invested long-only portfolios and sector constraints) using `solveQP`. The function `portfolioFrontier` is used to compute efficient frontiers with constraints. A number of variations (such as quadratic utility optimization, benchmark-relative optimization, and liability relative optimization) are briefly described. It is shown how to calculate implied returns and optimally combine forecasts with implied returns to obtain an estimate of mean returns. The chapter also discusses

Karush-Kuhn-Tucker conditions and the impact of constraints, and shows how to use the linear programming special case of the function `solveQP` to check for arbitrage opportunities.

Chapter 2. Chapter 2 introduces the SIMPLE modeling language component of NUOPT and shows how it may be used to solve general portfolio optimization problems that can not be handled by the special purpose functions `solveQP` and `portfolioFrontier` used in Chapter 1. The first part of the chapter provides the basics on how to use SIMPLE and how to solve some general function optimization problems, including a maximum likelihood estimate of a normal mixture model. Then its application to two non-quadratic utility functions is illustrated, as well as its application to multi-stage stochastic optimization. Finally, the use of some built-in S-PLUS optimization functions is illustrated on several simple finance problems (such as calculation of implied volatilities, fitting a credit loss distribution, and fitting a term structure model).

Chapter 3. This chapter on advanced issues in mean-variance optimization begins by treating the following non-standard problems: risk-budgeting constraints, min-max optimization with multiple benchmarks and risk regimes, and Pareto optimality for multiple benchmarks. Then several important portfolio optimization problems that require mixed integer programming (MIP) are presented, namely buy-in thresholds and cardinality constraints (e.g., finding optimal portfolios with the best k-out-of-n assets, round lot constraints, and tracking indices with a small number of stocks). Finally the chapter shows how to handle transaction cost constraints (such as turnover constraints, proportional costs, and fixed costs).

Chapter 4. This chapter introduces parametric and nonparametric bootstrap sampling in portfolio choice, with emphasis on the parametric approach assuming multivariate normality. It is shown that resampling when arbitrary short-selling is allowed recovers the Markowitz weights plus random noise that goes to zero as the resample size increases, whereas persistent bias is introduced in the case of long-only portfolios. Further exploration of the long-only case with a zero mean-return "lottery ticket" shows how volatility can induce bias in long-only portfolios, but with a trade-off due to increased risk associated with increased volatility. Here we discuss the deficiencies of portfolio construction via resampling and suggest that readers be wary of some advantages claimed for the approach. The chapter closes with a discussion of the use of a basic nonparametric bootstrap, as well as an increased precision double bootstrap, for assessing the uncertainty in Sharpe ratios and Sortino ratios. These are just two of many possible applications of the standard and double bootstrap in finance.

Chapter 5. This chapter discusses the use of scenario-based optimization of portfolios, with a view toward modeling non-normality of returns and enabling the use of utility functions and risk measures that are more suitable for the non-normal returns consistently encountered in asset returns. The chapter begins by showing how implied returns can be extracted when using a general utility function other than quadratic utility. Then we show a simple means of generating copulas and normal-mixture marginal distributions using S-PLUS. Subsequent sections show how to optimize portfolios with the following alternative risk measures, among others: mean absolute deviation, semi-variance, and shortfall probability. A particularly important section in this chapter discusses a desirable set of "coherence" properties of a risk measure, shows that conditional value-at-risk (CVaR) possesses these properties while standard deviation and value-at-risk (VaR) do not, and shows how to optimize portfolios with CVaR as a risk measure. The chapter concludes by showing how to value CDOs using scenario optimization.

Chapter 6. Here we introduce the basic ideas behind robust estimation, motivated by the fact that asset returns often contain outliers and use the S-PLUS Robust Library for our computations. Throughout we emphasize the use of robust methods in portfolio construction and choice as a diagnostic for revealing what outliers, if any, may be adversely influencing a classical mean-variance optimal portfolio. Upon being alerted to such outliers and carefully inspecting the data, the portfolio manager may often prefer the robust solution. We show how to compute robust estimates of mean returns, robust exponentially weighted moving average (EWMA) volatility estimates, robust betas and robust covariance matrix estimates, and illustrate their application to stock returns and hedge fund returns. Robust covariance matrix estimates are used to compute robust distances for automatic detection of multidimensional outliers in asset returns. For the case of portfolios whose asset returns have unequal histories, we show how to modify the classical normal distribution maximum-likelihood estimate to obtain robust estimates of the mean returns vector and covariance matrix. Robust efficient frontiers and Sharpe ratios are obtained by replacing the usual sample mean and covariance matrix with robust versions. The chapter briefly explores the use of one-dimensional outlier trimming in the context of CVaR portfolio optimization and concludes with a discussion of influence functions for portfolios.

Chapter 7. This chapter discusses modern Bayes modeling via the Gibbs sampler form of Markov Chain Monte Carlo (MCMC) for semi-conjugate normal distribution models as well as non-normal priors and likelihood models, as implemented in the S+Bayes Library. Empirical motivation is provided for the use of non-normal priors and likelihoods. The use of S+Bayes is first demonstrated with a simple mean-variance model for a single stock. We then use it to obtain Bayes estimates of alpha and beta in the single factor model and to illustrate Bayes estimation for the general linear model in a cross-sectional regression model. We show how to use the Gibbs sampler output to produce tailored posterior distributions of quantities of interest (such as mean returns, volatilities, and Sharpe ratios). The chapter shows how to compute Black-Litterman models with the usual conjugate normal model (for which a formula exists for the posterior mean and variance), with a semi-conjugate normal model via MCMC, and with t distribution priors and likelihood via MCMC. The chapter concludes by outlining one derivation of a Bayes-Stein estimator of the mean returns vector and shows how to compute it in S-PLUS.

Downloading the Software and Data

The software and data for this book may be downloaded from the Insightful Corporation web site using a web registration key as described below.

The S-PLUS Software Download

The S-PLUS for Windows and S+NUOPT software being provided by Insightful for this book expires 150 days after install. As of the publication of this book, the S+Bayes software is an unsupported library available free of charge from Insightful. To download and install the S-PLUS software, follow the instructions at http://www.insightful.com/support/splusbooks/martin05.asp. To access the web page, the reader must provide a password. Please contact Insightful Technical Support at `keys@insightful.com` to obtain the password and a trial license web registration key. **In order to activate S-PLUS for Windows and S+NUOPT, the reader must use the web registration key.**

S-PLUS Scripts and CRSP Data Download

To download the authors' S-PLUS scripts and the CRSP data sets in the files *scherer.martin.scripts.v1.zip* and *scherer.martin.crspdata.zip*, follow the instructions at http://www.insightful.com/support/splusbooks/martin05.asp. The first file contains approximately 100 S-PLUS scripts, and the second file contains the CRSP data. **The reader must use the web registration key obtained from Insightful Technical Support to download these files.**

The S-PLUS Scripts

As a caveat, we make no claims that the scripts provided with this book are of polished, professional code level. Readers should feel free to improve upon the scripts for their own use.

With the exception stated in the next paragraph, the scripts provided with this book are copyright © 2005 by Bernd Scherer and Douglas Martin. None of these scripts (in whole or part) may be redistributed in any form without the written permission of Scherer and Martin. Furthermore the scripts may not be translated or compiled into any other programming language, including, but not limited to, R, MATLAB, C, C++, and Java.

The script `multi.start.function.ssc`, which is not listed in the book but is included in the file *scherer.martin.scripts.v1.zip*, was written by Heiko Bailer and is in the public domain.

The CRSP Data

The CRSP data are provided with permission of the **Center for Research in Security Prices** (Graduate School of Business, The University of Chicago). The data were provided for educational use and only for the course program(s) for which this book is intended and used. The data may not be sold, transmitted to other institutions, or used for purposes outside the scope of this book. CRSP® data element names are trademarked, and the development of any product or service link to CRSP® data will require the permission of CRSP® (www.crsp.uchicago.edu).

The CRSP data zip file *scherer.martin.crspdata.zip* contains a number of CRSP data sets in S-PLUS `data.dump` format files. Relative price change returns for twenty stocks are contained in each of the following files:

`microcap.ts.sdd`	(Monthly returns, 1997–2001)
`smallcap.ts.sdd`	(Monthly returns, 1997–2001)
`midcap.ts.sdd`	(Monthly returns, 1997–2001)
`largecap.ts.sdd`	(Monthly returns, 1997–2001)

Each of the above files contains market returns (defined as the portfolio of market-cap-weighted AMEX, NYEX, and Nasdaq returns), and returns on the 90-day T-bill. In addition, the mid-cap returns file

`midcapD.ts.sdd` (Daily returns, 2000–2001)

contains the daily stock returns and market returns. We also include the following file containing monthly returns for three stocks from CRSP:

`returns.three.ts.sdd` (Monthly returns, 02/28/91–12/29/95)

We note that there are a few data sets appearing in examples in the book that are not distributed with the book. Readers are encouraged to substitute a CRSP data set or other data set of their choice in such cases.

Using the Scripts and Data

Under Microsoft Windows, we recommend using the scripts and data as follows. First, create an empty project folder for the scripts with a name of your choice, (e.g., *PortOpt*), and unzip the file *scherer.martin.scripts.v1.zip* in that folder. Next, create a project folder for the data sets (e.g., named *DataForPortOpt*), and attach it below the project folder for the scripts. Unzip the file *scherer.martin.crspdata.zip* in that folder. You should then run the script `load.returns.ssc` by opening it in S-PLUS and clicking on the **Run** button. This will load all the above data sets for the book, as well as the functions `panel.superpose.ts.` and `seriesPlot`, which are extended versions of similar functions in the S+FinMetrics package. Now you can run scripts in your project folder by clicking on a script to open it and clicking the **Run** button.

Acknowledgments

Special thanks go to Chris Green for his painstaking work in producing this book in both early stages and in the final stage of clean-up for delivery to Springer. Chris did so many things with great care, including code checks and improvements, careful editing, indexing, and final formatting in MS Word. We could not have delivered the book without his extensive help. Leonard Kannepell provided a tremendous amount of indispensable help in the production of book as delivered to Springer for copy editing, enduring many painful crashes of Microsoft Word in the process, and his help is deeply appreciated. Eric Aldrich provided appreciated editing on a portion of the book. Any errors not caught by these individuals are as always the sole responsibility of the authors.

We thank Heiko Bailer for the script `multi.start.function.ssc` in which he implemented the classical Stambaugh method for estimating the mean and covariance for unequal histories of returns, as well as a robust version of the Stambaugh method for dealing with outliers. Heiko produced the unequal histories examples in Sections 6.8 and 6.9. He also provided an improved version of the time series plotting function `seriesPlot`, and added the nice horizontal axes values for risk and return in Figure 6.48 that we wish we had used in other such plots in the book.

Alan Myers at the Chicago Center for Research in Security Prices kindly made the arrangement that allowed us to provide a sample of CRSP data for purchasers of the book, an aspect that we feel is very beneficial to the learning process. We are grateful to Insightful Corporation for its cooperation in making it possible to deliver the book with S-PLUS software to enrich the value of the book, and for hosting the authors' scripts and the CRSP data sets on their web site. John Kimmel at Springer was a constant source of help and support throughout the publication process, for which we are warmly appreciative.

Chapter 6 could not have been written without the research contributions to the S-PLUS Robust Library by Alfio Marazzi and Victor Yohai as primary consultants, and by Ricardo Maronna, David Rocke, Peter Rousseeuw, and Ruben Zamar, as well as the software development efforts of Kjell Konis, Matias Salibian-Barrera, and Jeff Wang. Chapter 7 could not have been written without the initial leadership of Yihui Zhan and the extensive research and development efforts of Alejandro Murua in producing the S+Bayes Library.

John Tukey encouraged Doug Martin to enter the field of statistics and facilitated this process by arranging Doug's initial consulting contract with the Bell Laboratories Mathematics and Statistics Research Center, an engagement that spanned ten years. Without these events S-PLUS would not exist today.

Bernd Scherer thanks his family for always encouraging his academic interests and in particular Katja Goebel for her patient understanding and unconditional support.

Finally, untold thanks to our families and friends for putting up with our time away from them during the many hours we devoted to working on this book.

Contents

List of Code Examples

1 Linear and Quadratic Programming

1.1 Linear Programming: Testing for Arbitrage

1.1.1 Arbitrage

In order to familiarize the reader with NUOPT for S-PLUS, we will start with the most prominent subjects in both finance and operations research and show how we can check for arbitrage in security returns using linear programming techniques.[1] Suppose all securities available to investors cost one monetary unit, but the returns (**R**) they offer to investors in different states of the world differ. Our model consists of n assets and m states of the world. Security returns can hence be summarized in an $m \times n$ matrix $\mathbf{S}$ of gross returns known to all and identical for all investors (i.e., the same information and no differential taxation),

$$\mathbf{S} = \begin{pmatrix} 1+R_{11} & \dots & 1+R_{1n} \\ \vdots & \ddots & \vdots \\ 1+R_{m1} & \cdots & 1+R_{mn} \end{pmatrix}. \tag{1.1}$$

Each row represents a different state of the world. Each column stands for a different asset. If the number of assets equals the number of states of the world, a market is called **complete**. We will later come back to this definition.

Suppose further that investors want to maximize their end-of-period wealth and always prefer more to less. Investors are assumed to be unrestricted (holdings do not have to sum to one and can be long and short) in purchasing a portfolio of securities, described by an $n \times 1$ vector $\mathbf{w}$, where element w_i denotes the percentage of holdings in security i. Arbitrage exists if investors can either

- extract money by setting up a portfolio that has no further obligation (i.e., zero cash flows in all states of the world, or **first-order arbitrage**), or
- purchase a portfolio at zero cost that will pay off a positive amount in at least one state of the world while paying out nothing in all other states (**second-order arbitrage**).

Suppose an investor searches for arbitrage. Formally, we can describe the problem that has to be solved as a simple linear program,

$$\min_{\mathbf{w}} c = \mathbf{w}^T \mathbf{I}, \tag{1.2}$$

subject to

$$\mathbf{Sw} \geq \mathbf{0}, \tag{1.3}$$

where the costs, c, of setting up an arbitrage portfolio (1.2) are minimized subject to the payoff constraint (1.3). In our notation, $\mathbf{0}$ denotes an $m \times 1$ vector of zeros, while $\mathbf{I}$ denotes an $n \times 1$ vector of ones. It is clear that a 100% cash portfolio would always satisfy (1.3), but it could not be purchased at zero cost. If (1.2) becomes negative we have been able to generate cash. Alternatively, if it becomes zero, trades have either been costless or did not take place at all ($\mathbf{w} = \mathbf{0}$). The "no trade solution" places an upper value of zero on our objective function. According to what we described above, we are now able to distinguish the three cases summarized in
Table 1.1.

Table 1.1 Arbitrage conditions

Case	Objective	State constraint
No arbitrage	$c = 0$	$\mathbf{Sw} = \mathbf{0}$
First-order arbitrage	$c = -\infty$	Feasible
Second-order arbitrage	$c = 0$	Feasible, not all constraints binding

We will now look at duality theory to extract the price of primitive securities that pay off one monetary unit in one state of the world and nothing in all other states. It is well-known from the theory of linear programming that the dual to (1.2)–(1.3) can be expressed using (1.4)–(1.5):

$$\max_{\mathbf{d}} \mathbf{d}^T \mathbf{0}, \tag{1.4}$$

$$\text{subject to } \mathbf{S}^T \mathbf{d} \leq \mathbf{I}. \tag{1.5}$$

We also know from the strong duality property that in the case of no arbitrage (where the primal problem is bounded and all state constraints are binding), both solutions will coincide (i.e., $\mathbf{w}^T \mathbf{I} = \mathbf{d}^T \mathbf{0} = 0$ and equality constraints will hold):

$$\mathbf{S}^T \mathbf{d} = \mathbf{I}. \tag{1.6}$$

This condition is well-known to finance students. It simply says that we can recover security prices (standardized to one in the current setting) by multiplying cash flows in each state of the world by their respective state prices.

1.1.2 First Steps with `solveQP()`

We are now equipped to solve our first little problem in NuOPT. First we need to generate asset returns for three states of the world. For simplicity, all returns are assumed to be drawn independently (with 8% expected return and 20% volatility)[2].

```
set.seed(10)      # Use to replicate random numbers
n <- 3
m <- 3
S <- 1+matrix((rmvnorm(m, mean=rep(0.08, n),
  cov=diag(rep((.2)^2,n)))), ncol=n)
S

         [,1]     [,2]      [,3]
[1,] 1.120444 1.138809 1.1339732
[2,] 1.193477 0.803357 0.9066128
[3,] 1.036637 1.217810 0.6883356
```

In NuOPT for S-Plus, the command

```
solveQP(objQ,objL,A,cLO,cUP,bLO,bUP,x0,isint,
  type = minimize,trace=T)
```

will generally solve linear and quadratic problems of the form $\frac{1}{2}\mathbf{x}^T \mathbf{Q}_{obj} \mathbf{x} + \mathbf{L}_{obj}^T \mathbf{x}$, subject to linear constraints $\mathbf{c}_{Lo} \leq \mathbf{A}\mathbf{x} \leq \mathbf{c}_{Up}$ and $\mathbf{b}_{Lo} \leq \mathbf{x} \leq \mathbf{b}_{Up}$, where inputs are defined as in Table 1.2. In order to solve our arbitrage model, we define the linear objective function with `objL=rep(1, ncol(S))`, set the lower state constraints to zero with `cLO=rep(0, nrow(S))`, and allow an unbounded upside with `cUP=rep(Inf, nrow(S))`.

```
module(nuopt) # load Nuopt module
solution <- solveQP(, objL=rep(1, ncol(S)), S,
  cLO=rep(0, nrow(S)), cUP=rep(Inf, nrow(S)),
  type=minimize)
```

The object `solution` contains all the output from NUOPT[3]. In order to see what output is available, type "`solution`" at the command prompt. This will list all the output contained in the variable `solution`. Individual outputs can be extracted from `solution` using the "`$`" operator; for example, `solution$variables$x$current`, `solution$objective`, etc. Next we want to test for first-order (`test.1`) and second-order arbitrages (`test.2`) using the function in Code 1.1. It returns 1 if an arbitrage opportunity exists and 0 otherwise.

Table 1.2 Inputs for `solveQP`

Input	Description	NUOPT notation
$\mathbf{Q}_{obj}$	Quadratic term in objective function. Matrix of dimension $n \times n$.	`objQ`
$\mathbf{L}_{obj}$	Linear part of the objective function. Vector of dimension $n \times 1$.	`objL`
$\mathbf{x}$	Decision variables. Vector of dimension $n \times 1$.	`x0`
$\mathbf{A}$	Coefficient matrix of linear constraints. Matrix of dimension $m \times n$.	`A`
$\mathbf{c}_{Lo}, \mathbf{c}_{Up}$	Lower, upper bounds for the linear constraints. Vectors of dimension $m \times 1$.	`cLO, cUP`
$\mathbf{b}_{Lo}, \mathbf{b}_{Up}$	Lower, upper bounds for the variables. Vectors of dimension $n \times 1$.	`bLO, bUP`
—	Logical $n \times 1$ vector (`T` if integer variable)	`isint`
—	Minimization or maximization of objective	`type=minimize` `type=maximize`
—	Logical value to indicate if execution trace is shown	`trace=T`

```
arbitrage.check <- function(solution)
{
  test.1 <- if(solution$objective < 0) 1 else 0
  payoffs <- S%*%matrix(
     round(solution$variables$x$current, digit=6),
     ncol=1)
  test.2 <- if (min(payoffs)>=0 &
     max(payoffs)>0)) 1 else 0
  if (max(test.1, test.2)==1) 1 else 0
}
```

Code 1.1 Function to Check for Arbitrage

As there has been no arbitrage for our random draw, we can employ the dual problem and use (1.6) to calculate state prices for our example above.

```
d <- solve(t(S), matrix(rep(1, nrow(S)), ncol=1))
d

                [,1]
[1,] 0.84982823642
[2,] 0.04002501517
[3,] 0.00004414494
```

1.1.3 Pitfalls in Scenario Generation

Our discussion so far has been very theoretical. How can we apply in practice what we just learned? Suppose we have 200 observations (either from a data bank or generated by simulation methods) for 100 assets.[4] How can we check for arbitrage opportunities? Code 1.2 will run n_{sim} simulations together with the associated arbitrage checks.

```
n <- 100
m <- 200
n.s <- 100    # number of simulations
count <- matrix(0, ncol=1, nrow=n.s) # storage

for(i in 1:n.s){
  S <- 1+matrix(rmvnorm(m, mean=rep(0.08, n),
     cov=diag(rep((.2)^2,n))), ncol=n)
  solution <- solveQP(, objL=rep(1, ncol(S)), A=S,
     cLO=rep(0, nrow(S)),cUP=rep(Inf, nrow(S)),,,,
     type=minimize, trace=F)
```

```
  count[i] <- arbitrage.check(solution)
}
hist(count)
```

Code 1.2 Arbitrage Check

The results of 100 runs for 100 assets and 200 states of the world are summarized in Figure 1.1. It shows a roughly 36% chance that our simulated scenarios contain an arbitrage situation. Hence we should take great care when using scenario optimization for a large number of assets, as an optimizer will take advantage of these arbitrage situations. This will result in unintuitive results that are overoptimistic in what they promise can be achieved.

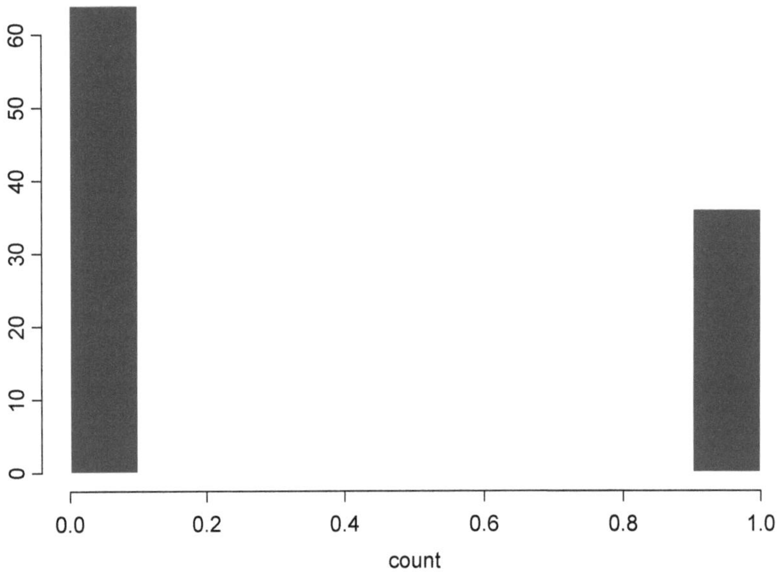

Figure 1.1 Arbitrage Opportunities

1.2 Quadratic Programming: Balancing Risk and Return

1.2.1 Classical Markowitz Optimization

Mean-variance-based portfolio construction lies at the heart of modern asset management.[5] It rests on the presumption that rational investors choose among risky assets purely on the basis of expected return and risk, with risk measured as variance.[6] Portfolios are considered mean-variance efficient if they minimize the variance for a given mean return or if they maximize the expected mean return for a given variance. Mean-variance efficiency rests on firm theoretical grounds if either[7]

- investors exhibit quadratic utility, in which case they ignore non-normality in the data, or
- returns are multivariate normal, in which case the specific utility function is irrelevant, as the set of efficient portfolios is the same for all investors (whatever attitudes toward risk they might have).

We start with an ordinary Markowitz optimization in which we want to minimize risk for a given target return. In order to perform this within NUOPT we need to set `objQ` equal to the covariance matrix $\mathbf{\Omega}$. While a covariance matrix is routinely used, we need to check whether it is positive semi-definite in order to ensure $\mathbf{w}^T\mathbf{\Omega w} \geq 0$. It is well-known from matrix algebra that this condition is met as long as all eigenvalues are non-negative and at least one eigenvalue is strictly positive. Violations arise if either we have fewer time periods than assets or we use an artificial covariance matrix.[8]

We can use the S-PLUS function `eigen()` to test this condition. However, as we know from the previous section, this does not rule out the existence of arbitrage when using $\mathbf{\Omega}$ to randomly draw scenarios. Now the matrix of linear constraints contains not only the full investment constraint, $\sum_i w_i = 1$, but also the minimum return constraint, $\mu_{\min} \leq \sum_i w_i \mu_i \leq \infty$, while upper and lower bounds on variables reflect short-selling constraints. Excess returns (nominal returns minus cash returns) expectations for asset i are given by μ_i and summarized in the $n \times 1$ vector $\boldsymbol{\mu}$. We describe the portfolio construction using the notation of Table 1.2, and Code 1.3 shows how to optimize the portfolio.[9]

Minimize

$$\begin{pmatrix} w_1 \\ \vdots \\ w_n \end{pmatrix}^T \begin{pmatrix} \sigma_{11} & \cdots & \sigma_{1n} \\ \vdots & \ddots & \vdots \\ \sigma_{m1} & \cdots & \sigma_{mn} \end{pmatrix} \begin{pmatrix} w_1 \\ \vdots \\ w_n \end{pmatrix} \tag{1.7}$$

subject to

$$\mathbf{c}_{Lo} = \begin{pmatrix} \mu_{\text{target}} \\ 1 \end{pmatrix} \leq \begin{pmatrix} \mu_1 & \cdots & \mu_n \\ 1 & \cdots & 1 \end{pmatrix} \begin{pmatrix} w_1 \\ \vdots \\ w_n \end{pmatrix} \leq \begin{pmatrix} \infty \\ 1 \end{pmatrix} = \mathbf{c}_{Up} \tag{1.8}$$

$$\mathbf{b}_{Lo} = \begin{pmatrix} 0 \\ \vdots \\ 0 \end{pmatrix} \leq \begin{pmatrix} w_1 \\ \vdots \\ w_n \end{pmatrix} \leq \begin{pmatrix} 1 \\ \vdots \\ 1 \end{pmatrix} = \mathbf{b}_{Up}. \tag{1.9}$$

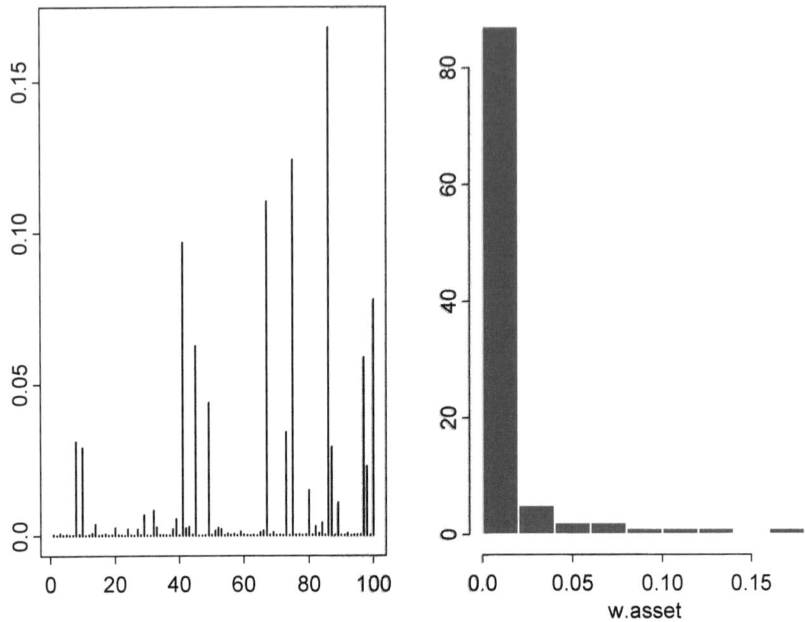

Figure 1.2 Optimal Portfolio Weights for 200 Observations

The results for an arbitrary simulation run are summarized in Figure 1.2. Note that the optimal solution in this case is known in advance without any calculation. If all assets return 10% on average and also exhibit equal volatilities and constant correlation (across assets), the minimum-risk solution for a 10% target return must be an equally weighted portfolio (1% weight for each asset). However, due to sampling error (meaning that we do not have enough data to correctly estimate our inputs), we get a very concentrated portfolio (80% of the assets drop out of the optimal solution even though we know they should be included with a weight of 1%). Repeating the same exercise for 10,000 drawings, we get closer to the optimal solution, but we are still quite significantly wrong, as shown in Figure 1.3. The reason for this lies in the high-correlation assumption we used in setting up the covariance matrix, as it is well-known that high correlations will increase the return sensitivity of the optimal solution. High correlation effectively means that assets are very close substitutes, with expected returns (notoriously difficult to estimate) becoming the distinguishing element. In the next section, we will show a simple way to safeguard against these solutions.

Step 1: Draw random return data to generate the necessary inputs.

```
n <- 100            # number of assets
m <- 200            # states of the world
```

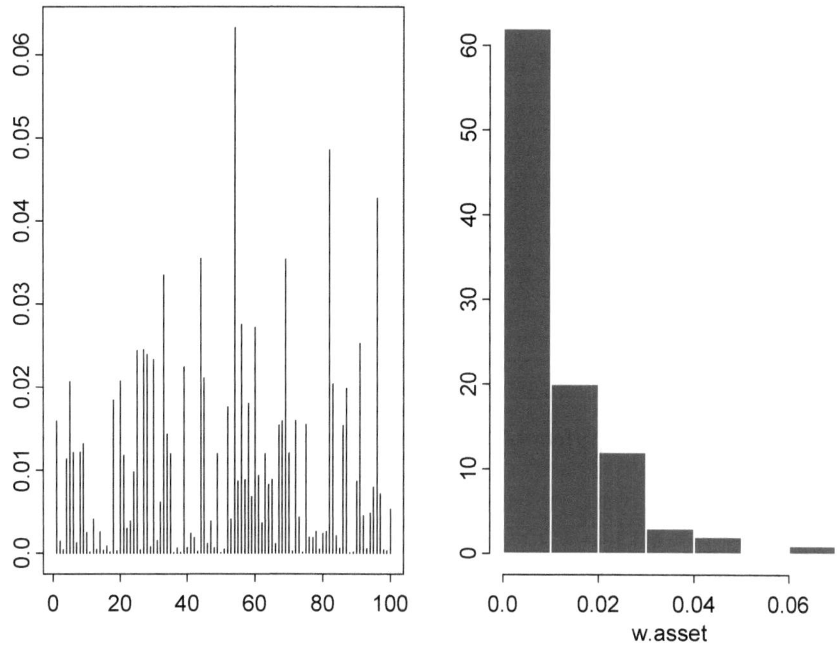

Figure 1.3 Optimal Portfolio Weights for 10,000 Observations

```
rho <- 0.7     # constant correlation
sigma <- 0.2   # constant volatility
mu <- 0.1      # same expected return for all assets
Cov <- matrix(rho*sigma*sigma, ncol=n, nrow=n)
diag(Cov) <- rep(sigma*sigma, n)
S <- 1+matrix(rmvnorm(m, mean=rep(mu, n),
  cov=Cov),ncol=n)
```

Step 2: Define Inputs

```
Cov <- var(S)         # calculate covariance matrix
mu <- apply(S, 2, mean)# calculate mean vector
mu.target <- 0.1        # define return target
A <- rbind(mu,1)        # set up A matrix

cLO <- c(mu.target,1)   # set up bounds on
cUP <- c(Inf,1)         # linear constraints
bLO <- rep(0, n)        # set up bounds on variables
bUP <- rep(1, n)
```

Step 3: Solve

```
# optimize
solution <- solveQP(objQ=Cov,,A,cLO,cUP,bLO,bUP,,)
```

Step 4: Plot solution

```
# address vector of portfolio weights
w.asset <- solution$variables$x$current
graphsheet()                # plot solution
par(mfrow=c(1,2))
plot(w.asset, type="n")
lines(w.asset,, type="h")
hist(w.asset)
```

Code 1.3 Portfolio Optimization

NuOPT for S-PLUS offers two additional functions to solve mean-variance problems. While `portfolioQPCov()` allows us to solve problems where $m < n$ (where the covariance matrix would not be of full rank, i.e., not positive semi-definite), `portfolioQPSparse()` speeds up calculations if constraints or objectives are sparse (contain many zeros). We refer the reader to the NuOPT manual to investigate both functions.

1.2.2 Implied Returns

Suppose we do not know the distribution from which asset returns have been drawn and instead use the statistical estimates as guidance. Suppose further that we have a strong belief in the efficiency of an equally weighted portfolio (i.e., we want to anchor the solution around this portfolio and we are confident about our risk estimate $\mathbf{\Omega}$). It would be natural to calculate what are called the implied returns of an equally weighted portfolio. These are the return forecasts that would return the equally weighted portfolio if they were used in a mean-variance optimization. To back out these returns, we start from the well-known mean-variance objective $U = \mathbf{w}^T\boldsymbol{\mu} - \frac{\lambda}{2}\mathbf{w}^T\mathbf{\Omega}\mathbf{w}$, where λ denotes a risk aversion coefficient that penalizes risk (measured as variance). Taking first derivatives, we get $\frac{dU}{d\mathbf{w}} = \boldsymbol{\mu} - \lambda\mathbf{\Omega}\mathbf{w} = 0$. Solving for the optimal weights, we get the well-known expression for optimal portfolio weights $\mathbf{w}^* = \lambda^{-1}\mathbf{\Omega}^{-1}\boldsymbol{\mu}$. Working backward, however, would lead us to $\boldsymbol{\mu}_{impl} = \lambda\mathbf{\Omega}\mathbf{w}$. Assuming that a given portfolio $\mathbf{w}$ is optimal and hence $\lambda = \mu/\sigma^2$ (the slope of the investor's utility curve equals the risk-return trade-off offered by a given portfolio, where μ and σ^2 reflect portfolio return and variance), we arrive at the expression for implied returns

$$\begin{aligned} \boldsymbol{\mu}_{impl} &= \left(\frac{\mu}{\sigma}\right)\frac{\boldsymbol{\Omega}\mathbf{w}}{\sigma} = \boldsymbol{\beta}\mu \\ \boldsymbol{\mu}_{impl} &= \left(\frac{\mu}{\sigma}\right)\mathbf{MCTR}, \end{aligned} \tag{1.10}$$

where we use $\frac{\boldsymbol{\Omega}\mathbf{w}}{\sigma^2} = \boldsymbol{\beta}$ and **MCTR** denotes an $n \times 1$ vector of marginal contributions to portfolio risk. Note that

$$\frac{d\sigma}{d\mathbf{w}} = \frac{d(\mathbf{w}\boldsymbol{\Omega}\mathbf{w})^{1/2}}{d\mathbf{w}} = \tfrac{1}{2}2\boldsymbol{\Omega}\mathbf{w}(\mathbf{w}\boldsymbol{\Omega}\mathbf{w})^{-\frac{1}{2}} = \frac{\boldsymbol{\Omega}\mathbf{w}}{\sigma} = \mathbf{MCTR}, \tag{1.11}$$

where the i-th element is given by

$$\frac{d\sigma}{dw_i} = \frac{w_i\sigma_{ii} + \sum_{j\neq i} w_j\sigma_{ij}}{\sigma}. \tag{1.12}$$

Code 1.4 illustrates how to compute implied returns in S-PLUS.

```
w.eq <- matrix(1/n, ncol=1, nrow=n)
# Subscript [1] to make numeric
vol.eq <- sqrt(t(w.eq)%*%Cov%*%w.eq)[1]
mu.impl <- (Cov%*%w.eq)/vol.eq^2*mu.target
mctr <- Cov%*%w.eq/vol.eq
plot(mctr, mu.impl,
  xlab="marginal contribution to risk",
  ylab="implied return")
```

Code 1.4 Implied Returns

For a portfolio to be optimal, the relationship between marginal contribution to portfolio risk and marginal return (implied return) has to be linear, with slope equal to the Sharpe ratio of this portfolio. This is shown in Figure 1.4. As our risk estimates are much better than our return estimates (lower sampling error), the dispersion of implied returns becomes very small. It ranges between 9% and 11%. We know that the implied returns will lead us back to a portfolio we would reckon to be efficient in the absence of further information. Deviations from this portfolio[10] only occur if we have strong enough conviction that our return forecasts do indeed carry information.

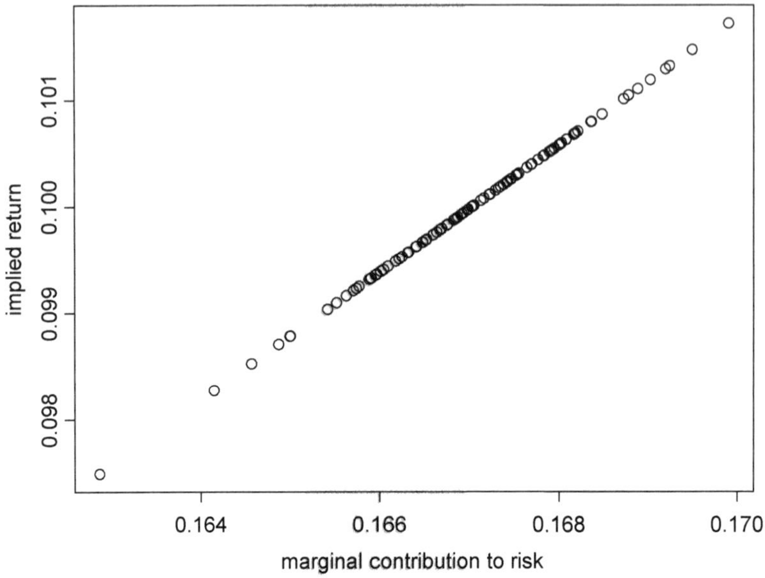

Figure 1.4 Relationship Between Implied Return and Marginal Risk for 200 Observations

1.2.3 Adding Forecasts

Suppose we are given an $f \times 1$ vector $\mathbf{f}$ that contains a set of f return forecasts together with an $f \times f$ covariance matrix of estimation errors $\mathbf{\Sigma}$.[11] The higher the conviction of a particular forecast, the lower the corresponding value on the main diagonal of $\mathbf{\Sigma}$ will be. Off-diagonal elements describe the covariance between forecast errors. In order to specify the forecasts, we need an additional $f \times n$ matrix $\mathbf{P}$. The meaning of this matrix can be illustrated by a couple of examples.

If the only return we forecast is on asset 1, $\mathbf{P}$ becomes a $1 \times n$ vector that contains one as the first element and zero otherwise: $\mathbf{P} = \begin{pmatrix} 1 & 0 & \cdots & 0 \end{pmatrix}$. Obviously, $\mathbf{P}$ becomes a diagonal matrix with ones on the main diagonal if we attempt to forecast all assets in the covariance matrix. Alternatively, we can express opinions on return differentials between various pairs of assets,

$$\mathbf{P}_{4\times n} = \begin{pmatrix} 1 & 0 & -1 & 0 & \cdots \\ 1 & -1 & 0 & 0 & \cdots \\ 0 & 0 & 0 & 1 & \cdots \\ \frac{1}{2} & -1 & \frac{1}{2} & 0 & \cdots \end{pmatrix}. \tag{1.13}$$

Note that we can also express more forecasts than assets ($f > n$). This framework offers a great deal of flexibility. We can, for example, express opinions such as: "I am 66% confident that the return differential between asset 1 and asset 3 (first row above) will range between 1% and 2%." Assuming normality, we should interpret this as a 1.5% outperformance on average with a standard deviation of 0.5% because we know from elementary statistics that 66% of all observations are 1 standard deviation away from the mean ($1\% \leq 1.5\% \pm 1 \cdot 0.5\% \leq 2\%$). The corresponding entries would be

$$\mathbf{f} = \begin{pmatrix} 1.5 \\ \vdots \end{pmatrix}, \; \mathbf{\Sigma} = \begin{pmatrix} 0.5^2 & \cdots \\ \vdots & \ddots \end{pmatrix}. \tag{1.14}$$

The optimal combination of explicit return forecasts and implicit returns is a straightforward matrix calculation:

$$\boldsymbol{\mu} = \boldsymbol{\mu}_{impl} + \mathbf{\Omega P}^T \left(\mathbf{P \Omega P}^T + \mathbf{\Sigma} \right)^{-1} \left(\mathbf{f} - \mathbf{P} \boldsymbol{\mu}_{impl} \right). \tag{1.15}$$

It is interesting to note that changes in implied returns are driven by both conviction (the natural enemy of diversification) and the covariance between assets via $\mathbf{\Omega P}^T$. A high-conviction signal on a single asset would hence also affect all assets, depending on their historical correlation. This makes intuitive sense: if we changed return expectations on the German bond market, for example, we might not want the French bond market unchanged. It mitigates the problem of correlation-inconsistent return forecasts, which are the main reason for poor optimization results.

Let us investigate, using Code 1.5, how the methodology introduced above works in the current setting. We assume that the volatility of forecast errors is five times the volatility of asset returns.

```
# This code needs the results of Code 1.4
n <- 100       # number of assets
m <- 10000     # states of the world
rho <- 0.7     # constant correlation
sigma <- 0.2   # constant volatility
mu <- 0.1      # same expected return for all assets
Cov<-matrix(rho*sigma*sigma, ncol=n, nrow=n)
diag(Cov)<-rep(sigma*sigma, n)
# set up matrices
P <- diag(rep(1,n))
S <- 1+matrix(rmvnorm(m, mean=rep(mu, n), cov=Cov),
  ncol=n)
Cov <- var(S)
```

```
f <- apply(S, 2, mean)
Cov.f <- 25*Cov
mu.ad <- mu.impl+Cov%*%t(P)%*%
  solve((P%*%Cov*t(P)+Cov.f))%*%(f-P%*%mu.impl)
# rerun optimization
A <- rbind(t(mu.ad),1)
solution <- solveQP(Cov,, A, cLO, cUP, bLO, bUP,,)
w.asset.ad <- solution$variables$x$current
graphsheet()
par(mfrow=c(1,2))
plot(w.asset.ad, type="n")
lines(w.asset.ad, type="h")
hist(w.asset.ad)
```

Code 1.5 Adding Forecasts

As a result of our return adjustments, we end up with much more diversified (in the sense of being closer to our anchor portfolio) holdings, as can be seen in Figure 1.5.

1.2.4 Variations of St. Markowitz

In this section, we want to provide you with a number of suggestions on how to use `solveQP()` for some nonstandard-looking portfolio optimization problems.[12] The according procedures are summarized in Table 1.3.

Utility Optimization. Suppose that, instead of minimizing risk for a given return expectation, we need to maximize utility (given as $U = \mathbf{w}^T\boldsymbol{\mu} - \frac{\lambda}{2}\mathbf{w}^T\boldsymbol{\Omega}\mathbf{w}$) subject to an arbitrary set of constraints. Since `solveQP` minimizes $\frac{1}{2}\mathbf{x}^T\mathbf{Q}_{obj}\mathbf{x} + \mathbf{L}_{obj}^T\mathbf{x}$, the adjustments we need to undertake are to multiply `objQ` by $-\lambda$, set `objL` to $\boldsymbol{\mu}$, and set `type` equal to `maximize`. The matrix of coefficients for the linear constraints now only contains the 100% investment constraint $A = (1 \ \cdots \ 1)$, as returns are already included in the objective. The upper and lower bounds on linear constraints are now the scalars $c_{Lo} = 1, c_{Up} = 1$.

Table 1.3: Adjustments in `solveQP` to Solve Nonstandard Problems

Problem	Adjustment
Utility Optimization $U = \mu - \frac{\lambda}{2}\sigma^2$	Multiply `objQ` by $-\lambda$ Set `objL` to $\boldsymbol{\mu}$ Change `type` to `maximize` $\mathbf{A} = (1 \ \cdots \ 1)$ $\mathbf{c}_{Lo} = 1, \mathbf{c}_{Up} = 1$ $\mathbf{b}_{Lo} = (0 \ \cdots \ 0)^T, \mathbf{b}_{Up} = (1 \ \cdots \ 1)^T$
Estimation Error $U = \mu - \frac{\lambda}{2}\theta\sigma^2$ $\theta = \frac{m+1}{m-n-2}$	Same as above Multiply `objQ` by $-\lambda\theta$
Liability Relative Optimization $(\mu - \mu_l) - \frac{\lambda}{2}\theta(\sigma^2 + \sigma_l^2 - 2\rho\sigma\sigma_l)$	Same as above Change `objQ` and `objL` $\boldsymbol{\mu} = \mathbf{T}\boldsymbol{\mu}^*, \boldsymbol{\Omega} = \mathbf{T}\boldsymbol{\Omega}^*\mathbf{T}^T$ $\mathbf{T} = \begin{pmatrix} 1 & 0 & & & -1 \\ 0 & 1 & 0 & & -1 \\ & 0 & \ddots & 0 & -1 \\ & & 0 & 1 & -1 \end{pmatrix}$
Benchmark-Relative Optimization $U = \mu_a - \frac{\lambda}{2}\theta\sigma_a^2$	Same as Estimation Error $\mathbf{b}_{Lo} = \begin{pmatrix} \mathbf{0}_{n\times1} \\ -\mathbf{w}_{b_{n\times1}} \end{pmatrix}_{2n\times1}^T$ $\mathbf{b}_{Up} = \begin{pmatrix} 1_{n\times1} \\ -\mathbf{w}_{b_{n\times1}} \end{pmatrix}_{2n\times1}^T$ $\mathbf{c}_{Lo} = 0, \mathbf{c}_{Up} = 0$
Dual Benchmark Optimization $U = U_1 + U_2$ $U_1 = \mu_{a_1} - \frac{\lambda}{2}\theta\sigma_{a_1}^2$ $U_2 = \mu_{a_2} - \frac{\lambda}{2}\theta\sigma_{a_2}^2$	Same as Benchmark-Relative Optimization $\varphi = \frac{\lambda_1}{\lambda_1 + \lambda_2}$ $\mathbf{w}_b = \varphi\mathbf{w}_{b_1} + (1-\varphi)\mathbf{w}_{b_1}$ $\lambda = \varphi\lambda_1 + (1-\varphi)\lambda_2$

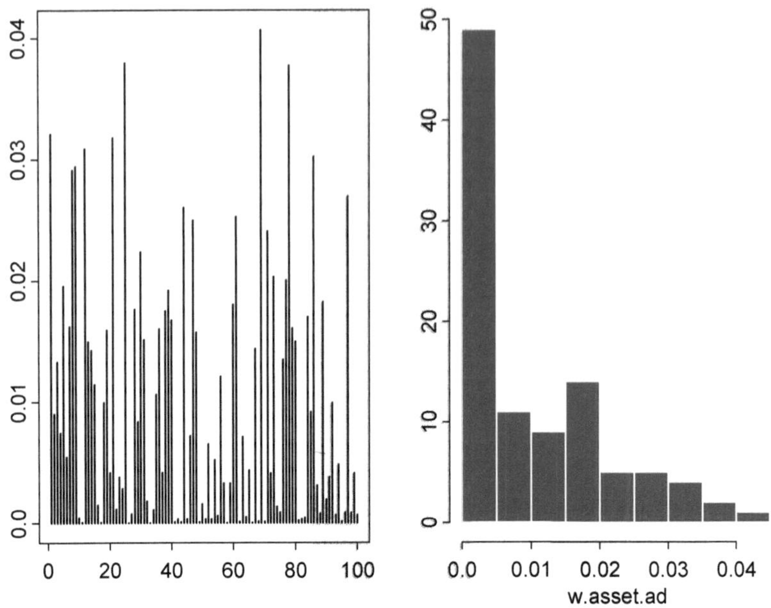

Figure 1.5 Optimization results for adjusted returns

Estimation Error. If inputs are measured with error and we only have an noninformative prior (all we know is that the inputs are measured with error) about the real parameters, the world has clearly become a riskier place. The optimal cure (according to Bayesian statistics) in this situation is to leverage the covariance matrix of asset returns by multiplying it by a scalar $\theta = \frac{m+1}{m-n-2} \geq 1$. [13] This leverage factor creates a new "pseudo-risk aversion", $\tilde{\lambda} = \lambda\theta$. Estimation error decreases if the number of observations rises relative to the number of assets. Note that expected returns remain unaffected by estimation error, as there is (by definition) no uncertainty about an expected value.

Asset Liability Management. Most institutional investors have some kind of liabilities. In fact, the only reason they need assets is because they face liabilities. How can we incorporate them into mean-variance analysis?[14] Assume we not only have time series of $n \times 1$ asset returns but also one time series of liability returns, and assume the means and covariance of the combined data set are summarized in the $(n+1) \times 1$ vector of mean returns, $\boldsymbol{\mu}^*$, as well as in the $(n+1) \times (n+1)$ covariance matrix $\boldsymbol{\Omega}^*$. Let us introduce an $n \times (n+1)$ transformation matrix, $\mathbf{T}$,

$$\mathbf{T} = \begin{pmatrix} 1 & 0 & & & -1 \\ 0 & 1 & 0 & & -1 \\ & 0 & \ddots & 0 & -1 \\ & & 0 & 1 & -1 \end{pmatrix}. \tag{1.16}$$

We can now transform mean returns as well as covariances into liability-relative mean returns and liability-relative covariances by calculating $\boldsymbol{\mu} = \mathbf{T}\boldsymbol{\mu}^*$ and $\boldsymbol{\Omega} = \mathbf{T}\boldsymbol{\Omega}^*\mathbf{T}^T$. For example, the element in the first row and first column of the covariance matrix is no longer the variance of returns for asset one but rather the variance of the return difference of a portfolio with a 100% long position in asset one and a 100% short position in the liabilities. The objective becomes to trade off relative returns versus relative risks $(\mu - \mu_l) - \frac{\lambda}{2}\theta\left(\sigma^2 + \sigma_l^2 - 2\rho\sigma\sigma_l\right)$, where μ_l represents the expected liability return, σ_l denotes the volatility of liability returns and ρ expresses the correlation between returns of the asset portfolio with liability returns.

Active Optimization. Active portfolio management attempts to beat a benchmark portfolio that serves as a yardstick for the skills of the active manager.[15] Assume that the active manager stays within the investment universe of the benchmark. All we need to do is to introduce the negative benchmark weights as the decision variable (effectively a short portfolio) into the objective function (1.7), constrain benchmark weights to their current holdings, and make sure that asset and benchmark weights now add up to zero. This will automatically change the calculated risk number from a measure of total portfolio volatility into a measure of active volatility, σ_a^2 (tracking error), and expected return into expected active return, μ_a. It should be clear by now that liability-relative optimization can also be set up as a benchmark-relative optimization.

Dual Benchmark Optimization. Finally, many investors have more than one benchmark.[16] They might want to optimize versus their liabilities, but they also want to keep up with their peers. While this sounds complicated, it actually is not. We can transform the problem of dual benchmark optimization (that is, to find one vector of portfolio weights that maximizes the utility from both subproblems) by creating a new benchmark that effectively is a mixture of both benchmarks, where the weight attached to a particular benchmark depends on the importance (risk aversion) of the respective benchmark risk.

1.3 Dual Variables and the Impact of Constraints

1.3.1 KKT Conditions and Portfolio Optimization

So far, we have seen what uncertainty in inputs can do to portfolio optimization. It is not surprising that practitioners have always been looking at ways to reduce the impact of estimation error on optimal portfolios and exclude solutions that might make mathematical sense but little investment sense. One of the most obvious safeguards against the unwarranted side effects of portfolio optimization is to impose constraints on the set of solutions. Constraints come in many forms: full investment constraints (the asset weights have to add up to 100%), non-negativity constraints (no short-selling allowed), or group constraints (groups of assets have to stay within bounds). In this section, we want to calculate the impact of constraints on portfolio construction by looking at how our actual forecasts are modified via constraints.

Suppose we are given a set of forecasts for all asset returns as summarized in the $f \times 1$ vector $\mathbf{f}$, where in this case $f = n$. Suppose further that we also know the $n \times 1$ vector of equilibrium returns $\boldsymbol{\mu}_{impl}$ (those returns that would yield a portfolio assumed to be efficient; i.e., a portfolio we would hold in the absence of any forecasting power). A Markowitz investor wants to solve the following constrained portfolio optimization problem: [17]

Maximize

$$\mathbf{w}^T\mathbf{f} - \tfrac{\lambda}{2}\mathbf{w}^T\boldsymbol{\Omega}\mathbf{w}, \tag{1.17}$$

subject to

$$\mathbf{c}_{Lo} \le \mathbf{A}\mathbf{w} \le \mathbf{c}_{Up}, \tag{1.18}$$

$$\mathbf{b}_{Lo} \le \mathbf{w} \le \mathbf{b}_{Up}. \tag{1.19}$$

The well-known KKT conditions for the problem described in (1.17)–(1.19) imply that

$$\mathbf{f} = \lambda\boldsymbol{\Omega}\mathbf{w} + \mathbf{A}^T\boldsymbol{\Lambda}_{\mathbf{A}} + \boldsymbol{\Lambda}_{\mathbf{w}}, \tag{1.20}$$

where $\boldsymbol{\Lambda}_{\mathbf{A}}$ denotes the vector of dual variables associated with the n_c group constraints and $\boldsymbol{\Lambda}_{\mathbf{w}}$ represents the vector of dual variables associated with the n asset-specific constraints.

Assume that we have found the optimal solution $\mathbf{w}^*$ to (1.17). We already know that the implied returns for this portfolio are given by $\mathbf{f}^* = \lambda\boldsymbol{\Omega}\mathbf{w}^*$. If none of the constraints were binding, all dual variables would be zero and we would get $\mathbf{f} = \mathbf{f}^*$. In this case, original forecasts and implied forecasts are the same. The constraints did not alter our forecasts (as they were not binding in the first place). However, this case is rare; constraints are often binding, particularly as they often have been imposed on the problem after a first unconstrained optimization yielded an "unpleasant" result (e.g., many assets received zero weight). As we can always express our forecast as $\mathbf{f} = \boldsymbol{\mu}_{impl} + \boldsymbol{\alpha}$ and $\mathbf{f} = \mathbf{f}^* + \boldsymbol{\eta}$, we can decompose the implied returns of the optimal solution to the constrained optimization above as

$$\mathbf{f}^* = \boldsymbol{\mu}_{impl} + \boldsymbol{\alpha} - \boldsymbol{\eta}, \tag{1.21}$$

where we can further decompose $\boldsymbol{\eta}$ into $\mathbf{A}^T\boldsymbol{\Lambda}_\mathbf{A} + \boldsymbol{\Lambda}_\mathbf{w}$. Now that we have set up the methodological foundations, we can proceed to a numerical example.

1.3.2 Can Constraints Safeguard Against Bad Research?

The objective of this section is to provide an illustration of the results above and a "how-to-do" it recipe for NUOPT. After the usual housekeeping operations

```
remove(ls())
module(nuopt)
```

we load the covariance data into S-PLUS and calculate $\boldsymbol{\Omega}$.

```
# define correlation matrix
corr <- matrix(data=
  c(  1, 0.4, 0.5, 0.5, 0.4, 0.1, 0.1, 0.1,
    0.4, 1.0, 0.3, 0.3, 0.1, 0.4, 0.1, 0.1,
    0.5, 0.3, 1.0, 0.7, 0.1, 0.1, 0.5, 0.1,
    0.5, 0.3, 0.7, 1.0, 0.1, 0.1, 0.1, 0.5,
    0.4, 0.1, 0.1, 0.1, 1.0, 0.0, 0.0, 0.0,
    0.1, 0.4, 0.1, 0.1, 0.0, 1.0, 0.0, 0.0,
    0.1, 0.1, 0.5, 0.1, 0.0, 0.0, 1.0, 0.2,
    0.1, 0.1, 0.1, 0.5, 0.0, 0.0, 0.2, 1.0),
   nrow=8, ncol=8)

# define diagonal volatility matrix
vol <- diag(c(17, 21, 22, 20, 8, 8, 8, 8))
# calculate variance covariance matrix
Cov <- vol %*% corr %*% vol
n <- ncol(Cov)
```

Next we can calculate $\boldsymbol{\mu}_{impl}$ assuming a weight vector $\mathbf{w}$ and an associated portfolio return of $\mu = 3.5\%$.

```
# create implied returns
w <- matrix(c(0.24, 0.18, 0.12, 0.06, 0.16, 0.12,
   0.08, 0.04), ncol=1)
mu.impl <- as.numeric(3.5 * (Cov %*% w)/ (t(w) %*%
  Cov %*% w)[1])
```

In Code 1.6, we assume an arbitrary set of group and asset constraints.

```
# set up group constraint
# full investment constraint
group <- c(1,1,1,1,1,1,1,1)

# assets 1-4 build one group
group <- rbind(group, c(1,1,1,1,0,0,0,0))

# assets 5,6,7,8 build a 2nd group
group <- rbind(group, c(0,0,0,0,1,1,1,1))

# define upper and lower weights
bUP <- c(0.39, 0.33, 0.27, 0.21, 0.31, 0.27, 0.23,
  0.19)
# asset constraints
bLO <- c(0.09, 0.03, 0.00, 0.00, 0.01, 0.00, 0.00,
  0.00)
# group constraints
cUP <- c(1, 0.8, 0.5)
cLO <- c(1, 0.4, 0.1)
```

Code 1.6 Setup for Portfolio Optimization

As a little exercise, the reader might want to calculate the minimum variance portfolio and the maximum return portfolio. The latter is quite useful if we want to trace out an efficient frontier in the presence of constraints. The solution is given in Code 1.7.

```
# This code needs the results of Code 1.6
# calculate minimum variance portfolio
w.min.var <- matrix(solveQP(2*Cov, , group, cLO,
     cUP, bLO, bUP, rep(0, n), type=minimize,
     trace=T)$variables$x$current,ncol=1)
mu.min <- t(w.min.var) %*% mu.impl
# calculate maximum return portfolio
mu.max <- solveQP( , mu.impl, group, cLO, cUP, bLO,
  bUP, rep(0, n), type=maximize, trace=T)$objective
```

Code 1.7 Minimum Variance and Maximum Return Portfolios

Let us now try to recover the portfolio weights we used in deriving the implied returns. A straightforward way is to use (1.17), but what value do we need to assume for λ? Recalling that for the optimal portfolio weights

$$\lambda = \mu / \sigma^2 = 3.5/98.1 = 0.0357,$$

we can obtain the optimal solution using `solveQP` as follows.

```
solveQP(-0.0357*Cov, as.numeric(mu.impl), group,
     cLO, cUP, bLO, bUP, as.numeric(w.min.var),
     type=maximize, trace=T)$variables$x$current
NUOPT 5.2.3a, Copyright (C) 1991-2001 Mathematical
  Systems Inc.
PROBLEM_NAME anon.QP
NUMBER_OF_VARIABLES 8
NUMBER_OF_FUNCTIONS 4
PROBLEM_TYPE MAXIMIZATION
METHOD LINE_SEARCH
<preprocess begin>..........<preprocess end>
<iteration begin>
 res=5.2e+000 .... 5.1e-005 . 3.8e-008
<iteration end>
STATUS OPTIMAL
VALUE_OF_OBJECTIVE 1.748819986
ITERATION_COUNT 7
FUNC_EVAL_COUNT 10
FACTORIZATION_COUNT 7
RESIDUAL 3.841505323e-008
ELAPSED_TIME(sec.) 0.06
        1         2         3          4
0.2398441 0.1798606 0.1198895 0.05996219

        5         6          7          8
0.1600725 0.1201176 0.08013711 0.04011639
```

This is very close to our original set of weights. If we were to increase the precision of NUOPT, we could recover the weights exactly. Suppose now the set of forecasts

```
f <- c(4.65,3.61,8.7,6.96,0.39,0.38,1.66,0.93)
```

are used in a portfolio optimization. Note that we keep λ the same so that we can compare both solutions while maintaining the same risk-return trade-off.

```
solution <- solveQP(-0.0357*Cov, as.numeric(f),
     group, cLO, cUP, bLO, bUP,
     as.numeric(w.min.var), type=maximize, trace=T)
```

The optimal solution $\mathbf{w}^*$ is recovered by

```
w.star <- matrix(solution$variables$x$current,
     ncol=1)
round(w.star, digits = 2)
```

```
      [,1]
[1,] 0.14
[2,] 0.03
[3,] 0.27
[4,] 0.19
[5,] 0.01
[6,] 0.05
[7,] 0.23
[8,] 0.07
```

Note that asset #1 hits its upper boundary while asset #8 hits its lower boundary. If it were unconstrained, the optimizer would have wanted to invest further in asset #3 while shorting asset #8.

We can now calculate the implied returns for the optimized portfolio:

```
0.0357 * Cov %*% w.star
```

```
          [,1]
[1,] 4.8057517
[2,] 3.7683019
[3,] 8.7089196
[4,] 7.1157412
[5,] 0.5997224
[6,] 0.5357518
[7,] 1.6077238
[8,] 1.0857500
```

How does this compare with our decomposition? That is, do we get the same results using (1.20)?

```
L.w <- matrix(solution$variables$x$dual, ncol=1)
L.A <- matrix(solution$constraints$"1"$dual,
     ncol=1)

f + t(group) %*% L.A + L.w
```

```
           [,1]
[1,] 4.8057517
[2,] 3.7683017
[3,] 8.7089183
[4,] 7.1157396
[5,] 0.5997223
[6,] 0.5357519
[7,] 1.6077225
[8,] 1.0857508
```

Finally, we want to quantify how much each constraint contributed to the difference between $\mathbf{f}$ and $\mathbf{f}^*$. First we look at which asset constraints have been binding:

```
round(L.w, digits = 2)
```

```
      [,1]
[1,]  0.00
[2,]  0.00
[3,] -0.15
[4,]  0.00
[5,]  0.05
[6,]  0.00
[7,] -0.21
[8,]  0.00
```

As we have already seen, constraints on asset #3 and asset #8 have been binding. This is reflected in our dual variables $\mathbf{\Lambda_w}$ above. We can interpret the entries as follows. Without a non-negativity constraint on asset #8, we would have needed a 0.22 higher return on this asset in order to obtain a non-negative weighting. The reverse is true for asset #3: without an upper constraint on asset #3, we would have needed to lower return expectations for asset #3 by 0.15 to obtain the same result. Next we look at $\mathbf{\Lambda_A}$.

```
round(L.A, digits = 2)
```

```
     [,1]
[1,] 0.16
[2,] 0.00
[3,] 0.00
```

A positive entry in the first element of $\mathbf{\Lambda_A}$ means that the full investment constraint has been binding. We can interpret a positive value as the need to increase the overall return by 0.47 to justify the full investment, which is currently imposed by the constraints. We can now come back to our starting question: are constraints a safeguard against bad research? If we compare our original alphas ($\boldsymbol{\alpha} = \mathbf{f} - \boldsymbol{\mu}_{impl}$) with the implied alphas of our constrained optimization ($\boldsymbol{\alpha}^* = \mathbf{f}^* - \boldsymbol{\mu}_{impl}$), we can observe a considerable change. For example, the alpha for asset #1 changed from –0.33 to 0.13.

What are the business implications of these findings? Well, if an asset manager spends a large amount of money on his or her research analysts, but the research information entering portfolios gets lost or changed by a set of constraints, the asset manager should rethink his or her resource allocation.

1.4 Analysis of the Efficient Frontier

In this section, we will go beyond calculating a single efficient portfolio and trace out a complete efficient frontier. The function `portfolioFrontier()` allows us to find the minimum risk portfolios for required returns ranging from the return of the minimum risk portfolio to the maximum return portfolio. We will use the same covariance matrix and restrictions as in the last section. For a start, we assume expected returns.

```
# This code needs the results of Codes 1.5-1.7 and
# intermediate results in the text
f <- c(3,4,5,6,0.25,0.5,0.75,1)
frontier.uc <- portfolioFrontier(Cov, f, wmin=-Inf,
  max.ret=max(mu.impl), n.ret=30)
graphsheet()
par(mfrow=c(1,2))
plot(frontier.uc$sd, frontier.uc$returns,
  xlab="Risk measured in standard deviation",
  ylab="Expected return", type="b")
title(main="Efficient frontier (unconstrained) ")
barplot(frontier.uc$weights)
title(main="Frontier portfolios")
```

Code 1.8 NUOPT Code for the Unconstrained Efficient Frontier

The efficient frontier with no restrictions (apart from the restriction that all positions have to add up to 100%) is shown in Figure 1.6; the S-PLUS code needed to produce it is given in Code 1.8. It has some interesting characteristics[18]:

- The maximum return portfolio is not constrained to a 5% return, as we excluded constraints (i.e., the optimizer can use leverage).

- Asset weights are either linearly rising or linearly falling with leverage, where long positions eventually change into short positions.
- All frontier portfolios can be expressed as a weighted combination of any two frontier portfolios.
- Leverage increases as we increase the return requirements.

Next we introduce a long-only constraint (no short-selling allowed) in Code 1.9.

```
# This code needs the results of Code 1.8
frontier.nss <- portfolioFrontier(Cov, f, wmin=0,
  max.ret=max(mu.impl), n.ret=30)
graphsheet()
par(mfrow=c(1,2))
plot(frontier.nss$sd, frontier.nss$returns,
  xlab="Risk measured in standard deviation",
  ylab="Expected return", type="b")
title(main="Efficient frontier (no short sales)")
barplot(frontier.nss$weights)
title(main="Frontier portfolios")
```

Code 1.9 Efficient Frontier, No Short-Selling

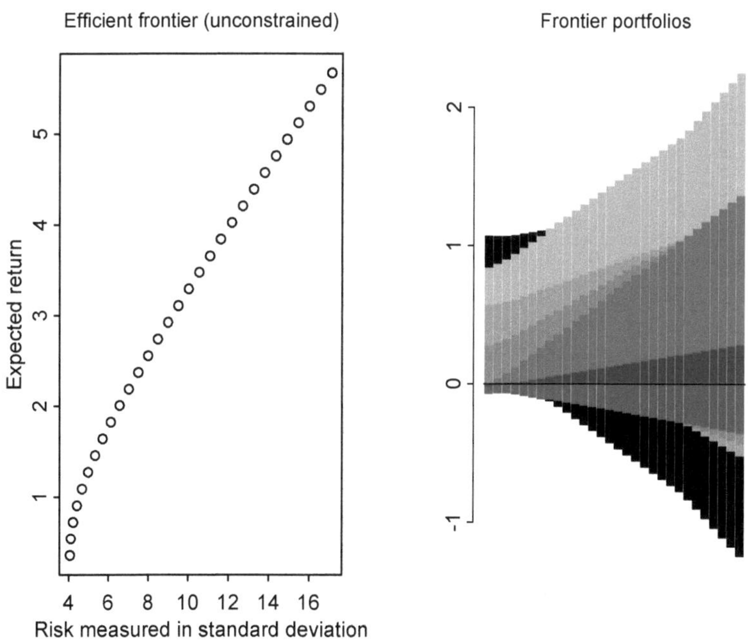

Figure 1.6 Efficient Frontier without Short-Selling Constraint

The results are displayed in Figure 1.7. We observe that the efficient frontier stretches to a maximum 6% return. Leverage has not been allowed in this example.

Finally, we can add all the constraints of the previous section.

```
wmax <- c(0.39,0.33,0.27,0.21,0.31,0.27,0.23,0.19)
wmin <- c(0.09,0.03,0.00,0.00,0.01,0.00,0.00,0.00)
grmat <-
     rbind(c(1,1,1,1,0,0,0,0),c(0,0,1,1,0,0,1,1))
grmax <- c(0.8, 0.5)
grmin <- c(0.4, 0.1)
```

However, in tracing out the efficient frontier, we need to specify the return on the maximum return portfolio, as it is not clear what the maximum return will be in the face of various interacting constraints. Code 1.10 illustrates how to do this.

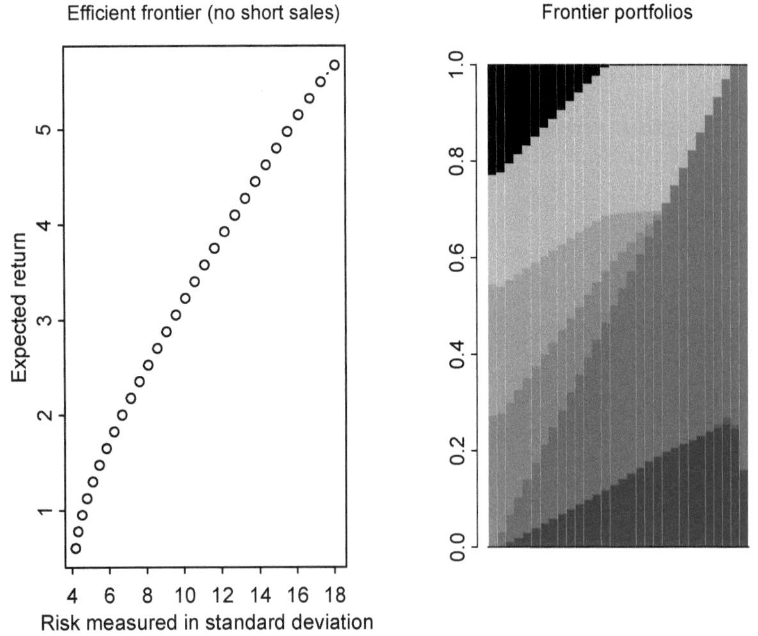

Figure 1.7 Efficient Frontier with Short-Selling Constraints

```
# This code needs the results of Code 1.9
mu.max <- solveQP( ,f,rbind(1,grmat),
  cLO=c(1,grmin),cUP=c(1,grmax),bLO=wmin,bUP=wmax,
  rep(0,n),type=maximize,trace=T)$objective
```

```
frontier.c <- portfolioFrontier(Cov,f,wmin=wmin,
  wmax=wmax, max.ret=mu.max, grmat=grmat,
  grmax=grmax, grmin=grmin, n.ret=30)
graphsheet()
par(mfrow=c(1,2))
plot(frontier.c$sd, frontier.c$returns,
  xlab="Risk measured in standard deviation",
  ylab="Expected return", type="b")
title(main="Efficient frontier (all constraints)")
barplot(frontier.c$weights)
title(main="Frontier portfolios")
```

Code 1.10 Efficient Frontier with Group Constraints

The results are summarized in Figure 1.8. Note that the frontier ranges up to the maximum return portfolio, which is considerably below the maximum single expected return within the investment universe. Portfolios for all frontiers look considerably constrained.

Finally, we compare the three different frontiers in Figure 1.9 (produced using Code 1.11). The use of constraints increases the riskiness of portfolios for every level of expected returns.

```
# This code needs the results of Codes 1.8-1.10
label<-c("unconstrained", "no short sales",
  "fully constrained")
frontier.df<-data.frame(x=c(frontier.uc$sd,
  frontier.nss$sd, frontier.c$sd),
  y=c(frontier.uc$returns, frontier.nss$returns,
  frontier.c$returns), type=ordered(rep(1:3,
  rep(30,3)), labels=label))
xyplot(y ~ x, groups=type, lwd=3, data=frontier.df,
  panel=panel.superpose, type="b", pch=5:7,
  key=list(x=0.95, y=0.05, corner=c(1,0),
text=list(label), lines=list(lwd=3, col=2:4,
  lty=2:4, type="b", pch=5:7)),
  xlab="Risk measured in standard deviation",
  ylab="Expected return")
```

Code 1.11 Comparison of Efficient Frontiers

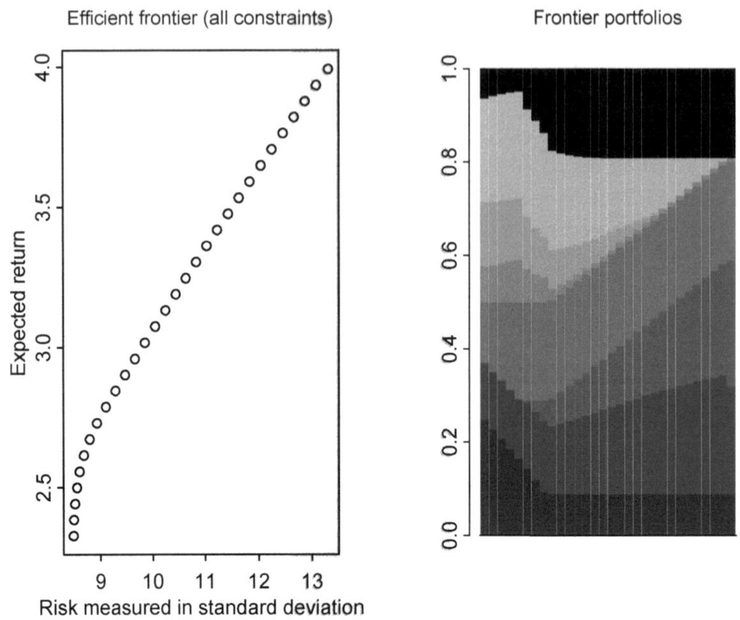

Figure 1.8 Efficient Frontier with All Constraints

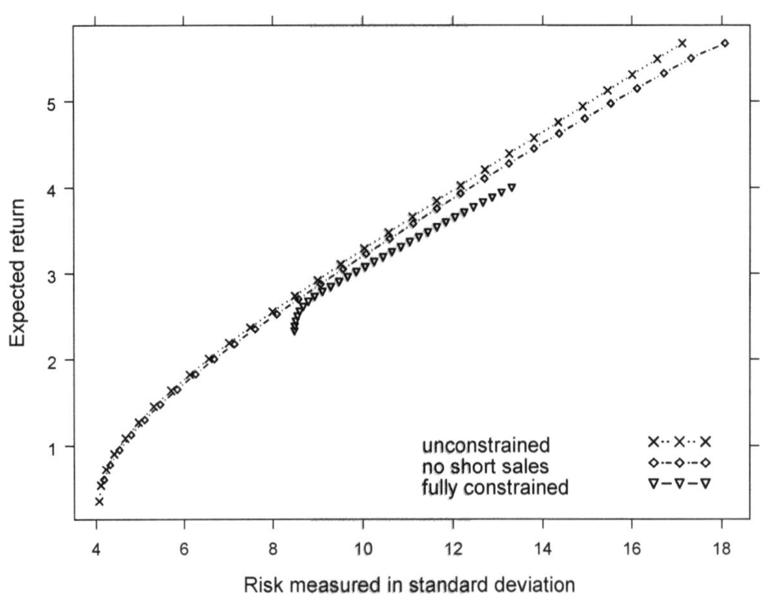

Figure 1.9 Comparison of Efficient Frontiers

We will finish this section with a puzzle to solve. If we repeat the exercise above, but use the implied returns as forecast instead, we get Figure 1.10. The constrained and unconstrained frontiers coincide most of the time. Can you explain this?

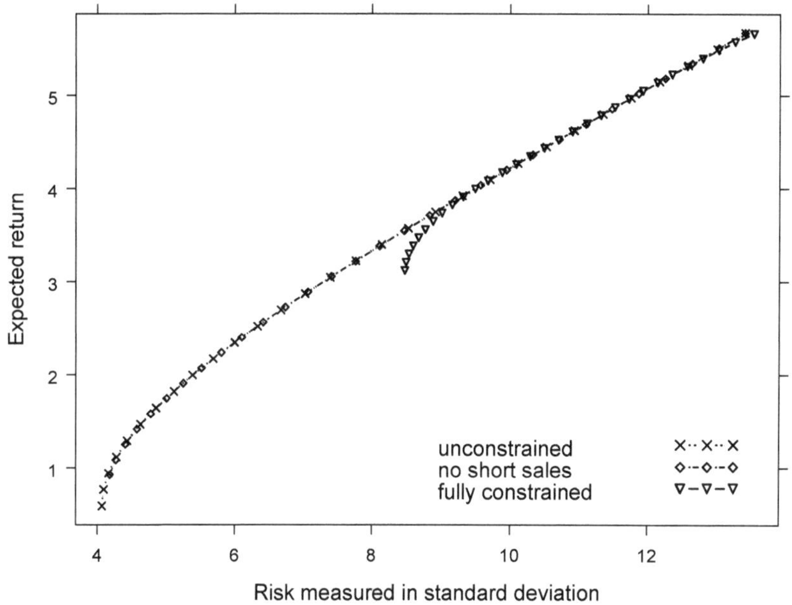

Figure 1.10 Comparison of Efficient Frontiers for Implied Returns

Exercises

1. Acquire some data on yield, convexity, and duration for five yield curve buckets[19] in the U.S. market, together with the same data for the index as a whole.

   (a) Use `solveQP` to find a portfolio that maximizes the relative yield versus the index but stays fully invested and matches duration as well as convexity.
   (b) Use S-PLUS to decompose the yield changes of all five buckets into three uncorrelated principal components and their associated loadings. Assume that these three loadings explain most of the variance. Is this a good assumption?
   (c) Calculate the tracking error of the portfolio you found under a). Where does the risk come from?
   (d) Use the mean-variance approach discussed in this chapter to eliminate active risk. What portfolio do you arrive at?

   In order to solve this exercise, you might find the following short digression useful. Using principal component analysis, we can decompose yield changes into three uncorrelated principal components (Δpc_i), which usually explain most of the variance of the underlying series apart from "odd" places such as Japan. Together with the associated loadings (b_{ij}), we can write

$$\Delta y_j = \sum_{i=1}^{3} b_{ij} \Delta pc_i .$$

   We need to calculate the matrix of factor loadings $\mathbf{b}$ and the covariance matrix of principal components $\mathbf{\Omega}_{\Delta pc}$. We can combine this information with the duration vector $\mathbf{D}$ to arrive at active risk:

$$\sigma^2 = (\mathbf{w} - \mathbf{w}_b)^T \mathbf{D}^T \mathbf{b} \mathbf{\Omega}_{\Delta pc} \mathbf{b}^T \mathbf{D} (\mathbf{w} - \mathbf{w}_b).$$

   Risk from the respective principal components can be calculated from $\frac{d\sigma_a}{d\mathbf{pc}} = \frac{(\mathbf{\Omega}_{\Delta pc} \mathbf{b}^T) \mathbf{D}(\mathbf{w} - \mathbf{w}_b)}{\sigma_a}$. Hint: See Scherer (2004).

2. Suppose you are given the following expected returns and covariance matrix of asset returns. Inputs have been calculated from ten annual observations.

$$\mathbf{\Omega} = \begin{pmatrix} 400 & 25 \\ 25 & 25 \end{pmatrix}, \quad \mu = \begin{pmatrix} 7 \\ 5 \end{pmatrix}.$$

(a) Calculate efficient frontiers with and without the inclusion of estimation error. How do the slopes change?

(b) Does the composition of the tangency portfolio change if you include a risk-free asset returning 3% per year? Explain. Hint: See Klein and Bawa (1976).

3. Assume the covariance between assets is described via $\mathbf{\Omega}_{aa}$ ($k \times k$ matrix, where k equals the number of assets), while the covariance between assets and liabilities is described in $\mathbf{\Omega}_{al}$ ($k \times 1$ vector of covariances). If we denote the funding ratio (defined as assets divided by liabilities) by f, we can generally express total (surplus) risk (σ^2) as

$$\sigma^2 = \begin{bmatrix} \mathbf{w}_a \\ -f \end{bmatrix}^T \begin{bmatrix} \mathbf{\Omega}_{aa} & \mathbf{\Omega}_{al} \\ \mathbf{\Omega}_{al}^T & \sigma_l^2 \end{bmatrix} \begin{bmatrix} \mathbf{w}_a \\ -f \end{bmatrix}. \tag{1.22}$$

Expanding expression (1.22), we get

$$\sigma^2 = \mathbf{w}_a^T \mathbf{\Omega}_{aa} \mathbf{w}_a - \tfrac{1}{f} 2 \mathbf{w}_a^T \mathbf{\Omega}_{al} + \tfrac{1}{f^2} \sigma_l^2 .$$

Suppose we now want to find the liability-mimicking portfolio (i.e., the asset weights that minimize asset-liability risk subject to a full investment constraint). Find the solution to

$$\min_{\mathbf{w}_a} \mathbf{w}_a^T \mathbf{\Omega}_{aa} \mathbf{w}_a - \tfrac{1}{f} 2 \mathbf{w}_a^T \mathbf{\Omega}_{al} + \tfrac{1}{f^2} \sigma_l^2 + \lambda \left(1 - \mathbf{w}_a^T \mathbf{I}\right);$$

i.e., describe how to arrive at

$$\mathbf{w}_a = (1-\theta) \frac{\mathbf{\Omega}_{aa}^{-1} \mathbf{I}}{\mathbf{I}^T \mathbf{\Omega}_{aa}^{-1} \mathbf{I}} + \tfrac{1}{f} \mathbf{\Omega}_{aa}^{-1} \mathbf{\Omega}_{al}, \tag{1.23}$$

where $\theta = \tfrac{1}{f} \mathbf{I}^T \mathbf{\Omega}_{aa}^{-1} \mathbf{\Omega}_{al}$. Interpret the result.

Endnotes

[1] See Pliska (1997) on mathematical finance and Hillier and Lieberman (1995) on mathematical programming in operations research.
[2] The reader can exactly reproduce results in this book that use simulated random numbers by using the command `set.seed(10)` as we have done here. If the user does not use `set.seed` at all or uses `set.seed(n)` with a value other than `n=10`, somewhat different results will be obtained due to the generation of different random numbers than we have used for the examples.
[3] Note that some `S` matrices will give rise to an objective of $-\infty$ and cause `solveQP` to print an error message. This is not wrong—it indicates the existence of a first-order arbitrage opportunity. See
Table 1.1.
[4] This idea is adopted from Dert (1995).
[5] See Markowitz (2000) for a reissue of his work with some interesting chapters on computer implementation added.
[6] Note also that we sometimes take risk to mean standard deviation or volatility.
[7] See Huang and Litzenberger (1998).
[8] When estimating covariance matrices, we need to balance sampling error (we do not have enough data, particularly in the presence of many assets) and specification error (imposing a factor structure on the covariance matrix will reduce sampling error at the expense of increased specification error).
[9] Note that we will use μ to characterize expectation on either the risk premium or the total return. Which interpretation is meant will be clear from the context. We will use $\hat{\mu}$ for sample estimates (either from history or from a drawn set of scenarios).
[10] Normally a capitalization-weighted "market" portfolio would play the "anchor" role here.
[11] See Satchell and Scowcroft (2000) for an excellent review on Bayesian portfolio choice.
[12] See Scherer (2004) for a more detailed treatment of these topics and the relevant literature.
[13] See Stambaugh (1997).
[14] For an excellent review of mean-variance-based single-period ALM, see Leibowitz et al. (1995). While solution methods for one-period methods are well-developed and widely available, the asset-liability problem is often inherently path-dependent, a feature that is not dealt with in these models.
[15] Lee (2000) provides an excellent overview on all relevant aspects of tactical asset allocation.
[16] Wang (1999) describes how many problems within multiple benchmark optimization can be dealt with using a simple mean-variance optimizer.
[17] This and the next section draw heavily on Grinold and Easton (1998). We also use the same data in order to encourage the reader to reproduce all results from their article.
[18] See Ingersoll (1987) or Grinold and Kahn (2000).

[19] A bucket is a collection of assets with a certain maturity range. For example, we could have a bucket of 1–3 year government bonds.

2 General Optimization with SIMPLE

2.1 Indexing Parameters and Variables

So far, we have learned how to use NUOPT functions to solve a variety of application problems. NUOPT for S-PLUS also offers a powerful and intuitive modeling language called SIMPLE that allows us to formulate complicated models to be passed on to the numerical optimization package of NUOPT. In the next few sections, we will introduce the reader to SIMPLE with the use of a fully worked-out application example. Note that this is meant to be an introduction; many applications covering different aspects of SIMPLE are treated throughout the rest of the book.

Suppose we invest our starting wealth of one monetary unit in a portfolio with n assets. After one period, we want to minimize the average shortfall over m scenarios of our final wealth, W, below some minimum wealth requirement, $W_{\min}$:

$$\frac{1}{m}\sum_{s=1}^{m}\max\left(W_{\min}-W_s,0\right).$$

Minimization of this equation takes place subject to the set of constraints

$$\frac{1}{m}\sum_{s=1}^{m}W_s \geq W_{\text{target}}, \tag{2.1}$$

$$\sum_{i=1}^{n}w_i = 1, \tag{2.2}$$

$$w_i \geq 0, \tag{2.3}$$

where $W_s = \sum_{i=1}^{n} w_i (1 + R_{is})$ for all scenarios $s = 1, \cdots, m$. Equation (2.1) states that average wealth should be above a specified target wealth, W_{target}, while (2.2) and (2.3) represent the usual full investment and non-negativity constraints. In order to define an optimization problem, we usually start with parameters and variables. Parameters are those objects that are treated as constants by NUOPT. Parameters could be the coefficients in an equation, the elements of a variance-covariance matrix or the returns in a scenario matrix. In the example above, the parameters that require definition are

- the elements of the $m \times n$ scenario matrix of asset returns **S**,
- the target wealth W_{target}, and
- the minimum wealth $W_{\min}$.

Apart from parameters, the problem in (2.1)–(2.3) also includes yet unknown quantities called variables. These are the n asset weights. Note that we have not yet addressed W_s. It will be treated in the next section, as we could set it up as a variable or as an expression.

Notice that all the parameters and variables in our optimization problem are indexed (i.e., they have subscripts). This is necessary to know exactly how variables and parameters interact. In SIMPLE, we need to do the same. We can define variables and parameters using the direct or indirect methods. Suppose we are given a scenario matrix of asset returns **S** for six scenarios and four assets:

```
m <- 6
S <- 1 + matrix(rmvnorm(m,
  mean = c(0.02, 0.06, 0.08, 0.12),
  cov = diag(c(0.02, 0.05, 0.1, 0.2))),ncol = 4)
S
numeric matrix: 6 rows, 4 columns.

          [,1]      [,2]      [,3]     [,4]
[1,] 1.1517432 1.1100860 0.8626316 1.071162
[2,] 0.8787766 0.9618646 0.6678107 1.009150
[3,] 0.9797007 1.0657226 1.1382760 1.066186
[4,] 1.1029785 1.0891477 0.7423621 1.047908
[5,] 1.1350944 1.2769152 1.3875332 1.278555
[6,] 1.1688834 1.3251761 1.0179170 1.175775
```

The number of columns equals the number of assets. Let us define the set of subscripts needed for our **variables** using the **direct method**. It is called direct because it forces us to explicitly define the index set.

```
n <- Set(1:ncol(S))       # set n contains integers
                          # from 1 to 4
i <- Element(set = n)  # i defined as elements of n
w <- Variable(index = i)  # w carries subscripts i
w
 1 2 3 4
 0 0 0 0
attr(, "indexes"):
[1] "i"
```

However, we can also define variables using the **indirect method**. The advantage of the indirect method is that variables are automatically assigned starting values.

```
w.start <- as.array(c(rep(1/ncol(S), ncol(S)))
# note: NuOPT expects arrays
n <- Set()
i <- Element(set = n)
w <- Variable(w.start, index = i)
w
    1    2    3    4
 0.25 0.25 0.25 0.25
attr(, "indexes"):
[1] "i"
```

It is easy to see that the number of elements specified for `w.start` defines the number of variables, their subscripts, and their starting values.

After defining our set of variables, we need to assign the correct subscripts to our **parameters,** too. Let's forget for a moment that we have already defined all variables. We need to define two sets, the first for subscripts on assets and the second for subscripts on scenarios.

```
n <- Set(1:ncol(S))
m <- Set(1:nrow(S))
# elements in set of asset subscripts
i <- Element(set=n)
# elements in set of scenario subscripts
s <- Element(set=m)
> n
{ 1 2 3 4 }
> m
{ 1 2 3 4 5 6 }
```

The SIMPLE command `dprod()` offers an intuitive way to define parameters that carry two indices (such as our scenario matrix).[1]

```
S1 <- Parameter(index = dprod(s, i))
S1
  1 2 3 4
1 0 0 0 0
2 0 0 0 0
3 0 0 0 0
4 0 0 0 0
5 0 0 0 0
6 0 0 0 0
attr(, "indexes"):
[1] "s" "i"
```

Note that we have not yet defined the parameters. If instead we type

```
S2 <- Parameter(S, index = dprod(s, i))
S2

          1         2         3        4
1 1.1517432 1.1100860 0.8626316 1.071162
2 0.8787766 0.9618646 0.6678107 1.009150
3 0.9797007 1.0657226 1.1382760 1.066186
4 1.1029785 1.0891477 0.7423621 1.047908
5 1.1350944 1.2769152 1.3875332 1.278555
6 1.1688834 1.3251761 1.0179170 1.175775

attr(, "indexes"):
[1] "s" "i"
```

we automatically assign the correct index to each row and column of our scenario matrix. After this has been done (after the index has been set), we could also define variables (asset weights):

```
w <- Variable(index=i)
```

Parameters that also require definition are minimum wealth and target wealth. Note that neither parameter needs to be indexed. Therefore we can use the simplest definition of parameters SIMPLE can offer.

```
Wealth.target <- Parameter(1.07)
Wealth.min <- Parameter(1.00)
```

For later convenience, we will also define the average return an asset offers across all scenarios as a parameter:

```
# calculate mean for dimension 2 (columns) of S
mu.bar <- apply(S, 2, mean)
# define parameter
mu.bar <- Parameter(as.array(mu.bar), index=i)
```

Now that we have set up variables and parameters, we need to build the model in (2.1)–(2.3) using arithmetic operations and expressions.

2.1.1 Arithmetic Operations and Expressions

Models typically involve the calculation of expressions. In our example, an expression that needs to be calculated is $W_s = \sum_{i=1}^{n} w_i (1 + R_{si})$. The syntax is virtually the same as for parameters and variables:

```
# wealth is defined(indexed) in each scenario
W <- Expression(index=s)
```

However, the calculation of an expression is yet new. Again SIMPLE is very intuitive, as it allows us to directly translate summations or products into code. Table 2.1 exhibits some examples from a portfolio context. In our example above, we still need to specify the defined expression for final portfolio wealth.

```
W[s]~Sum(w[i]*R[s,i],i)
```

Table 2.1 SIMPLE Definitions

Name	Mathematical Formula	SIMPLE formula
Weight on asset subgroup	$\sum_{i=3}^{n} w_i$	`Sum(w[i],i,i>2)`
Wealth for each scenario	$\sum_{i=1}^{n} w_i (1 + R_{si})$	`Sum(w[i]*R[s,i],i)`
Grand mean of all scenarios	$\frac{1}{nm}\sum_{s=1}^{m}\sum_{i=1}^{n}(1 + R_{si})$	`Sum(Sum(R[s,i],i),s)/(n*m)`

In terms of our objective (2.1), it is necessary to distinguish between cases where final wealth is higher or lower than minimum wealth. Combining `Expression` and yet another function `ife()`, we can achieve this. The use of `ife(condition, Expression1, Expression2)` allows calculating an expression contingent on whether a condition is met:

```
# define positive and negative wealth
W.pos <- Expression(index=s)
W.neg <- Expression(index=s)
# calculate expression
W.pos[s] ~ ife(W[s]>=Wealth.min, W[s], 0)
W.neg[s] ~ ife(W[s]<=Wealth.min, W[s], 0)
```

We can later use `W.neg[s]`, which effectively is $\max(W_s - W_{\min}, 0)$, within our objective function to solve our simple scenario-optimization model.

2.1.2 Constraints and Objectives

Constraints can be used in many forms.[2] The obvious use is to constrain asset weights in very much the same way the reader might write them down on paper.

```
W[i]>=0     # non-negativity constraint on all
            # assets
W[2]>=0.2   # minimum weight constraint on asset 2
Sum(w[i], i>2, i<6)>=0.5  # assets 3 to 5 must sum
                          # up to at least 50%
Sum(w[i],i)==1     # full investment constraint
```

Alternatively, we can also use constraints to implicitly define variables. Coming back to our original example, it might be useful to define a set of variables that equal $\max(W_s - W_{\min}, 0)$ in all scenarios.

```
up <- Variable(index=s)    # index new variables
dn <- Variable(index=s)
# implicitly define deviations with the use of
# constraints
up[s]-dn[s] == Sum(w[i]*S2[s,i],s)-Wealth.min
up[s] >= 0
dn[s] >= 0
```

We stop our discussion of constraints here, as later chapters will discuss more complicated modeling situations, such as the use of constraints in mixed integer problems, in more detail.

Nothing can be optimal without an **objective**. SIMPLE allows objectives to be specified in a very natural way. It only requires declaring the name of the objective and whether it is a maximization or a minimization problem. In this model, the objective is a risk measure (average shortfall) we want to minimize.

```
risk <- Objective(type="minimize")
```

Specification of this risk measure evolves in very much the same way as before.

```
risk ~ Sum(dn[s],s)/nrow(S)
```

We have now learned all the details necessary to solve our simple scenario-optimization model outlined at the beginning of this section.

2.1.3 Build, Control, and Solve Models

Remember that SIMPLE allows us to set up an optimization model that is then passed to the numerical algorithms available in NUOPT. How is this done? The first step is to build a complete model from the ingredients above by writing a function that contains the relevant commands, as in Code 2.1.

```
scenario.model <- function(S, Wealth.min,
  Wealth.target)
{
  # define subscripts
  n <- Set(1:ncol(S))
  m <- Set(1:nrow(S))
  i <- Element(set=n)
  s <- Element(set=m)

  # parameters
  S2 <- Parameter(S, index=dprod(s,i))
  mu.bar <- apply(S, 2, mean)
  mu.bar <- Parameter(as.array(mu.bar), index=i)
  Wealth.target <- Parameter(Wealth.target)
  Wealth.min <- Parameter(Wealth.min)

  # variables
  w <- Variable(index=i)
  up <- Variable(index=s)
  dn <- Variable(index=s)
  up[s]-dn[s] == Sum(w[i]*R[s,i],i)-Wealth.min
  up[s] >= 0
  dn[s] >= 0
```

```
  # objective
  risk <- Objective(type="minimize")
  risk ~ Sum(dn[s],s)/nrow(S)

  # constraints
  Sum(mu.bar[i]*w[i],i) == Wealth.target
  Sum(w[i],i) == 1
  w[i] >= 0
}
```

Code 2.1 Scenario Modeling in SIMPLE

Next we need to expand the model into a system of equations that can be solved by NUOPT using the `System` command:

```
> scenario.system <- System(scenario.model, S,
  Wealth.min, Wealth.target)
Evaluating
     scenario.model(S,Wealth.min,Wealth.target)
  ... ok!
Expanding (1/7)(2/7)(3/7)(4/7)(5/7)(6/7)(7/7)ok!
```

Note that the fact that our expansion went well is not a guarantee that the NUOPT solver will be able to come up with a solution or that the solution found is indeed a global maximum or minimum. In order to see what has been passed on to NUOPT, we can view the complete system using the `show` command:

```
> show(scenario.system)
1-1 : -up[1]+dn[1]+1.15174*w[1]+1.11009*w[2]
  +0.862632*w[3]+1.07116*w[4]-1 == 0
1-2 : 0.878777*w[1]+0.961865*w[2]+0.667811*w[3]
  +1.00915*w[4]-up[2]+dn[2]-1 == 0
1-3 : 0.979701*w[1]+1.06572*w[2]+1.13828*w[3]
  +1.06619*w[4]-up[3]+dn[3]-1 == 0
1-4 : 1.10298*w[1]+1.08915*w[2]+0.742362*w[3]
  +1.04791*w[4]-up[4]+dn[4]-1 == 0
1-5 : 1.13509*w[1]+1.27692*w[2]+1.38753*w[3]
  +1.27856*w[4]-up[5]+dn[5]-1 == 0
1-6 : 1.16888*w[1]+1.32518*w[2]+1.01792*w[3]
  +1.17577*w[4]-up[6]+dn[6]-1 == 0
2-1 : up[1] >= 0
2-2 : up[2] >= 0
2-3 : up[3] >= 0
2-4 : up[4] >= 0
```

```
2-5 : up[5] >= 0
2-6 : up[6] >= 0
3-1 : dn[1] >= 0
3-2 : dn[2] >= 0
3-3 : dn[3] >= 0
3-4 : dn[4] >= 0
3-5 : dn[5] >= 0
3-6 : dn[6] >= 0
4-1 : 1.06953*w[1]+1.13815*w[2]+0.969422*w[3]
  +1.10812*w[4] == 1.07
5-1 : w[1]+w[2]+w[3]+w[4] == 1
6-1 : w[1] >= 0
6-2 : w[2] >= 0
6-3 : w[3] >= 0
6-4 : w[4] >= 0
risk<objective>:0.166667*dn[1]+0.166667*dn[2]
  +0.166667*dn[3]+0.1666670*dn[4]+ 0.166667*dn[5]
  +0.166667*dn[6] (minimize)
```

While this is only informative for a small set of scenarios, it can be a useful way of checking the model for different optimization problems. Finally we solve the optimization problem using the `solve` command:

```
solution <- solve(scenario.system, trace = T)
weight <- matrix(round(solution$variable$w$current,
  digit = 5) * 100, ncol = 1)
weight
       [,1]
[1,] 25.750
[2,]  0.000
[3,] 20.321
[4,] 53.929
```

As we already know from our previous discussion, use of a small number of scenarios is completely inappropriate, as it might offer arbitrage opportunities and misrepresent the underlying distributions (sampling error). For what follows, we draw 1000 scenarios for four assets. In order to get a complete picture of investment opportunities, we want to trace out an efficient frontier (i.e., the geometric location of the minimal average shortfall, relative to a specified minimum wealth, for each given wealth target). This requires a function (given in Code 2.2) that returns portfolio weights and associated risks for a single optimization run (i.e., for a given wealth target).

```
portfolio <- function(S, Wealth.min, Wealth.target)
{
  scenario.system <- System(scenario.model, S,
     Wealth.min, Wealth.target)
  solution <- solve(scenario.system, trace=T)
  weight <-
     matrix(round(solution$variable$w$current,
     digit=5)*100, ncol=1)
  risk <- solution$objective
  return(weight,risk)
}
```

Code 2.2 Portfolio Weights Function

Finally, we need a function (given in Code 2.3) that returns an efficient frontier, ranging from a minimum to maximum wealth target.

```
scenario.frontier <- function(S, Wealth.min,
  Wealth.target, n.pf)
{
  # contains risk return results
  Risk <- matrix(0, ncol=1, nrow=n.pf)
  # n.pf denotes number of frontier portfolios
  Return <- matrix(0, ncol=1, nrow=n.pf)

  # define wealth targets
  mu.max <- max(apply(S, 2, mean))
  mu.min <- min(apply(S, 2, mean))
  mu.range<-seq(mu.min, mu.max,
     (mu.max-mu.min)/(n.pf-1))

  # contains asset weights
  weight <- matrix(0, ncol=1, nrow=ncol(S))
  for(i in 1:n.pf){
     x <- portfolio(S, Wealth.min,
        Wealth.target=mu.range[i])
     Risk[i,1] <- x$risk
     Return[i,1] <- mu.range[i]
     weight <- cbind(weight,x$weight)
  }

  # plots frontier and frontier portfolios
  graphsheet()
  par(mfrow=c(1,2))
  plot(Risk, Return, type="b")
  title("Scenario Frontier")
```

```
  barplot(weight[,-1])
  title("Frontier Portfolios")
  list("optimal.weights" = weight, "Risk"=Risk,
     "Return"=Return)
}
```

Code 2.3 Scenario Model

We can now run a frontier analysis plotting target wealth versus average wealth shortfall:

```
scenario.frontier(S, Wealth.min, Wealth.target,
  n.pf=5)
```

As inputs, we require a scenario matrix, minimum and target wealth, and the number of frontier portfolios. The results are shown in Figure 2.1. Each dot represents a different portfolio with portfolio weights shown in the bar chart to the right. Optimal solutions appear to be diversified (intermediate return portfolios contain all four assets). Minimum and maximum wealth target portfolios are fully invested in the minimum and maximum return assets, respectively. This ends our brief discussion of SIMPLE; many more examples will be given in the following chapters.

2.2 Function Optimization

So far, we have always used SIMPLE within a portfolio context. However, we can also apply it to a straightforward optimization problem. This will show the reader the flexibility and generality of SIMPLE. We start with maximizing a nonlinear function of two variables,

$$-10x_1^2 + 4x_1 + \frac{1}{3}x_2^2 + 20x_2, \tag{2.4}$$

under nonlinear constraints $x_1^2 + x_2^2 \leq 16$, $x_1 x_2 \geq 3$, $x_1 \geq 0$, $x_2 \geq 0$.[3] The necessary code is given in Code 2.4. Note that as the problem is small we neither defined the parameters nor indexed the variables.

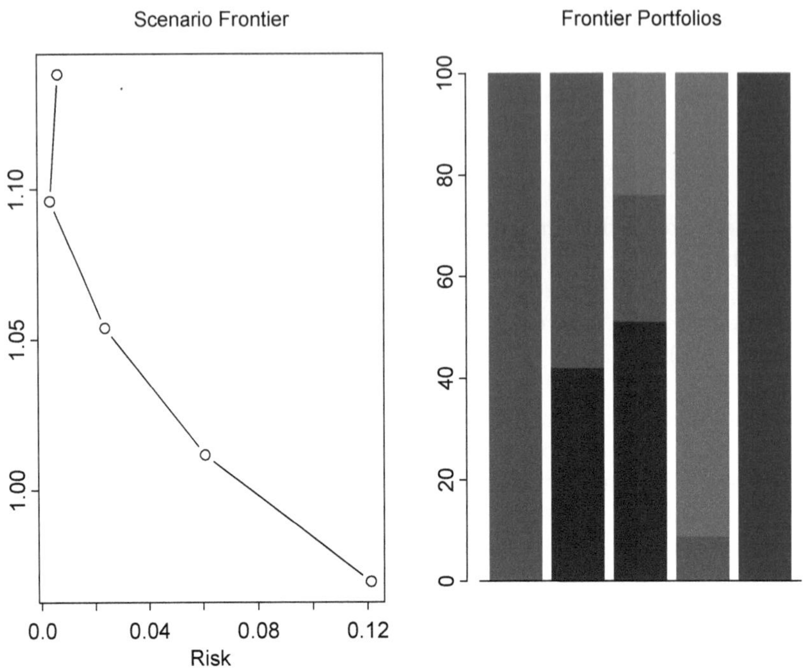

Figure 2.1 Scenario Frontier and Underlying Frontier Portfolio

```
> model.1 <- function(){
  x1 <- Variable()                    # define variables
                                      # with no index
  x2 <- Variable()
  f <- Objective(type="maximize")    # define
     objective
  f ~ -10*x1^2 + 4*x1 + (1/3)*x2^2 + 20*x2
  x1^2+x2^2 <= 16                     # set constraint
  x1*x2 >= 3
  x1 >= 0
  x2 >= 0
}

> # transform model into system of equations
> system.model.1 <- System(model.1)
> # check model
> show(system.model.1)
1-1 : x1*x1+x2*x2 <= 16
2-1 : x1*x2 >= 3
3-1 : x1 >= 0
4-1 : x2 >= 0
```

```
f<objective>: -10*x1*x1+0.333333*x2*x2+4*x1+
  20*x2 (maximize)

> # solve model for x1 and x2
> x <- solve(system.model.1)

> x$variables$x1$current
[1] 0.7640694
> x$variables$x2$current
[1] 3.926346
```

Code 2.4 Maximizing a Nonlinear Function of Two Variables

However, it should be noted that the existence of a solution does not guarantee that we have found a global maximum (minimum). Suppose we want to maximize instead

$$f(x,y) = \cos(x+y)\cos(3x-y) + \cos(x-y)\sin(x+3y) + 5\exp\left(-\frac{x^2+y^2}{8}\right) \quad (2.5)$$

with no constraints imposed. You should be able to write the necessary short piece of code yourself by now. In contrast with the previous example, we have used starting values to initialize the optimization:

```
> model.2 <- function(x.start, y.start){
  x <- Variable(x.start)
  y <- Variable(y.start)
  f <- Objective(type="maximize")
  f ~ cos(x+y)*cos(3*x-y) + cos(x-y)*sin(x+3*y) +
     5*exp(-(x^2+y^2)/8)
}

> system.model.2 <- System(model.2, x.start=20,
   y.start=20)
> # trace=F limits output to what is required
> solve(system.model.2, trace=F)$objective
  current
 1.799038
```

In order to see whether this is a global maximum, we can plot (2.5) within the range of –5 to +5.

```
x <- seq(-5, 5,length=50) #define range for x and y
y <- seq(-5, 5,length=50)
```

```
# define function
f <- function(x,y){
  cos(x+y)*cos(3*x-y)+cos(x-y)*sin(x+3*y)+
  5*exp(-(x^2+y^2)/8)
}

# plot function
z <- outer(x,y,f)
persp(x,y,z)
contour(x,y,z, nlevels=10, xlab="x", ylab="y")
```

The objective value of 1.79 is not a global maximum, as we can see from Figure 2.2 and Figure 2.3. For a different set of starting values, we get the optimal solution shown in Code 2.5.

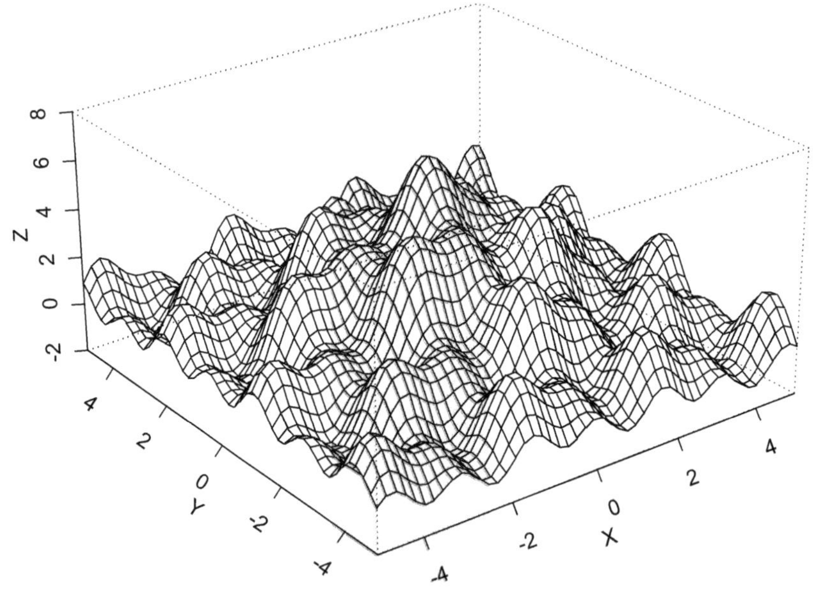

Figure 2.2 2D Plot for $\cos(x+y)\cos(3x-y)+\cos(x-y)\sin(x+3y)$
$+5\exp\left(-\frac{x^2+y^2}{8}\right)$

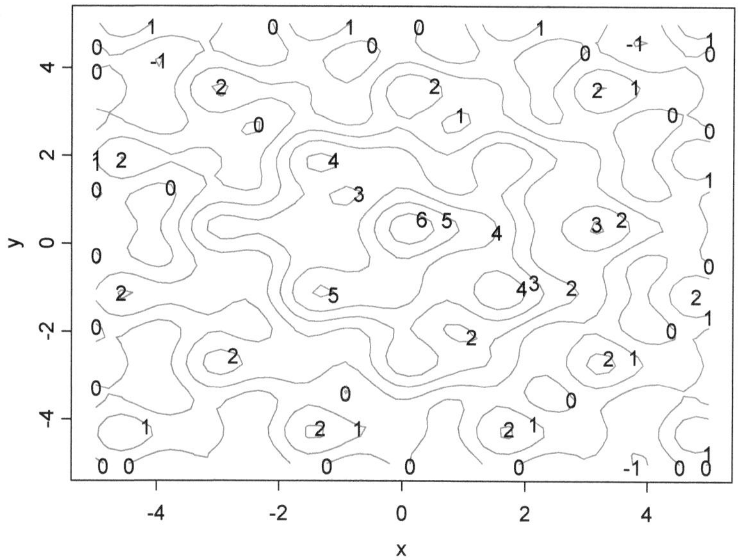

Figure 2.3 Contour Plot for $\cos(x+y)\cos(3x-y)+\cos(x-y)\sin(x+3y)$ $+5\exp\left(-\frac{x^2+y^2}{8}\right)$

```
> system.model.2 <- System(model.2,x.start=3.5,
  y.start=4)
> solve(system.model.2, trace=F)$objective
  current
 6.703374
```

Code 2.5 Function Optimization

Objective functions like that of (2.5) are called nonconvex. Their optimization will always require some heuristics. Without prior knowledge about "good" starting values, it is recommended to randomly search the space of admissible variable combinations and record the corresponding objective values in an attempt to find the global maximum. However, be aware that functions such as

$$1-\exp\left(-\frac{1}{x^2+y^2}\right)$$

contain many solutions with the same maximum; see Figure 2.4 and Figure 2.5.

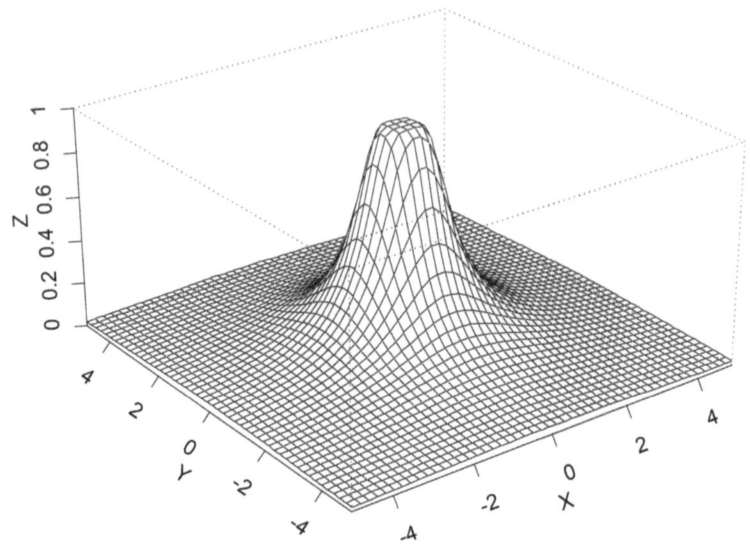

Figure 2.4 2D Plot for $1 - \exp\left(-\frac{1}{x^2 + y^2}\right)$

2.3 Maximum Likelihood Optimization

Models that can be understood intuitively and fit the data well have the greatest chance of receiving attention from practitioners. As market participants tend to think in regimes (e.g., periods of different volatility) it is natural to model a distribution as a combination of two (or more) normal distributions, with each distribution representing a different regime. Shifts from one regime to another take place randomly with a given probability. These mixtures of normal distributions have found great interest in finance, as they can model skewness as well as kurtosis and fit the data much better than a non-normal alternative. In this section, we will show how to simulate, estimate, test, and apply a mixture-of-normals model of asset returns (see Figure 2.6 using S-PLUS and NUOPT. In order to allow the reader to reproduce the calculations of this section, we will generate a data set by drawing a total of 15,000 samples from two normal distributions, each with a mean return of 10%: one with a volatility of 40%, representing a high-volatility regime, and one with a volatility of 15%, representing a low-volatility regime. The probability of a draw from the high-volatility regime is $\frac{1}{3}$. This results in the typical fat-tailed distribution that is characteristic of so many financial time series.

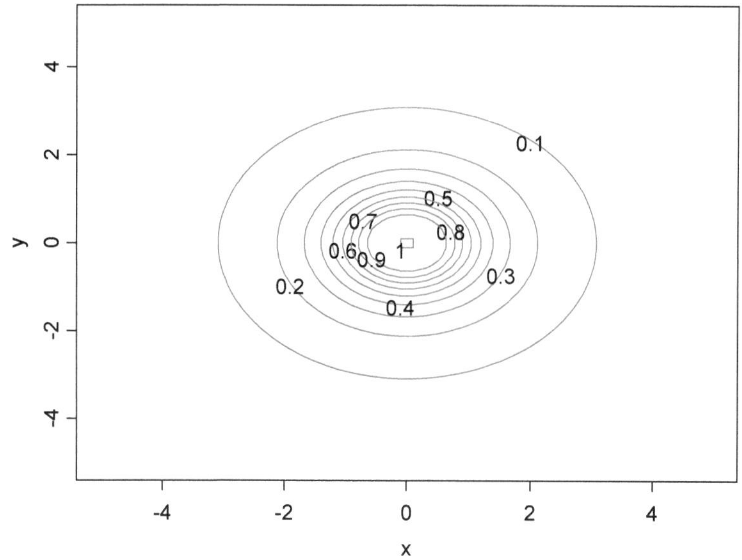

Figure 2.5 Contour Plot for $1 - \exp\left(-\frac{1}{x^2+y^2}\right)$

```
data <- c(rnorm(5000, 0.1, 0.4),
  rnorm(10000, 0.1, 0.15))
hist(data)
```

The likelihood function for a mixture of the two normals with densities

$$f_i = \frac{1}{\sqrt{2\pi}\sigma_i}\exp\left\{-\frac{1}{2}\left(\frac{R_s-\mu_i}{\sigma_i}\right)^2\right\}, \qquad i=1,2, \tag{2.6}$$

is usually written in its log form:

$$\log(L) = \sum_{s=1}^{m}\log(pf_1 + (1-p)f_2). \tag{2.7}$$

Maximum likelihood estimation then becomes the maximization of the objective (2.7) with respect to the variables p (probability of drawing from distribution #1), μ_1, σ_1 (mean and standard deviation of returns for distribution #1), and μ_2, σ_2. The return data R_s for all $s = 1, \cdots, m$ scenarios represent parameters (as they are fixed once the sample is drawn). We need to impose non-negativity constraints on standard deviations and on p:

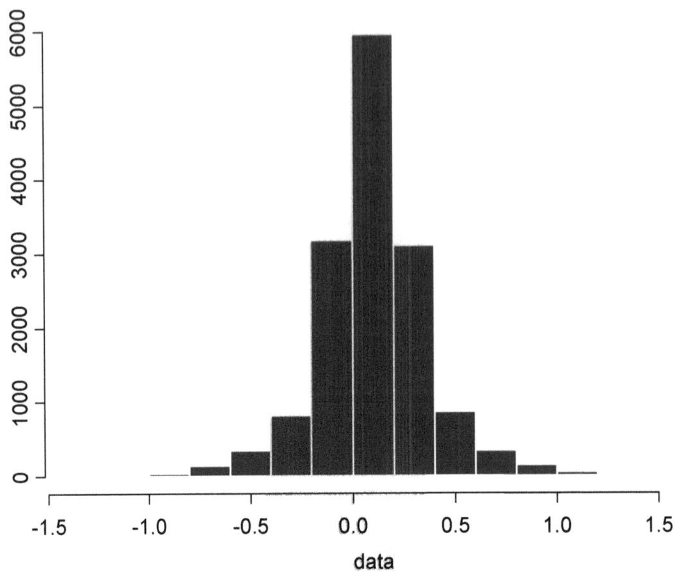

Figure 2.6 Mixture of Two Normals

```
MoN.model <- function(data){
  # parameter
  m <- Set(1:length(data))
  s <- Element(set=m)
  R <- Parameter(as.array(data), index=s)
  # variable
  I <- Set(1:5)
  i <- Element(set=I)
  p <- Variable(index=i)
  # objective
  logL <- Objective(type="maximize")
  logL ~ Sum(log(p[1]/(sqrt(2*pi)*p[3])*
     exp(-0.5*((R[s]-p[2])/p[3])^2)
     +(1-p[1])/(sqrt(2*pi)*p[5])*
     exp(-0.5*((R[s]-p[4])/p[5])^2)),s)

  # constraints
  p[1]>=0
  p[3]>=0
  p[5]>=0
}
# solve system
MoN.system <- System(MoN.model, data=data)
```

```
solution <- solve(MoN.system, trace=F)
# get parameters and likelihood function
p <- solution$variables$p$current
> p

         1         2         3          4         5
 0.3461775 0.1066749 0.3978774 0.09752759 0.1471847
attr(, "indexes"):
[1] "i"
```

We have been able to almost exactly recover our assumed parameters when simulating the data set.[4] This is due to the high number of observations and the consequent low sampling error. However if you need assurance that our estimated distribution is significantly different from its (single) normal alternative, we can employ a likelihood ratio test (with obvious significance), as illustrated in Code 2.6.

```
LogL.uc <- sum(log(p[1]*dnorm(data, p[2], p[3])+
  (1-p[1])*dnorm(data, p[4], p[5])))
LogL.c <- sum(log(dnorm(data, mean(data),
  stdev(data))))
LR.test <- -2*(LogL.c-LogL.uc)
LR.test
[1] 2242.879
```

Code 2.6 Maximum Likelihood Optimization and Regime Probabilities

So far, we have estimated the unconditional probability p (33%) that a given return is drawn from a hectic (i.e., high-volatility period). However, using Bayes' rule, we can also calculate the conditional probability that a given observation has been drawn from the hectic distribution

$$prob(hectic|R_s, p, \mu_1, \sigma_1, \mu_2, \sigma_2) = \frac{pf(R_s|\mu_1, \sigma_1)}{pf(R_s|\mu_1, \sigma_1) + (1-p)f(R_s|\mu_2, \sigma_2)} \tag{2.8}$$

where $f(R_s \mid \mu_i, \sigma_i)$ denotes the usual marginal density. We can calculate (2.8) for every data point, as shown in Figure 2.7. The probability that a given data point has been drawn from the hectic regime is highest when the observation is extreme (positive as well as negative). For small return realizations, the reverse is true. This calculation is useful, as we can use it to estimate the correlation between two assets conditional on the first asset experiencing a hectic regime.[5]

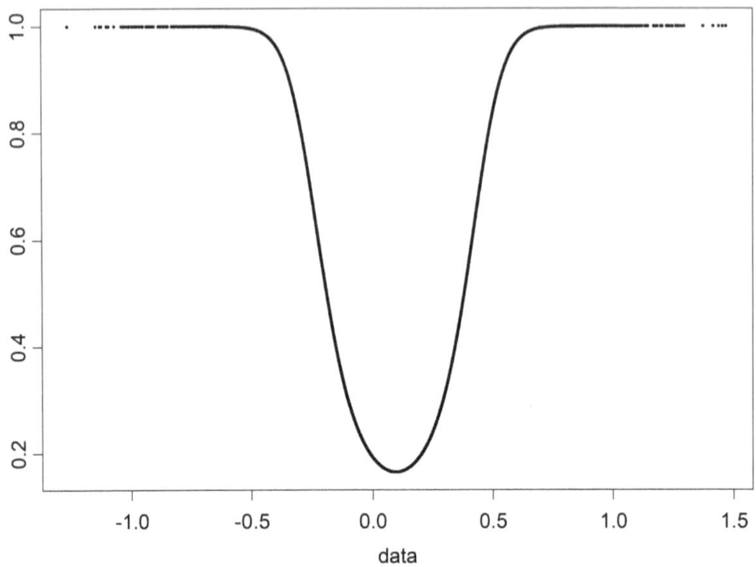

Figure 2.7 Conditional Probability

2.4 Utility Optimization

2.4.1 Semi-quadratic Utility Maximization

A great deal of research has been put into determining whether some types of objectives in portfolio optimization are compatible with the behavior of utility-maximizing agents. The utility function (2.9) is particularly interesting, as it has been shown that the decision-making under the mean-semi-variance objective (see Chapter 3) is fully compatible with the maximization of expected utility from

$$U_s = \begin{cases} R_{ps} & \text{for } R_{ps} \geq \tau \\ R_{ps} - \kappa(\tau - R_{ps})^2 & \text{for } R_{ps} < \tau, \end{cases} \tag{2.9}$$

where $R_{ps} = \sum_{i=1}^{n} w_i (1 + R_{is})$ equals the portfolio return and κ can be interpreted as a risk aversion parameter. It is also called **semi-quadratic**, as the quadratic part is only defined for $R_{ps} \leq \tau$. Figure 2.8 shows a semi-quadratic utility function for $\kappa = 4$. The expected utility maximization of (2.9) allows us to introduce (in Code 2.7) the use of `ife()` and `Expression()` in defining an objective function that is piecewise-defined.

```
# generate scenarios for asset.1 to asset.4
scenarios <- matrix(rmvnorm(1000,
  mean=c(0.02, 0.04, 0.05, 0.08),
  cov=diag(c(0.02, 0.04, 0.1, 0.2))), ncol=4)

FUT.model <- function(scenarios, k, g, h)
{
  # number of observations and assets
  n.obs <- nrow(scenarios)
  n.assets <- ncol(scenarios)

  # NuOPT set up
  asset <- Set()
  period <- Set(1:n.obs)
  j <- Element(set=asset)
  t <- Element(set=period)

  # define parameters
  S <- Parameter(scenarios, index=dprod(t,j))
  K <- Parameter(k)
  G <- Parameter(g)
  H <- Parameter(h)

  # define "x" variable (weights)
  x <- Variable(index=j)
  R <- Expression(index=t)
  R[t] ~ Sum(x[j]*S[t,j],j)
  u <- Expression(index=t)
  u[t] ~ ife(R[t]>=G, R[t], R[t]-K*(R[t]-G)^H)

  # define utility measure
  utility <- Objective(type="maximize")
  utility ~ Sum(u[t],t)/n.obs

  # constraints (add up)
  Sum(x[j],j) == 1
  # constraints (non-negativity)
  x[j] >= 0
}

# run model
```

```
FUT.system <- System(FUT.model, scenarios, k=1,
     g=0, h=2)
solution <- solve(FUT.system, trace=T)
weight <- matrix(round(solution$variable$x$current,
  digit=4)*100, ncol=1)
```

Code 2.7 Semi-quadratic Utility Optimization

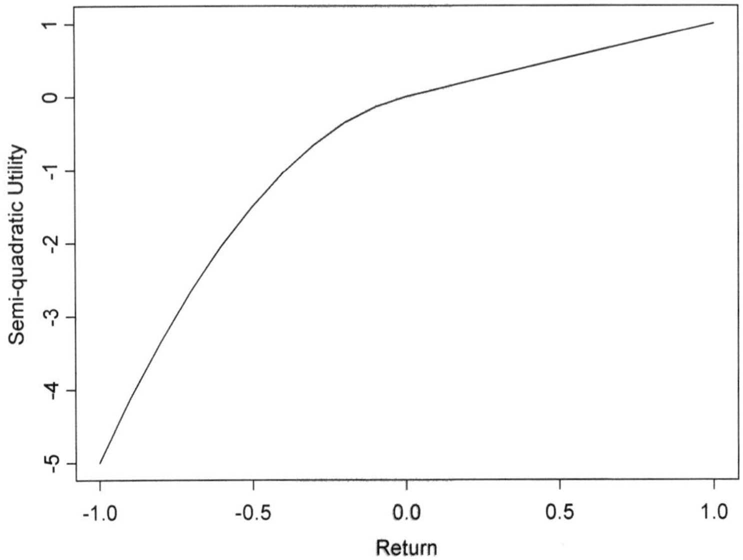

Figure 2.8 Semi-quadratic Utility

We can repeat this optimization for risk aversion parameters ranging from 1 to 10 (this is left to the reader as an exercise) to arrive at the matrix of portfolio weights for different risk aversions.

```
> weight
       [,1]  [,2]  [,3]  [,4]  [,5]  [,6]  [,7]
[1,]  0.01  0.09 14.89 23.18 28.07 31.38 33.74
[2,] 44.76 52.60 46.16 42.50 40.35 38.87 37.80
[3,] 18.30 21.25 19.00 17.51 16.63 16.05 15.63
[4,] 36.92 26.06 19.94 16.80 14.95 13.70 12.82

      [,8]  [,9] [,10]
[1,] 35.48 36.83 37.95
[2,] 37.00 36.38 35.85
[3,] 15.34 15.12 14.93
[4,] 12.18 11.68 11.27
```

```
> barplot(weight,
    legend=paste("asset.",sep="",1:4),
    names=paste("", sep="", 1:10),
    xlab="risk aversion", ylab="weight")
```

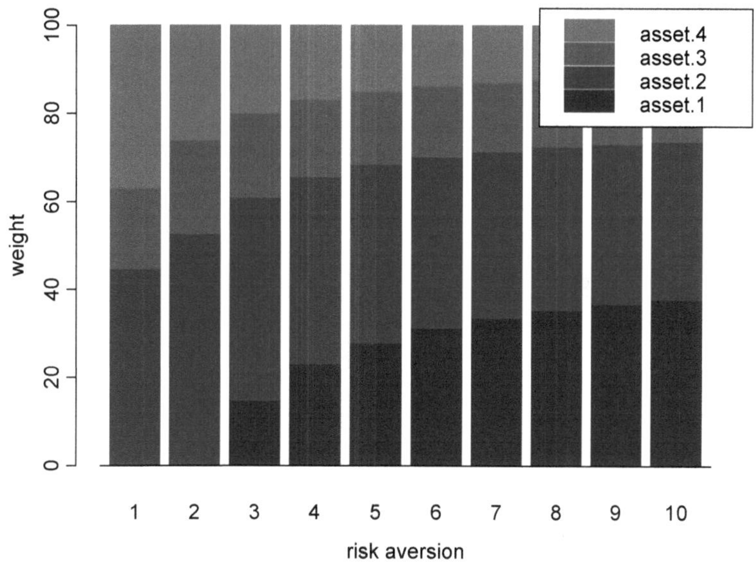

Figure 2.9 Optimal Portfolios for Semi-quadratic Utility

Figure 2.9 plots the resulting weights in a bar chart. As risk aversion rises, our portfolio optimization will reduce the weightings in the riskier assets. The reader might want to show that high risk aversions will recover the minimum variance portfolio. Why is this the case? Haven't we explicitly stated non-mean-variance objectives? The answer is that our return generation has been drawing returns from a normal distribution; hence the set of all mean-variance-efficient portfolios will also be the optimal set of portfolios for non-mean-variance preferences.

2.4.2 Utility Optimization using Piecewise Linear Approximation

In many optimization applications, it is useful to linearize a nonlinear objective function to speed up calculations or to involve linear solver technology, which is widely available and well-developed. In this section, we will show how to do this within NUOPT. Suppose we assume a standard CRRA (constant relative risk aversion) utility function that the expresses utility in scenario s as

$$U_s = \begin{cases} \frac{1}{1-\gamma}\left(\sum_{i=1}^{n} w_i\,(1+R_{is})\right)^{1-\gamma} & \text{for } \gamma \geq 0 \\ \log\left(\sum_{i=1}^{n} w_i\,(1+R_{is})\right) & \text{for } \gamma = 1, \end{cases} \tag{2.10}$$

where γ reflects risk aversion. The higher γ, the higher the risk aversion. Values between 3 and 5 are assumed to be realistic for decision makers. Figure 2.10 shows (2.10) for $\gamma = 5$.

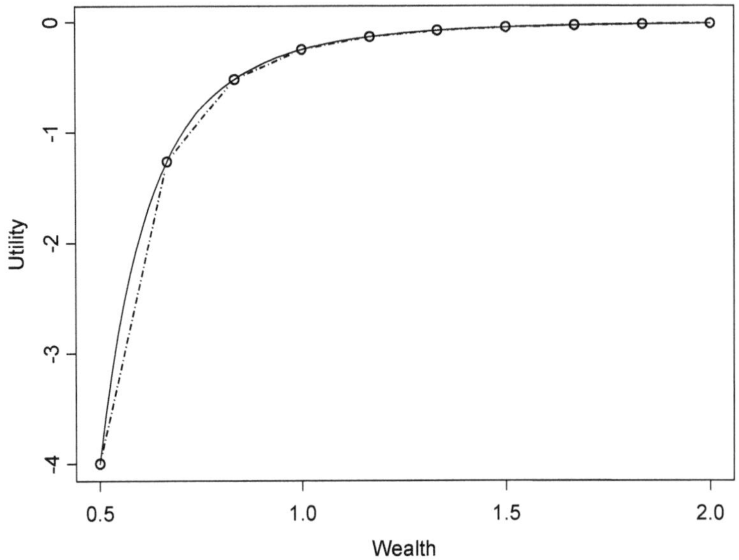

Figure 2.10 Piecewise Linear Approximation of CRRA Utility Function

```
risk.aversion <- 5
wealth <- seq(0.5, 2, length=100)
Utility <- 1/(1-risk.aversion)*
  wealth^(1-risk.aversion)
wealth.grid <- seq(0.5, 2, length=10)
utility.grid <- 1/(1-risk.aversion)*
  wealth.grid^(1-risk.aversion)
plot(wealth, Utility, type="l")
lines(wealth.grid, utility.grid, type="b", col=5)
```

Notice that higher risk aversion is marked by greater curvature in the utility function. Also, for large levels of wealth, the utility function becomes extremely flat. Hence, for large wealth levels, a linear approximation works relatively well, while for small levels of wealth it becomes more critical (i.e., more steps are needed).

The utility function above is concave. We can therefore interpret the decreasing slopes of the approximating line segments as decreasing marginal utility. The notion of concavity is important, as it will allow us to use separate variables for each line segment. Maximizing expected utility amounts to finding

$$\max_{\mathbf{w}} \frac{1}{m} \sum_{s=1}^{m} U_s \tag{2.11}$$

subject to the usual non-negativity and group constraints. Before we present the final program, we will describe the basic idea behind our approach. Equation (2.11) is approximated with $j = 1, \cdots, k$ line segments. The slope of each line segment (marginal utility) is denoted by a_j. We also need to define wealth for each line segment j and each scenario $s = 1, \cdots, m,$ leaving us with $m \cdot k$ wealth variables. Utility in state s is hence given by

$$U_s = \sum_{j=1}^{k} a_j W_{s,j}. \tag{2.12}$$

Note that marginal utilities and total utility (given as the product of marginal utility and wealth) are defined for their respective line segments

$$\begin{array}{lll} a_1 W_{1,s} & \text{for} & 0 \le W_{1,s} \le \overline{W}_1 \\ a_2 W_{2,s} & \text{for} & \overline{W}_1 \le W_{2,s} \le \overline{W}_2 \\ a_3 W_{3,s} & \text{for} & \overline{W}_2 \le W_{3,s} \le \overline{W}_3 \\ \ldots & & \end{array}$$

where $\overline{W}_j$ reflects the boundary of the respective line segment. Next we need to link wealth with portfolio allocation using

$$\sum_{i=1}^{n} w_i \left(1 + R_{i,s}\right) = \sum_{j=1}^{k} W_{s,j}. \tag{2.13}$$

It might immediately come to mind that we need to impose constraints on $W_{s,j}$ that guarantee, for example, that

$$W_{2,s} = 0 \text{ for } W_{1,s} < \overline{W}_1. \tag{2.14}$$

Constraints of this format, however, would not qualify for a linear program. Fortunately we do not need to impose these constraints, as our utility function is concave. This assures that smaller wealth variables are used first, as they have the largest impact on the objective function due to having the largest slopes. Expected utility can now be expressed as

$$\frac{1}{m}\sum_{s=1}^{m}\left(\sum_{j=1}^{k} a_j W_{s,j}\right). \tag{2.15}$$

The corresponding code is given in Code 2.8.

```
utility.model <- function(S, risk.aversion,
  wealth.grid)
{
  # check data for missing values
  if(any(is.na(S))==T)
     stop("no missing data are allowed")

  # check for number of observations and assets
  n.obs <- nrow(S)
  n.assets <- ncol(S)
  n.grids <- length(wealth.grid)

  # slope
  utility.grid <- 1/(1-risk.aversion)*
     wealth.grid^(1-risk.aversion)
  slope <- rep(0,(n.grids-1))
  for(i in 1:(n.grids-1)){
     slope[i] <- (utility.grid[i+1]-
        utility.grid[i])/(wealth.grid[i+1]-
        wealth.grid[i])
  }

  # NuOPT set up
  asset <- Set()
  period <- Set()
  grid <- Set()
  j <- Element(set=asset)
  t <- Element(set=period)
  k <- Element(set=grid)

  # define parameters
  R <- Parameter(S, index=dprod(t,j))
  slope <- Parameter(as.array(slope), index=k)
  bounds <- Parameter(as.array(wealth.grid[-1]),
     index=k)
```

```
  # define variable weights
  asset.weight <- Variable(index=j)

  # define wealth dummies
  wealth.dummy <- Variable(index=dprod(t,k))

  # equalize utility and end-of-period wealth
  1 + Sum(R[t,j]*asset.weight[j],j) ==
     Sum(wealth.dummy[t,k],k)

  # define risk measure
  utility <- Objective(type="maximize")
  utility ~ Sum(Sum(wealth.dummy[t,k]*
     slope[k],k),t)/(n.obs-1)

  # constraints (add up)
  Sum(asset.weight[j],j) == 1
  wealth.dummy[t,k] <= bounds[k]
  wealth.dummy[t,k]>= 0
  asset.weight[j] >= 0
}

utility.system <- System(utility.model, S,
  risk.aversion, wealth.grid)
show(utility.system)
solution <- solve(utility.system, trace=T)
```

Code 2.8 Piecewise Linearization of Utility Function

Effectively we changed a nonlinear program with a small number of variables (portfolio weights) into a linear program with a large number of variables (portfolio weights plus wealth variables).

2.5 Multistage Stochastic Programming

2.5.1 Sample Problem: Financial Planning

Dynamic stochastic programming is a highly specialized and technical field in the application of optimization techniques to financial problems.[6] In order to introduce readers to the core concepts of stochastic programming for asset allocation problems, we will use the financial planning example introduced in Birge and Louveaux (1997) and Brandimarte (2002).

Preferences. Suppose we are equipped with initial wealth W_0 and need to meet a future liability L at the end of time $T = t_3$. In order to achieve this, we can invest in stocks and bonds. Preferences are modeled with a piecewise linear utility function (to avoid nonlinearity in the resulting mathematical program). If our final period wealth exactly meets the liabilities, we enjoy zero utility. Downside deviations are penalized with downside costs c_d, while upside deviations provide us with rewards c_u. Needless to say, the disutility from downside deviations is larger than the utility from upside deviations $(c_d > c_u)$.

Scenario Tree. In contrast with all other applications in this book, we allow for intermediate decisions. (We will later see single-period models that only allow one decision at the start of an investment period.) Apart from today, we can reallocate assets at times t_1 and t_2. At t_3, our time horizon ends and all we can do is watch the outcome of our decisions made in t_2. Typically, we describe uncertainty in multistage stochastic programming in a (nonrecombining) scenario tree as described in Figure 2.11. The root of a scenario tree reflects the current date where we look for an optimal decision. Note that although we allow for future decisions (in fact we choose today knowing that we can decide again later, contingent on what has happened), this does not mean that we will implement decisions at later stages as we travel through the scenario tree. In fact, a scenario tree is solved on a rolling basis. Each complete path from the root of the tree to the leaf (for example, the sequence of nodes $0 \to 2 \to 6 \to 13$) is called a scenario. Each scenario is a realization of a random variable. In the tree depicted in Figure 2.11, all scenarios are equally probable ($p_s = \frac{1}{8}$ for all s). Uncertainty is revealed as we move along the path. While at the start (t_0) we don't know which of the eight scenarios will be realized at t_3, we know considerably more at time t_1. If we arrive at node 1, we can say with certainty that we are in one of the scenarios $w_1, w_2, w_3,$ or w_4, but we don't yet know which. For optimization purposes, this means that all decisions taken at node 1 must not be arrived at with the knowledge of which scenario will eventually come true. This information is simply not available at time t_1. Otherwise, decisions would optimize for the known future scenario, discarding the effect of a decision on all other scenarios (which are not relevant, as they are known not to become true).

In order to keep things transparent, we have assumed that each node has two descendants with equal probability on each path. Hence each node has exactly one ancestor.

Optimization Model. We formulate the optimization model in its most direct form, called the **split-variable formulation**. Assume a_{it}^s is the amount (not weight) invested in asset i at the beginning of period t in scenario s. From the assumptions above, it is clear we need to choose allocations for two assets

($I = 2$, stocks and bonds) at three points in time (t_0, t_1, t_2) for eight scenarios each. This amounts to $48 = 2 \cdot 3 \cdot 8$ variables. Equally, R_{it}^{s} denotes the return of asset i in scenario $s = 1, \cdots, S$ (if the return of an asset is 5%, then R_{it}^{s} is 1.05), where $S = 8$ in the current example. Again this means 48 return realizations. We can now start to formulate our simple asset-liability model. The investor's objective is to maximize utility arising from period t_3,

$$\max \sum_{s} p_s \left(c_u \cdot surplus_s^{+} - c_d \cdot surplus_s^{-} \right) \tag{2.16}$$

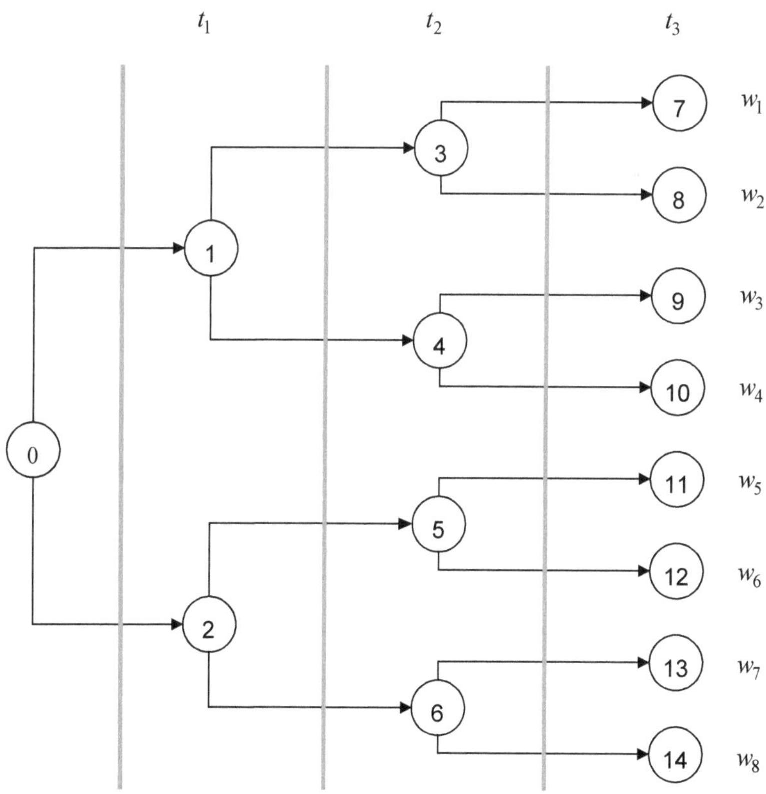

Figure 2.11 Event Tree for Multistage Stochastic Programming

where $surplus_s^{+}, surplus_s^{-}$ are variables restricted by non-negativity constraints, namely,

$$surplus_s^{+}, surplus_s^{-} \geq 0,$$

defining the surplus in state s. For any state, either $surplus_S^+$ (surplus) or $surplus_S^-$ (shortfall) is positive.

$$\sum_{i=1}^{I} R_{it_3}^s a_{it_2}^s = L + surplus_s^+ - surplus_s^- \quad \forall s \in S. \tag{2.17}$$

At times t_1 and t_2, wealth accumulates over time according to

$$\sum_{i=1}^{I} R_{i,t}^s a_{i,t-1}^s = \sum_{i=1}^{I} a_{i,t}^s \quad \forall s \in S, t = t_1, t_2. \tag{2.18}$$

This ensures that we can only invest what we have earned over the last period. In addition to these intertemporal budget constraints, we need to add the well-known budget constraint (to every single scenario) at time t_0

$$\sum_{i=1}^{I} a_{it_0}^s = W_0 \quad \forall s \tag{2.19}$$

with the usual non-negativity constraint on invested capital

$$a_{it}^s \geq 0 \quad \forall s \forall t. \tag{2.20}$$

So far, we can perfectly adjust to the scenario tree by making optimal decisions for every scenario, which in fact allows us to look ahead to the most favorable state of the world. As this will overstate the value of the objective function, we need to enforce **nonanticipativity**: at each node, we require that allocations be the same for all scenarios that are still undistinguishable. At node 0, all allocations have to be the same across all scenarios:

$$a_{it_0}^{s=1} = a_{it_0}^{s=2} = \cdots = a_{it_0}^{s=8}. \tag{2.21}$$

At nodes 1 and 2, we already know more about which path in the scenario tree we are on. However, we still need to restrict allocations to be the same for all paths crossing through nodes 1 and 2:

$$\begin{aligned} a_{it_1}^{s=1} &= a_{it_1}^{s=2} = \cdots = a_{it_1}^{s=4}, \\ a_{it_1}^{s=5} &= a_{it_1}^{s=6} = \cdots = a_{it_1}^{s=8}. \end{aligned} \tag{2.22}$$

Again the same logic applies to decisions taken in t_2. Although we know considerably more in node 3, we cannot anticipate the node at which the stochastic process will arrive:

$$\begin{aligned} a_{it_2}^{s=1} &= a_{it_2}^{s=2}, \\ a_{it_2}^{s=3} &= a_{it_2}^{s=4}, \\ a_{it_2}^{s=5} &= a_{it_2}^{s=6}, \\ a_{it_2}^{s=7} &= a_{it_2}^{s=8}. \end{aligned} \tag{2.23}$$

While this all looks a bit messy, we will show in the next section how we can easily implement this model within SIMPLE. In fact, all we have done is to set up a linear program with a large number of variables that are linked by nonanticipativity constraints.

2.5.2 Solving a Multistage Stochastic Program with NuOPT

In order to solve the problem laid out in the previous section, we need to specify the scenario tree first. The specification of the scenario tree is critical to the stochastic programming approach. In essence, a scenario tree tries to approximate a continuous (multivariate) distribution with a small amount of discrete scenarios. If the number of scenarios is too small or not representative of the continuous distribution, we will end up with a solution that is largely affected by estimation (specification) error. Making the number of scenarios large only partially helps: while the discrete approximation will become better, the number of variables will rise, making solutions computationally very expensive or infeasible.[7] Suppose we specify two return realizations (binomial tree) for each asset. Either equities outperform bonds $R_{equity} = 1.25 > R_{bonds} = 1.14$ or vice versa $R_{equity} = 1.06 > R_{bonds} = 1.12$. The optimization model in the above section can be formulated shown in Code 2.9.

```
DSP.model <- function(s.eq, s.bd, u.c, l.c, W.init,
  Liab)
{
  scenarios <- Set()
  steps <- Set()
  h <- Element(set=steps)
  s <- Element(set=scenarios)
  s.eq <- Parameter(s.eq, index=dprod(s,h))
  s.bd <- Parameter(s.bd, index=dprod(s,h))
  u.c <- Parameter(u.c)
  l.c <- Parameter(l.c)
  Liab <- Parameter(Liab)
```

```
W.init <- Parameter(W.init)
w.eq <- Variable(index=dprod(s,h))
w.bd <- Variable(index=dprod(s,h))
upper <- Variable(index=s)
lower <- Variable(index=s)
w.eq[s,1]+w.bd[s,1] == W.init
w.eq[s,1]*s.eq[s,1] + w.bd[s,1]*s.bd[s,1] ==
   w.eq[s,2]+w.bd[s,2]
w.eq[s,2]*s.eq[s,2] + w.bd[s,2]*s.bd[s,2] ==
   w.eq[s,3]+w.bd[s,3]
w.eq[s,3]*s.eq[s,3] + w.bd[s,3]*s.bd[s,3] ==
   Liab+upper[s]-lower[s]
w.eq[s,h]>=0
w.bd[s,h]>=0
upper[s]>=0
lower[s]>=0
w.eq[1,1]==w.eq[2,1]
w.eq[2,1]==w.eq[3,1]
w.eq[3,1]==w.eq[4,1]
w.eq[4,1]==w.eq[5,1]
w.eq[5,1]==w.eq[6,1]
w.eq[6,1]==w.eq[7,1]
w.eq[7,1]==w.eq[8,1]
w.eq[1,2]==w.eq[2,2]
w.eq[2,2]==w.eq[3,2]
w.eq[3,2]==w.eq[4,2]
w.eq[5,2]==w.eq[6,2]
w.eq[6,2]==w.eq[7,2]
w.eq[7,2]==w.eq[8,2]
w.eq[1,3]==w.eq[2,3]
w.eq[3,3]==w.eq[4,3]
w.eq[5,3]==w.eq[6,3]
w.eq[7,3]==w.eq[8,3]
w.bd[1,1]==w.bd[2,1]
w.bd[2,1]==w.bd[3,1]
w.bd[3,1]==w.bd[4,1]
w.bd[4,1]==w.bd[5,1]
w.bd[5,1]==w.bd[6,1]
w.bd[6,1]==w.bd[7,1]
w.bd[7,1]==w.bd[8,1]
w.bd[1,2]==w.bd[2,2]
w.bd[2,2]==w.bd[3,2]
w.bd[3,2]==w.bd[4,2]
w.bd[5,2]==w.bd[6,2]
w.bd[6,2]==w.bd[7,2]
```

```
  w.bd[7,2]==w.bd[8,2]
  w.bd[1,3]==w.bd[2,3]
  w.bd[3,3]==w.bd[4,3]
  w.bd[5,3]==w.bd[6,3]
  w.bd[7,3]==w.bd[8,3]
  utility <- Objective(type="maximize")
  utility ~ Sum(u.c*upper[s]-l.c*lower[s],s)
}

s.eq <- matrix(cbind(c(rep(1.25,4),rep(1.06,4)),
  rep(c(1.25, 1.25, 1.06,1.06),2),
  rep(c(1.25, 1.06),2)), ncol=3, nrow=8)
s.bd <- matrix(cbind(c(rep(1.14,4),rep(1.12,4)),
  rep(c(1.14, 1.14, 1.12,1.12),2),
  rep(c(1.14, 1.12),2)), ncol=3, nrow=8)
u.c <- 1
l.c <- 4

W.init <- 55
Liab <- 80

DSP.system <- System(DSP.model, s.eq, s.bd, u.c,
  l.c, W.init, Liab)
solution <- solve(DSP.system, trace=T)
```

Code 2.9 Stochastic Multiperiod Optimization

The solution is given below. We see that the nonanticipativity constraints force allocations to be the same for indistinguishable states of the world.

```
solution
$variables:
$w.eq:
         1        2            3
1 41.47927 65.09458 8.383990e+001
2 41.47927 65.09458 8.383990e+001
3 41.47927 65.09458 7.041860e-011
4 41.47927 65.09458 7.041860e-011
5 41.47927 36.74322 7.041443e-011
6 41.47927 36.74322 7.041443e-011
7 41.47927 36.74322 6.400000e+001
8 41.47927 36.74322 6.400000e+001
attr(, "indexes"):
[1] "s" "h"

$w.bd:
```

```
         1         2            3
1 13.52073  2.168138 7.857768e-011
2 13.52073  2.168138 7.857768e-011
3 13.52073  2.168138 7.142857e+001
4 13.52073  2.168138 7.142857e+001
5 13.52073 22.368029 7.142857e+001
6 13.52073 22.368029 7.142857e+001
7 13.52073 22.368029 5.604172e-011
8 13.52073 22.368029 5.604172e-011
attr(, "indexes"):
[1] "s" "h"

$upper:
        1        2        3            4        5
 24.79988 8.870299 1.428571 1.114117e-012 1.428571
            6             7             8
1.114083e-012 1.079866e-012 6.548158e-013

attr(, "indexes"):
[1] "s"

$lower:
             1             2             3
 6.548158e-013 6.548159e-013 6.548167e-013

             4             5
 1.588384e-012 6.548167e-013
             6             7     8
 1.588446e-012 1.663613e-012 12.16
attr(, "indexes"):
[1] "s"

$objective:
[1] -12.11268
```

2.5.3 An Alternative Formulation

The formulation above was intuitive but unfortunately requires us to use many variables (too many if the number of scenarios or variables becomes larger). Alternatively, we can write the financial planning problem using what is called the **compact formulation**. First let us distinguish between the root node $n = 0$, decision nodes $n \in D = \{1, \cdots, 6\}$, and end nodes $n \in E = \{7, 8, \cdots, 14\}$. As each node (apart from node 0) has exactly one ancestor (nonrecombining tree), we can identify every ancestor with a deterministic function $f(n)$. For example, $f(n=4)=1$ (i.e., the unique ancestor of node 4 is given by node 1). For each end node, we define $surplus_S^+, surplus_S^- > 0$ and the objective also remains the same:

$$\max \sum_s p_s \left(c_u \cdot surplus_s^+ - c_d \cdot surplus_s^- \right). \tag{2.24}$$

The decisions at node 0 need to satisfy the budget constraint $\sum_{i=1}^{I} a_{it_0}^s = W_0$. For all other decision nodes, we require

$$\sum_{i=1}^{I} R_{i,n} a_{i,f(n)} = \sum_{i=1}^{I} a_{i,n} \quad \forall n \in D. \tag{2.25}$$

We cannot use more wealth than the wealth generated by moving from the ancestor of n to n itself. Also, for all end nodes we demand

$$\sum_{i=1}^{I} R_{i,n} a_{i,f(n)} = L + surplus_s^+ - surplus_s^- \quad \forall n \in E. \tag{2.26}$$

The model is completed with $a_{i,n} \geq 0$. Apart from the usual dummy variables (defining nonzero surplus levels), we only require $14 = 2 \cdot 7$ variables. The formulation in SIMPLE is left to readers as an exercise.[8]

2.6 Optimization within S-PLUS

2.6.1 Optimization with One Variable

While NUOPT offers powerful optimization routines, the standard version of S-PLUS already comes with some built-in functions. We will review these functions using some simple finance-related examples.

A Simple Root Finding Problem. For continuous functions of one variable, S-PLUS offers the functions `uniroot` (find a zero) and `optimize` (find a local minimum). Suppose we need to find the internal rate of return of a savings plan that invested 1000 dollars every year for 10 years. The final wealth was 18,000 dollars. Mathematically, we need to solve for

$$-\frac{1000}{(1+y)^1} - \frac{1000}{(1+y)^2} - \cdots - \frac{1000}{(1+y)^{10}} + \frac{18000}{(1+y)^{11}} = 0.$$

We first write a function that calculates the internal rate of return for a series of equally spaced cash flows.

```
IRR <- function(cash.flow)
{
  pv <- function(x, cash.flow){
      sum(cash.flow*
         (1/(1+x))^(1:length(cash.flow)-1))
  }
  irr <- uniroot(pv, c(-0.99, 0.99),
      cash.flow=cash.flow)$root
}
```

In the example above, our cash flows need to be expressed by `cash.flow <- c(rep(-1000,10),18000)`. If we call `IRR(cash.flow)`, we get 0.1046 (about 10.5%). Note that we could have also used the function `polyroot(z)`, as the problem of finding the appropriate internal rate of return is indeed a polynomial:

$$1000a+1000a^2+\cdots 1000a^{10}-18000a^{11}=0, a=(1+y)^{-1}. \qquad (2.27)$$

The reader is encouraged to try this function.

Implied Volatility. Another typical root-finding problem in finance is to find the implied volatility of an option from quoted prices. The **implied volatility** ($\sigma_{implied}$) is defined as the volatility that equalizes model price (under the assumption that we use the correct model) and quoted price. We hence need to find

$$C_{market}-C_{Black\text{-}Scholes}\left(\sigma_{implied}\right)=0. \qquad (2.28)$$

Suppose a one year, at the money European call option trades at 16%. The one year interest rate is 3%. We first code a function to generate Black-Scholes model prices,

$$\begin{aligned} C_{Black\text{-}Scholes} &= S\cdot N(d)-\exp(-rT)\cdot X\cdot N\left(d-\sigma\sqrt{T}\right) \\ d &= \frac{1}{\sigma\sqrt{T}}\left(\log\left(\frac{S}{X}\right)+\left(r+\tfrac{1}{2}\sigma^2\right)T\right). \end{aligned} \qquad (2.29)$$

```
Black.Scholes.Call <- function(S,X,r,Time,sigma)
{
  d1 <- (log(S/X)+(r+0.5*sigma^2)*Time) /
     (sigma*sqrt(Time) )
```

```
  d2 <- d1-sigma*sqrt(Time)
  premium <- S*pnorm(d1)-exp(-r*Time)*X*pnorm(d2)
  list("premium"=premium)
}
```

Now we can define and solve the root-finding problem using Code 2.10.

```
f <- function(sigma,premium2){
  Black.Scholes.Call(S,X,r,Time,sigma)$premium -
  premium2
}

iv <- uniroot(f,c(0,1),keep.xy=T,premium2=0.16)
iv$root
```

Code 2.10 Root-Finding Problems

The implied volatility for the example above is 37%. It needed six evaluations (find out with `iv$nf`) to arrive at this result. Alternatively, we could have directly applied Newton's method.[9] The updating formula for volatility can be expressed as

$$\sigma_{s+1} = \sigma_s + \left(\tfrac{dC}{d\sigma}\right)^{-1}\left[C_{market} - C_{Black\text{-}Scholes}\left(\sigma_s\right)\right], \tag{2.30}$$

where

$$\frac{dC}{d\sigma} = S\sqrt{T}\tfrac{1}{\sqrt{2\pi}}\exp\left(-\tfrac{1}{2}x^2\right) \tag{2.31}$$

is also called "vega" (the option sensitivity to changes in volatility). We suggest extending the option code by two additional lines to allow for clean programming.

```
vega <- 1/(sqrt(2*pi)*exp((-d1^2)/2))*sqrt(Time)
list("premium"=premium, "vega"=vega)
```

Additionally we need to specify an initial estimate for σ as well as the maximum number of iterations and a convergence threshold. Having done this, we can code the following function that allows us to find implied volatilities via Newton's method:

```
sigma <- 0.2
max.it <- 10
tol <- 0.000000001
```

```
for(i in 1:max.it)
{
  diff <- premium -
     Black.Scholes.Call(S,X,r,Time,sigma)$premium
  vega <- Black.Scholes.Call(S,X,r,Time,sigma)$vega
  sigma <- sigma+diff/vega
  if(abs(diff) < tol) break
}
sigma
```

The code above will loop through a maximum number of iterations (`max.it`) but will stop as soon as the difference between the actual premium and model premium is sufficiently small (`tol`).

Portfolio Optimization and the Log-normality Assumption. Suppose now an investor is planning for her retirement and needs to decide how to split up her wealth between equities (or any other risky portfolio) and cash. Rather than going through the mathematics of optimal asset allocation, we will apply scenario optimization in conjunction with the standardized binomial density. The standardized binomial density can be regarded as the discrete version of the continuous standard normal density (at equally spaced points). The random variable z_i is evaluated at points $i = 0,1,\cdots,n,$ where $z_i = (2i-n)/\sqrt{n}$. The associated probability $b(z_i)$ amounts to $(n!/i!(n-i)!)(\frac{1}{2})^n$. In S-PLUS, we can easily generate z_i and $b(z_i)$:

```
n <- 100
z <- (2*(0:n)-n)/sqrt(n)
b.z <- dbinom(0:n, n, 0.5)
```

Note that z_i has mean zero and variance 1 (and skewness of zero and kurtosis of 3). In the first instance, we generate returns that are log-normally distributed $R_i = \exp(\mu + \sigma z_i) - 1$. Note that the expected return and variance of R_i can be calculated from $E(R) = \sum_{i=1}^{n} \exp(\mu + \sigma z_i) b(z_i)$ and $Var(R) = \sum_{i=0}^{n} (R_i - E(R))^2 b(z_i)$. We can also find a discrete version of the normal distribution with exactly the same expected return and variance:

```
R.lz <- exp(0.1+0.2*z)-1
mean <- sum(b.z*R.lz)
sig  <- sqrt(sum(b.z*(R.lz-mean)^2))
R.z  <- mean+sig*z
```

However, both distributions will exhibit different shapes. While the normal distribution is symmetric (offering the same odds for large negative and positive deviations from the mean return), the log-normal distribution shows fewer large negative returns and more large positive returns than the normal distribution. The reason for this is the positive skewness of the log-normal distribution. While the log-normal distribution is a correct representation of total returns (unleveraged losses should never exceed 100%), the normal distribution is not. We are accustomed to applying the normal distribution because of its additivity (weighted means yield portfolio mean), which unfortunately does not carry over to the lognormal distribution. Now we want to ask how allocations differ if we use the correct log-normal distribution rather than the normal distribution.[10] Suppose our investor aims at maximizing expected (power) utility. She hence wants to maximize

$$E(U)=\frac{1}{1-\gamma}E\left(W_i^{1-\gamma}\right)=\frac{1}{1-\gamma}\sum_{i=0}^{n}[1+wR_i+(1-w)r]b(z_i), \quad (2.32)$$

where $r = 0.03$ denotes the risk-free rate and γ expresses the degree of risk aversion. For normally distributed returns, we calculate utility from Code 2.11.

```
utility <- function(w, R=R.z)
{
  wealth <- 1+w*R+(1-w)*0.03
  sum(1/(1-g)*(wealth)^(1-g)*b.z)
}
```

Code 2.11 Edgeworth Expansion and Portfolio Optimization

The optimal portfolio allocation can now be found from `optimize(utility, c(0,1), maximum=T)$maximum` and amounts to 47.89% when $\gamma = 4$. For lognormally distributed returns, instead we find the optimal allocation to be 56.59%, an almost 9% difference. It becomes clear from this example that the choice of distribution is not trivial.

2.6.2 General Optimization

Credit Risk. The assessment of credit risk attracts increased interest by banks, regulators, and institutional investors. At its heart is the so-called **loss distribution** (i.e., the probability distribution of credit-related losses) illustrated in Figure 2.12. All risk quantities (expected loss, unexpected loss, economic capital, etc.) can be derived from it. We know from standard corporate finance that holding a corporate bond is equivalent to holding a long position in government bonds (assumed to be free of credit risk) and a short position in out-of-the-money put options on the underlying assets of the firm (written to shareholders in exchange for a premium that allows higher coupons for corporate bond returns). The return due to changes in the fate of the firm (the market value of underlying assets) is highly asymmetric. There is a high likelihood of zero losses (the put expires worthless) and a small likelihood of very large losses (the put ends in the money and is exercised by shareholders; i.e., the corporation defaults). Candidates for these kinds of asymmetric distributions are the general Beta distribution, the Weibull distribution, and the Gamma distribution, among others. Suppose credit losses $\tilde{L}$ follow a Gamma distribution and we observe losses L_i for $i = 1, \cdots, n$. How do we fit the distribution to the data? We suggest using maximum likelihood methods; that is, we work out the likelihood function for the Gamma $G(\alpha, \beta)$ distribution,

$$f(L) = \frac{\beta^{\alpha}}{\Gamma(\alpha)} L^{\alpha-1} \exp(-\beta L), \quad 0 < L < \infty. \tag{2.33}$$

Note that $E(L) = \alpha/\beta$ and $\text{Var}(L) = \alpha/\beta^2$.

$$l(\alpha, \beta | L) = \prod_{i=1}^{n} f(L_i | \alpha, \beta) = \frac{\beta^{n\alpha}}{\Gamma(\alpha)^n} \left(\prod_{i=1}^{n} L_i \right)^{\alpha-1} \exp\left(-\beta \Sigma_{i=1}^{n} L_i\right). \tag{2.34}$$

Now solve for the distribution parameters (α, β) that give the drawn data sample the maximum likelihood. Calculating first and second derivatives of the Gamma likelihood is a tedious algebraic exercise. Fortunately, S-PLUS offers a fast and reliable alternative. The function `nlminb` (find local minimum for smooth functions subject to box constraints) offers the most general optimization routine in S-PLUS.[11]

First we define the (log) likelihood function for the sampled data by summing over the log densities:

```
log.L <- function(x)   {
  e <- log(dGamma(LOSSES, x[1], x[2]))
  -sum(e)
}
```

Note that we put a minus sign in front of the sum, as `nlminb` is designed to minimize functions. Next we sample 1000 draws from a hypothetical distribution and call the optimization routine using Code 2.12.

```
> LOSSES <- rGamma(1000, 1, 5)
> hist(LOSSES, probability=T, xlab="LOSSES",
     main="GAMMA DISTRIBUTION OF CREDIT LOSSES")
> result <- nlminb(start=c(1,1), objective=log.L)
```

```
> result$parameters
[1] 1.044975 5.228908
```

Code 2.12 Fitting a Loss Distribution

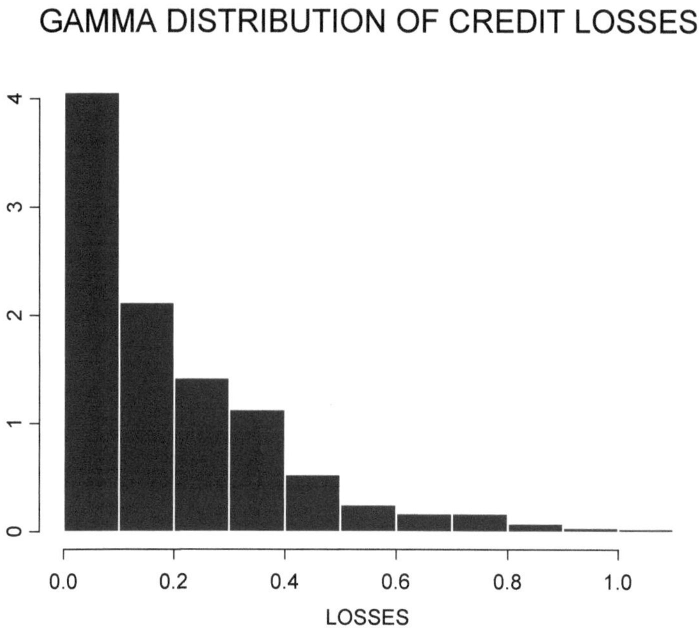

Figure 2.12 Loss Distribution

All we need to provide is a set of starting values (`start=c(1,1)`) and an objective (`objective=log.L`). The result differs from the simulated distribution due to sampling error. Note that we did not need a vector of first derivatives (gradient) or matrix of second derivatives (Hessian), as S-PLUS will approximate them with finite differences instead. Of course, if the derivatives are known (or have been calculated using `deriv`), they can be supplied. For more details, see the S-PLUS manual.

Term Structure Calibration. Modern academic finance generates a variety of term structure models; it can be difficult, even for the most ambitious researcher, to stay current with all of them. While some models require complex numerical techniques, parsimonious and intuitive models are more appealing to practitioners. One model of this sort (a similar version is used by the French Central Bank to calculate monthly available zero coupon rates) is described below. It postulates that the zero curve has the form[12]

$$R(T,\theta) = level - spread\left(\frac{1-\exp(-aT)}{aT}\right) + curvature\left(\frac{(1-\exp(-aT))^2}{4aT}\right) \quad (2.35)$$

$$\theta = (level, spread, curvature, a).$$

If maturity goes to infinity, all we are left with is the *level* term. We can hence interpret it as the long-term interest rate. If maturity goes to zero, the *spread* converges to $level - R(T,\theta)$ and can hence be regarded as a long-short spread. The two remaining parameters are *curvature*, describing how much the curve bends up or down, and the scale parameter, a, which can be interpreted as the strength of the mean reversion. Assuming our yield curve model is correct, we create a data set by introducing random error to our parameterized model (see Figure 2.13):

```
level <- 4
spread <- 1
curvature <- 10
a <- 1
Time <- seq(1,30,0.25)
zero <- (level-spread*((1-exp(-a*Time))/(a*Time)) +
  curvature*((1-exp(-a*Time))^2/(4*a*Time))) +
  rnorm(length(Time))/50
plot(Time, zero,ylab="ZERO RATE",
  xlab="TIME TO MATURITY")
```

In order to fit the model to the simulated data, we minimize the squared difference between model yield and simulated yield,

$$\min_{\theta} \sum_{i=1}^{n} [R(T) - R(T,\theta)]^2. \quad (2.36)$$

We summarize the minimization above in a special function called `term.fit`. This function is then minimized with respect to its parameter vector (see Code 2.13).

```
term.fit <- function(x)   {
  zero.fit <- (x[1]-x[2]*
     ((1-exp(-x[4]*Time))/(x[4]*Time))+
     x[3]*((1-exp(-x[4]*Time))^2/(4*x[4]*Time)))
     e <- sum((zero.fit-zero)^2)
  zero.fit
}
```

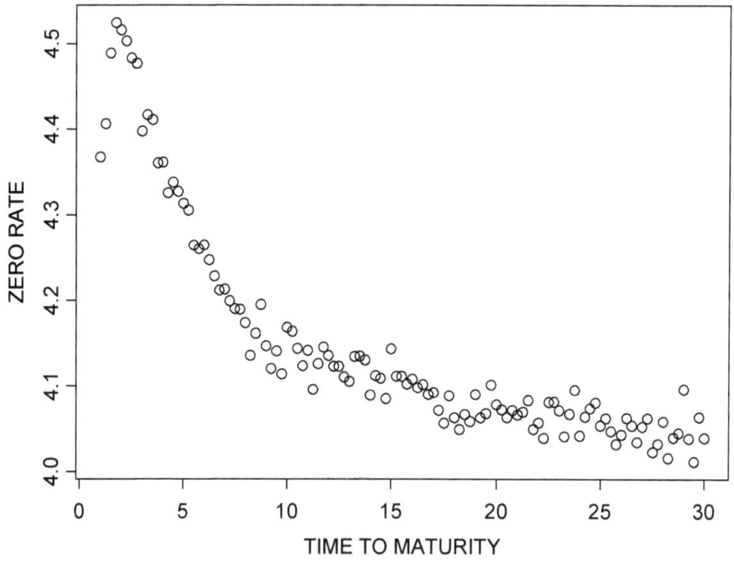

Figure 2.13 Simulated Yield Curve Data

```
result <- nlminb(start=c(1,1,1,0.4),
  objective=term.fit,
  lower=c(0.0001,0.0001,0.0001,0.0001),
  upper=c(10, 5, 40, 1))
x <- result$parameters
zero.fit <- (x[1]-x[2]*
  ((1-exp(-x[4]*Time))/(x[4]*Time)) +
  x[3]*((1-exp(-x[4]*Time))^2/(4*x[4]*Time)))
plot(Time, zero, pch=1, ylab="ZERO RATE",
  xlab="TIME TO MATURITY")
points(Time, zero.fit, type="l")
```

Code 2.13 Term Structure Fitting

The result is shown in Figure 2.14. With the principle above in mind, we could also fit this particular zero curve to any coupon bond curve, as the functional form for the zero curve is known and any coupon bond can be modeled as a portfolio of associated zero coupon bonds.

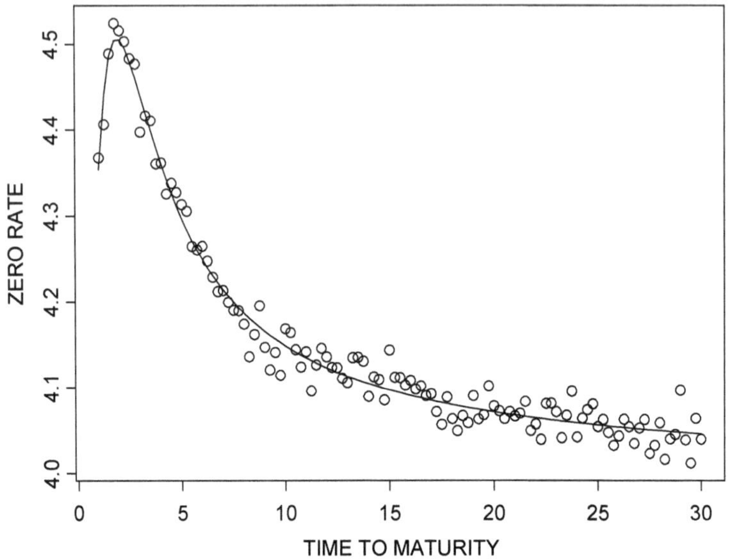

Figure 2.14 Simulated and Fitted Yield Curve Data

Exercises

1. Repeat the scenario optimization example, but solely use `Expressions` to define end-of-period wealth. Do not include variables other than asset weights.

2. Minimize the function: $f(x,y) = (x - y^2)(x - 4y^2)$. Check your solution. What is going on?

3. Use SIMPLE to program an active portfolio optimization using the covariance approach.

4. Use the program you wrote in Exercise 3 to trace out an active frontier without and with short-selling constraints. How do the frontiers differ? When does the benchmark matter in portfolio construction? Hint: See Grinold and Kahn (2000).

Endnotes

---

[1] We have used different names for the SIMPLE versions of S to avoid conflicts—SIMPLE objects do not work like regular S-PLUS objects.
[2] We always need the "$\leq$" in order to get closed constraint sets.
[3] The reader is encouraged to check the convexity of the objective function by calculating the matrix of 2nd derivatives (the Hessian matrix) and checking that it is positive semi-definite.
[4] Note that the solution to the SIMPLE system is not unique, as the roles of the two distributions can be interchanged (i.e., if $(p_1, p_2, p_3, p_4, p_5)$ is a solution, then so is $(1 - p_1, p_4, p_5, p_2, p_3)$).
[5] See Kim and Finger (2000) for more details on how to use a mixture of normals for stress testing and risk management.
[6] See Ziemba and Mulvey (1998) or Scherer (2004).
[7] Scenario generation is a complex field in its own right. Interested readers are referred to Ziemba (2003) for a nice review.
[8] Readers are encouraged to send in their solutions. The best program will be printed in the next edition, and the developer will get a single copy of all further editions of this book for free.
[9] See Judd (1998) for an economics-related textbook on numerical techniques.
[10] See the excellent exposition by Campbell and Viceira (2002).
[11] As `nlminb` encapsulates other functions such as `ms()` or `nlmin()` we will not discuss these.
[12] See El Karoui et al. (1998). It has been named the extended Vasicek model.

3 Advanced Issues in Mean-Variance Optimization

3.1 Nonstandard Implementations

3.1.1 Risk Budgeting Constraints

Risk budgeting is an increasingly trendy topic. The reason for this is clear: if a pension fund manager gets asked whether his plan is engaging in risk budgeting, it is hard to say "we do not budget risks," as this almost sounds negligent. In addition, as institutional investors become disappointed with the economic value traditional portfolio optimization results have provided, they become more willing to embrace a "budgeting" framework that allows them to plan and spend risk budgets rather than blindly following optimization results. Promoters of risk budgeting would like to "separate risk budgeting and VaR measurement from classic investment risk practices, such as asset allocation."[1] Others would argue that "we should regard a risk budget as an extension of mean-variance optimization that enables us to decouple a portfolio's allocation from fixed monetary values"[2] and hence "VaR and the risk capital budgeting metaphor pour old wine into new casks."[3] We very much agree with the latter two comments and argue that the merit of risk budgeting does not come from an increase in intellectual insight but rather from a more accessible way of decomposing and presenting investment risks.

We define risk budgeting as a process that reviews any critical assumption for the successful meeting of prespecified investment targets. It derives the appropriate trade-off between risks and returns associated with investment decisions. In a mean-variance world, this defaults to Markowitz portfolio optimization, where results are shown not only in terms of weights and monetary allocations but also in terms of risk contributions. A view that regards risk budgeting as a route to enforce diversification at least implicitly treats traditional Markowitz results with great suspicion. In this sense, risk budgeting can be viewed as a heuristic safeguard against estimation error.

Risk budgeting needs a decomposition of investment risk (i.e., we have to look for an additive description of investment risk). Standard portfolio theory, however, tells us that volatility is not additive. We start with the observation that portfolio volatility is a linear homogeneous function of portfolio weights. This allows us to rewrite portfolio volatility as the weighted sum of marginal risk contributions.

$$\sigma = \left(\sum_i \sum_j w_i w_j \sigma_{ij}\right)^{\frac{1}{2}} = \sum_i w_i \frac{d\sigma}{dw_i}. \tag{3.1}$$

To facilitate interpretation we divide by σ and arrive at risk budgets that sum to 100% (i.e., we can now attribute x% of total volatility as arising from asset i and its interactions with other assets)

$$1 = \sum_i risk\ budget_i = \sum_i \frac{w_i}{\sigma} \frac{d\sigma}{dw_i}. \tag{3.2}$$

Applying risk budgeting constraints in portfolio optimization allows us to limit the maximum and minimum risk contributions arising from individual positions. Some examples of risk budgeting are the following:

- No individual stock position must contribute more than 5% to total portfolio risk.
- Each allocation decision should contribute at least 5% to total portfolio risk.
- Market-timing decisions must not contribute more than 10% to total tracking error.
- Hedge funds are limited to a 10% risk contribution.
- None of my multiple managers must consume more than 15% of the total active risk budget.

The portfolio optimization problem as such changes slightly. Limits on risk budgets are, for example, expressed by $risk\ budget^{upper}$,

$$\begin{aligned}
&\min_{\mathbf{w}} \sum_i \sum_j w_i w_j \sigma_{ij} \quad \text{subject to} \\
&\sum_i w_i \mu_i = \mu \\
&\sum_i w_i = 1 \\
&risk\ budget^{lower} \le \frac{w_i}{\sigma} \frac{d\sigma}{dw_i} \le risk\ budget^{upper} \\
&w_i \ge 0.
\end{aligned} \tag{3.3}$$

We can transform this into S-PLUS code. All we need to enforce a given risk budget for an individual asset is to add an additional line of code to the usual mean-variance-based portfolio optimization, as in Code 3.1. Note that `min.rb` and `max.rb` denote the minimum and maximum risk budgets.

```
mv.rb <- function(Cov, mu.bar, mu.target, min.rb,
  max.rb)
{
  asset <- Set()
  j <- Element(set=asset)
  i <- Element(set=asset)
  Cov <- Parameter(Cov, index=dprod(i,j))
  min.rb <- Parameter(min.rb)
  max.rb <- Parameter(max.rb)
  w <- Variable(index=j)
  risk <- Objective(type="minimize")
  risk ~ Sum(Cov[i,j]*w[j]*w[i],i,j)
  mu.target <- Parameter(changeable=T)
  mu.bar <- Parameter(as.array(mu.bar),index=j)
  Sum(mu.bar[j]*w[j],j) >= mu.target
  Sum(w[j],j) == 1
  w[j] >= 0
  w[i]*Sum(Cov[i,j]*w[j],j)/risk <= max.rb
  w[i]*Sum(Cov[i,j]*w[j],j)/risk >= min.rb
}

# solve model
rb <- System(mv.rb, Cov, mu.bar, mu.target, min.rb,
  max.rb)
solution <- solve(rb)

# extract solution and calculate risk budgets
w <- matrix(solution$variables$w$current)
risk <- c(t(w)%*%Cov%*%w)
risk.budgets <- w*Cov%*%w/risk
```

Code 3.1 Optimization and Risk Budgeting Constraints

We recommend running the code above with some sample data to appreciate the mechanics of risk budgeting. Note again that risk budgeting constraints will enforce diversification at the expense of return generation. The resulting portfolios will lie below the unconstrained efficient frontier.

3.1.2 Min/Max Approach to Multiple Benchmarks and Rival Risk Regimes

Investors typically have many objectives. They want to maximize the total return of assets per unit of risk, but at the same time they do not want to depart too much from what their peers are doing or what their liability profile looks like. To address this issue, we need to be able to optimize against multiple benchmarks. One complication here is that decision makers often fail to agree on the probability ordering of alternative risk regimes. Hence we also require a technique that allows us to incorporate more than one estimate of the variance-covariance matrix. Suppose we face $s = 1,\cdots,S$ risk regimes, reflected in the associated covariance matrices $\mathbf{\Omega}_s$ and $b = 1,\cdots,B$ benchmark portfolios. We assume there are n assets available for investment; our positions in these assets are summarized in the vector of asset holdings, $\mathbf{w}$. We assume an investor seeks protection against the risk of adopting an investment strategy based on the wrong benchmark and/or the wrong risk regime. The optimization problem becomes[4]

$$\max_{\mathbf{w},\mathbf{w}\geq 0}\left(\mathbf{w}^T\boldsymbol{\mu} - \lambda \max_{s,b}\left[\left(\mathbf{w}-\mathbf{w}_b\right)^T \mathbf{\Omega}_s \left(\mathbf{w}-\mathbf{w}_b\right)\right]\right), \tag{3.4}$$

where λ reflects the decision maker's risk aversion. We can reformulate (3.4) in a way digestible to solvers of constrained quadratic programs:

$$\begin{aligned}
&\max_{\mathbf{w},\sigma^2_{\max}} \left(\mu - \lambda\sigma^2_{\max}\right)\\
&\left(\mathbf{w}-\mathbf{w}_b\right)^T \mathbf{\Omega}_s \left(\mathbf{w}-\mathbf{w}_b\right) \leq \sigma^2_{\max}\\
&\mathbf{w}^T\mathbf{I} = 1\\
&\mathbf{w}^T\boldsymbol{\mu} = \mu\\
&\mathbf{w} \geq 0.
\end{aligned} \tag{3.5}$$

Note that this defaults to standard Markowitz optimization for $s = 1$, $b = 1$. Before we present an algorithm to solve (3.5), it will help to suggest a method to arrive at rival risk regimes. It is well-known that correlations break down in times of market meltdowns (when portfolio managers need them most). We will not attempt to forecast the change in input parameters. However, we look for a tool to evaluate the diversifying properties of assets in rival risk regimes. As supervisory boards become more and more concerned about short-term performance, investors often do not have the luxury to bet on average correlation or average volatility. In order to come up with correlation and volatility estimates for what we define as "normal" and "hectic" times, we have to define exactly what we mean by “hectic” times.[5] Returns from “hectic” times are identified due to their statistical distance from the mean vector as given in

$$\left(\mathbf{R}_m - \bar{\boldsymbol{\mu}}\right)^T \bar{\boldsymbol{\Omega}}^{-1} \left(\mathbf{R}_m - \bar{\boldsymbol{\mu}}\right) = \mathbf{d}_m^T \bar{\boldsymbol{\Omega}}^{-1} \mathbf{d}_m, \tag{3.6}$$

where $\mathbf{d}_m$ reflects the distance vector at time m, $\mathbf{R}_m$ is a vector of return observations for n currencies at time m, $\bar{\boldsymbol{\mu}}$ denotes a vector of average currency returns, and $\bar{\boldsymbol{\Omega}}$ is the unconditional covariance matrix (over all m observations). For each cross section of stock returns we calculate (3.6) and compare it with the critical value $\chi^2_{0.95}(n)$. If we define an unusual observation as the outer 5% of a distribution (alternatively one might call it an outlier) and we look at five return series, our cutoff distance is 11.07. In the distance measure given in (3.6), return distances are weighted by the inverse of the covariance matrix. This takes into account volatilities (the same deviation from the mean might be significant for low-volatility series but not necessarily for high-volatility series) as well as correlations (returns of differing signs for two highly correlated series might be more unusual than for series with negative correlation). The mechanics of the algorithm described above will split up the data set into two subsets for “normal” and “hectic” times. Although it is clear from our discussion that in theory outliers are not necessarily associated with down markets, in practice they often are. As soon as a new data series is added to an existing data set, “hectic” and “normal” periods may well change. S-PLUS code for this methodology is provided in Code 3.2.

```
hectic.vs.normal <- function(datamatrix,
     percentage)
{
  series.names <- names(datamatrix)
  covar <- var(datamatrix)
  mean <-
     as.matrix(apply(datamatrix,2,mean),ncol=1)
  distance <-
     matrix(0,ncol=1,nrow=nrow(datamatrix))

  for(i in 1:nrow(datamatrix))  {
     distance[i] <- (datamatrix[i,]-mean) %*%
        solve(covar) %*% (t(datamatrix[i,])-mean)
  }

  normal <- matrix(
     datamatrix[distance<=qchisq(percentage,
        ncol(datamatrix))],ncol=ncol(datamatrix))
```

```
  stdev.normal <- apply(normal,2,stdev)
  names(stdev.normal) <- series.names

  hectic <- matrix(
     datamatrix[distance>qchisq(percentage,
        ncol(datamatrix))],ncol=ncol(datamatrix))
  stdev.hectic <- apply(hectic,2,stdev)
  names(stdev.hectic) <- series.names

  cor.normal <- cor(normal)
  cor.hectic <- cor(hectic)
  cov.normal <- var(normal)
  cov.hectic <- var(hectic)

  dimnames(cor.normal) <- list(series.names,
     series.names)
  dimnames(cor.hectic) <- list(series.names,
     series.names)

  dimnames(cov.normal) <- list(series.names,
     series.names)
  dimnames(cov.hectic) <- list(series.names,
     series.names)

  list("normal.correlation" = cor.normal,
       "hectic.correlation" = cor.hectic,
       "stdev.normal" = stdev.normal,
       "stdev.hectic" = stdev.hectic,
       "normal.covariance" = cov.normal,
       "hectic.covariance" = cov.hectic)
}
```

Code 3.2 Covariance Estimates in Good and Bad Times

Now we can turn to the S-PLUS code (Code 3.3) necessary to solve (3.5). To simplify matters, we focus on the case of two risk regimes and two benchmarks.

```
model.db.rr <- function(mu.bar, Cov.1, Cov.2,
     bench.1, bench.2, mu.target)
{
  asset <- Set()
  i <- Element(set=asset)
  j <- Element(set=asset)
  Q1 <- Parameter(Cov.1, index = dprod(i, j))
  Q2 <- Parameter(Cov.2, index = dprod(i, j))
  b1 <- Parameter(list(1:length(mu.bar), bench.1),
```

```
      index=i)
  b2 <- Parameter(list(1:length(mu.bar), bench.2),
      index=i)
  mu.bar <- Parameter(list(1:length(mu.bar),
      mu.bar),
      index=i)
  mu.target <- Parameter(mu.target)
  w <- Variable(index = i)
  te.max <- Variable()

  sigma.1 <- Expression(index = i)
  sigma.2 <- Expression(index = i)
  sigma.3 <- Expression(index = i)
  sigma.4 <- Expression(index = i)

  sigma.1[j] ~ Sum((w[i]-b1[i]) * Q1[i,j], i)
  sigma.2[j] ~ Sum((w[i]-b1[i]) * Q2[i,j], i)
  sigma.3[j] ~ Sum((w[i]-b2[i]) * Q1[i,j], i)
  sigma.4[j] ~ Sum((w[i]-b2[i]) * Q2[i,j], i)
  te.1 <- Expression()
  te.2 <- Expression()
  te.3 <- Expression()
  te.4 <- Expression()

  te.1 ~ Sum((w[i]-b1[i])*sigma.1[i],i)
  te.2 ~ Sum((w[i]-b1[i])*sigma.2[i],i)
  te.3 ~ Sum((w[i]-b2[i])*sigma.3[i],i)
  te.4 ~ Sum((w[i]-b2[i])*sigma.4[i],i)
  te.1 <= te.max
  te.2 <= te.max
  te.3 <= te.max
  te.4 <= te.max

  te <- Objective(minimize)
  te ~ te.max
  w[i] >= 0
  Sum(w[i], i) == 1
  Sum(mu.bar[i]*w[i], i) == mu.target
}
```

Code 3.3 Dual Benchmark, Dual Risk Optimization

Upon a cursory inspection of the program, readers will realize one peculiarity. Remember that SIMPLE requires us to formulate mathematical expressions that involve matrices as summations. We have therefore used the reformulation

$$\mathbf{w}^T \boldsymbol{\Omega} \mathbf{w} = \sum_i \sum_j w_i w_j \sigma_{ij} = \sum_j w_j \left(\sum_i w_i \sigma_{ij} \right). \tag{3.7}$$

Experimenting with the optimization code above, readers will learn that the results are (not surprisingly) dominated by the worst-case risk scenario. Hence the solutions where we use both regimes are virtually identical to the solutions where we only use the hectic regimes. The model is converted into a system readable by NuOPT and solved below using the usual commands:

```
sys.model.db.rr <- System(model.db.rr, mu.bar,
     Cov.1, Cov.2, bench.1, bench.2, mu.target)
solution <- solve(sys.model.db.rr)
```

In order to trace out an efficient frontier, we need to build a function around the code above that repeatedly optimizes portfolios for changing target returns. This is left to the reader as an exercise.

3.1.3 Multiple Benchmarks and Pareto Optimality

So far, we have not allowed for varying risk preferences, which would allow us to attach different weights to the various subproblems. We address this issue now.[6] Suppose an investor wants to maximize the minimum risk-adjusted performance under various risk regimes for benchmarks as well as risk aversions. How would we formulate this? In mathematical terms, we can express this as

$$\max_{\mathbf{w}, \mathbf{w} \geq 0} \left(\min_{s,b} \left[(\mathbf{w} - \mathbf{w}_b)^T \boldsymbol{\mu} - \lambda_{s,b} (\mathbf{w} - \mathbf{w}_b)^T \boldsymbol{\Omega}_s (\mathbf{w} - \mathbf{w}_b) \right] \right). \tag{3.8}$$

Essentially, Equation (3.8) poses the problem of maximizing the minimum risk-adjusted performance across alternative benchmarks, risk regimes, and the associated risk aversion coefficients. This is equivalent to maximizing the minimum utility (assuming mean-variance preferences). The resulting solution is Pareto optimal in the sense that we cannot increase utility any further without pushing utility from another subproblem below the minimum utility. Note that (3.8) differs from the conventional treatment of multiple benchmark problems,[7]

$$\max_{\mathbf{w}, \mathbf{w} \geq 0} \left(\mathbf{w}^T \boldsymbol{\mu} - \lambda_1 (\mathbf{w} - \mathbf{w}_1)^T \boldsymbol{\Omega} (\mathbf{w} - \mathbf{w}_1) - \lambda_2 (\mathbf{w} - \mathbf{w}_2)^T \boldsymbol{\Omega} (\mathbf{w} - \mathbf{w}_2) - \cdots \right), \tag{3.9}$$

by the presence of more than one risk regime as well as the min/max rule. S-PLUS code that allows us to solve (3.8) is given below in Code 3.4.

```
model.pareto <- function(mu.bar, Cov.1, Cov.2,
  bench.1, bench.2, lambda.1, lambda.2)
{
  asset <- Set()
  i <- Element(set=asset)
  j <- Element(set=asset)

  Q1 <- Parameter(Cov.1, index = dprod(i, j))
  Q2 <- Parameter(Cov.2, index = dprod(i, j))
  b1 <- Parameter(list(1:length(mu.bar), bench.1),
     index=i)
  b2 <- Parameter(list(1:length(mu.bar), bench.2),
     index=i)
  lambda.1 <- Parameter(lambda.1)
  lambda.2 <- Parameter(lambda.2)
  mu.bar <- Parameter(list(1:length(mu.bar),
     mu.bar),
     index=i)

  w <- Variable(index = i)
  U.min <- Variable()

  sigma.1 <- Expression(index = i)
  sigma.2 <- Expression(index = i)
  sigma.3 <- Expression(index = i)
  sigma.4 <- Expression(index = i)
  sigma.1[j] ~ Sum((w[i]-b1[i]) * Q1[i,j], i)
  sigma.2[j] ~ Sum((w[i]-b1[i]) * Q2[i,j], i)
  sigma.3[j] ~ Sum((w[i]-b2[i]) * Q1[i,j], i)
  sigma.4[j] ~ Sum((w[i]-b2[i]) * Q2[i,j], i)

  U.1 <- Expression()
  U.2 <- Expression()
  U.3 <- Expression()
  U.4 <- Expression()

  U.1 ~ Sum(mu.bar[i]*(w[i]-b1[i]), i)-
     lambda.1*Sum((w[i]-b1[i])*sigma.1[i],i)
  U.2 ~ Sum(mu.bar[i]*(w[i]-b1[i]), i)-
     lambda.1*Sum((w[i]-b1[i])*sigma.2[i],i)
  U.3 ~ Sum(mu.bar[i]*(w[i]-b2[i]), i)-
     lambda.2*Sum((w[i]-b2[i])*sigma.3[i],i)
  U.4 ~ Sum(mu.bar[i]*(w[i]-b2[i]), i)-
     lambda.2*Sum((w[i]-b2[i])*sigma.4[i],i)
```

```
    U.1 >= U.min
    U.2 >= U.min
    U.3 >= U.min
    U.4 >= U.min

    U <- Objective(maximize)
    U ~ U.min

    w[i] >= 0
    Sum(w[i], i) == 1
}

system.pareto <- System(model.pareto, mu.bar,
        Cov.1, Cov.2, bench.1, bench.2, lambda.1=0,
        lambda.2=30)
solve(system.pareto)
```

Code 3.4 Dual Benchmarks and Pareto Optimality

An example application of Code 3.4 is included in the code accompanying the book.

3.2 Portfolio Construction and Mixed-Integer Programming

3.2.1 Mixed-Integer Programming in SIMPLE

If you have ever tried to sell 22.345673 stock index futures, you might appreciate why we have to restrict ourselves to integer variables in some applications. (By integer variables we mean variables that can take on only integer values.) In portfolio construction, the most frequently used form of integer variable is the binary integer variable, that is, an integer variable that can assume either 1 or 0. NUOPT for S-PLUS, however, also allows us to model general integer variables (where every integer value is allowed). The method used to solve problems that contain integer variables is called **integer programming**. NUOPT allows us to handle problems that contain integer variables only (pure integer problems) as well as problems that contain both integer and other variables (mixed integer problems). Users can define integer variables for linear as well as quadratic programming problems in `solveQP()` by supplying an additional vector `isint` to `solveQP()`. This vector must only include logical values; its "true" entries indicate integer variables (i.e., `x[isint==T]` represents integer variables, while `x[isint==F]` represents continuous variables). For more general problems, we can also set up integer variables in SIMPLE in very much the same way as we set up continuous variables before:

```
x <- IntegerVariable(index=i, type=binary)
```

The use of integer variables allows us to model portfolio construction in a much more realistic setting. Some examples might help the reader appreciate the power of integer programming within a portfolio construction framework.[8]

- **Lower bound on weights.** Portfolio managers (and their clients) often hate small active positions (deviations from benchmark holdings) that they argue show little impact on total performance. Hence we could enforce this by adding the constraint that if an active position is established it needs to be at least of *x*%. This is by its very nature a go/no-go decision.
- **Upper bound on number of assets.** Portfolio diversification is helpful but too much of it might increase transaction costs as well as monitoring costs. We could hence limit the number of assets to a specified maximum in order to arrive at a manageable portfolio. Assets are counted as 0 or 1, depending on whether they are in or out of the solution set.
- **Fixed (and piecewise linear) transaction costs.** Sometimes it is useful to model transaction costs that contain both a fixed amount (ticket costs independent of trade size) as well as a volume-dependent cost. As fixed costs apply for any trade irrespective of size, this is another obvious go/no-go decision.
- **Threshold-based risk measures.** Some investors perceive risks as downside deviations below a specified threshold only. In the case of scenario optimization and non-normally distributed variables, every threshold violation is counted as 0 or 1.

The modeling of these problems is not very widespread as very little commercial software has previously been available. NUOPT allows us to solve these problems within S-PLUS at the greatest convenience for users, as will be shown below. Additionally, all the powerful analytics of S+FinMetrics can easily be integrated into our computations.

3.2.2 Buy-in Thresholds

Suppose we need to determine optimal portfolio weights w_i. However, we also want to restrict the number of assets we invest in. Note that a 0.0001% weight will contribute the same amount to the count of assets as a 10% weight. If we introduce a new binary variable δ_i that assumes 1 if an asset is included in the optimal solution and 0 otherwise, we are able to model the inclusion/exclusion of individual assets.

$$\delta_i = \begin{cases} 1 & \text{if asset } i \text{ is selected} \\ 0 & \text{otherwise.} \end{cases} \tag{3.10}$$

Remember that we want to count how many assets enter the optimal solution. As an asset is either in or out of the optimal solution, we can express this using

$$\begin{aligned} & w_i \leq \delta_i \cdot \text{large number} \\ & \delta_i = 0,1. \end{aligned} \tag{3.11}$$

Equation (3.11) operates like a switch. If an asset is included (even at a tiny size), the inequality is only satisfied for $\delta_i = 1$. As soon as the asset leaves the solution set, Equation (3.11) only holds for $\delta_i = 0$. Computationally, what we call here a "large number" should not be made "too" large. We can extend this logic to model typical buy-in thresholds as summarized in Table 3.1.

After these preliminaries we are ready to tackle our first portfolio construction problem using integer variables.

Table 3.1 Formulation of Buy-In Thresholds

Type	Formula
Either in or out	$w_i \leq \delta_i \cdot \text{large number},\ \delta_i = 0,1$
Either in-between or out	$\delta_i w_i^{\min} \leq w_i \leq \delta_i \cdot w_i^{\max},\ \delta_i = 0,1$
Either out or above	$w_i \leq \delta_i \cdot \text{large number},\ w_i^{\max} \delta_i \leq w_i, \delta_i = 0,1$
Cardinality constraint	$\sum_i \delta_i = \#\text{assets}$

3.2.3 Buy-in Thresholds and Cardinality Constraints

It is well-known that diversifying into a broader universe of assets has merits as well as limits. Positions may become very small, monitoring and research costs may rise, and diversification benefits may fade away as the number of assets becomes large. Investors might hence want to solve the following mixed integer quadratic program.

$$\begin{aligned}
&\min \sum_i \sum_j w_i w_j \sigma_{ij} \quad \text{subject to} \\
&\sum_i w_i \mu_i = \mu \\
&\sum_i w_i = 1 \\
&\delta_i w_i^{\min} \le w_i \le \delta_i \cdot w_i^{\max} \\
&\sum_i \delta_i = \#\text{assets} \\
&w_i \ge 0 \\
&\delta_i \in \{0,1\}.
\end{aligned} \tag{3.12}$$

Problem (3.12) can easily be transformed into the SIMPLE code shown in Code 3.5.

```
MV.model.card <- function(S, mu.target)
{
  if(any(is.na(S))==T)
     stop("no missing data are allowed")
  n <- Set(1:ncol(S))
  m <- Set(1:nrow(S))

  n.obs <- nrow(S)
  i <- Element(set=n)
  s <- Element(set=m)

  mu.bar <- apply(S, 2, mean)
  S <- Parameter(S, index=dprod(s,i))
  mu.bar <- Parameter(as.array(mu.bar), index=i)
  mu.target <- Parameter(as.numeric(mu.target),
     changeable=T)

  w <- Variable(index=i)
  dummy <- IntegerVariable(index=i, type=binary)

  w[i] <= 5*dummy[i]
  Sum(dummy[i], i) <= 2
  dev <- Variable(index=s)
  Sum((S[s,i]-mu.bar[i])*w[i],i) == dev[s]

  risk <- Objective(type="minimize")
  risk ~ 1/n.obs*Sum(dev[s]^2,s)

  Sum(mu.bar[i]*w[i],i) == mu.target
```

```
    Sum(w[i],i) == 1
    w[i] >= 0
  }
```

Code 3.5 Transformation of Buy-In Threshold Problem (3.12) into SIMPLE Code

In order to obtain a better understanding of the mechanics of (3.12), we will apply Code 3.5 to a simulated data set.

```
S <- matrix(rmvnorm(50,mean=c(0.02,0.04,0.05,0.08),
  cov=diag(rep(0.2,4))),ncol=4)
```

This code can now be used to trace out an efficient frontier and plot both risk/return combinations and the underlying portfolios. In our simulated data set, we have assumed a universe of four assets. Portfolios are restricted to contain at most two assets.

```
MV.pf.card <- function(S, mu.target)
{
  call(MV.model.card)
  MV.system.card <-
     System(MV.model.card,S,mu.target)
  solution <- solve(MV.system.card, trace=T)
  weight <-
     matrix(round(solution$variable$w$current,
     digit=5)*100, ncol=1)
  risk <- solution$objective
  return(weight,risk)
}
```

For a given set of scenarios `MV.frontier()` will generate an efficient frontier with 200 risk/return points (see Code 3.6). The results are shown in Figure 3.1.

```
MV.frontier.card <- function(S, n.pf)
{
  call(MV.pf.card)
  Risk <- matrix(0, ncol=1, nrow=n.pf)
  Return <- matrix(0, ncol=1, nrow=n.pf)
  n.obs <- nrow(S)
  equal.mean.scenarios <- S - matrix(
     rep(apply(S,2,mean),n.obs),nrow=n.obs,byrow=T)
  x <-
     MV.pf.card(S=equal.mean.scenarios,mu.target=0)
```

```
  mu.min <- t(x$weight) %*% apply(S, 2, mean)/100
  weight <- x$weight
  Risk[1,1] <- x$risk
  Return[1,1] <- mu.min
  mu.max <- max(apply(S, 2, mean))
  mu.range <- seq(mu.min, mu.max,
     (mu.max-mu.min)/(n.pf-1))
  for(i in 2:n.pf){
     x <- MV.pf.card(S, mu.target=mu.range[i])
     Risk[i,1] <- x$risk
     Return[i,1] <- mu.range[i]
     weight <- cbind(weight,x$weight)
  }
  graphsheet()
  par(mfrow=c(1,2))
  plot(Risk,Return, type="b")
  title("Mean-Variance Frontier with Cardinality
     Constraints")
  barplot(weight)
  title("Frontier Portfolios")
  list("optimal.weights" = weight)
}
x <- MV.frontier.card(S, n.pf=200)
```

Code 3.6 Optimization with Cardinality Constraints

At first sight, this is a very odd-looking frontier. Figure 3.1 is best understood if we think of the frontier as the envelope for all possible pairwise combinations (cardinality constraint of 2) within the four-asset universe (six combinations). Note that tracing out the frontier by stepwise increasing return requirements will lead to inefficient parts. In reality, the frontier becomes discontinuous, as it makes little sense to invest in dominated portfolios. Needless to say, the cardinality-constrained portfolio plots below an efficient frontier without cardinality constraints. The difference tends to be largest for the minimum variance portfolio (where diversification normally requires many assets to be included as long as the assets have similar risk characteristics), and it tends to be zero for the maximum return portfolio (which tends to be concentrated in a single asset, the maximum return asset).

We can also use the logic above to model round lots (stocks can only be purchased in blocks). As before, we model the holdings in the i-th asset as

$$w_i = \delta_i block_i,$$

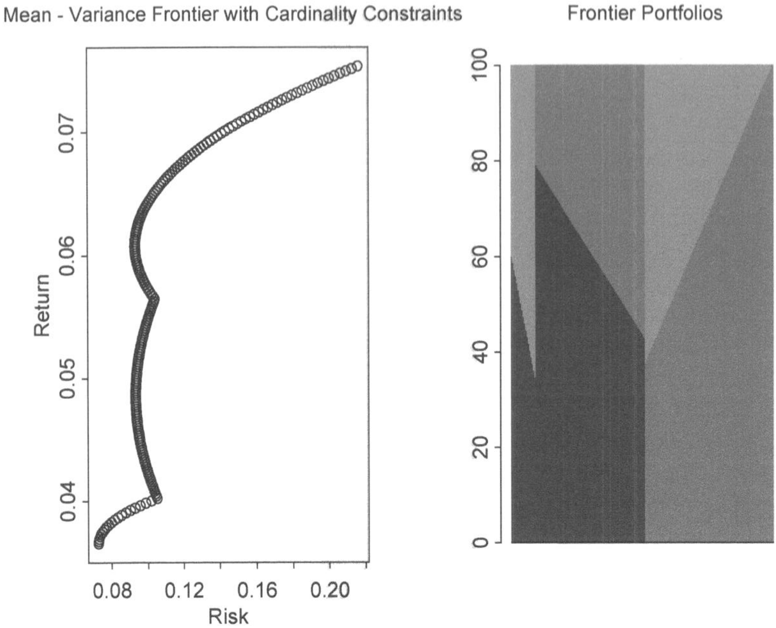

Figure 3.1 Efficient Frontier for Cardinality Constrained Portfolios

where $block_i = \frac{\text{round lot transaction for asset } i}{\text{total wealth}}$. Note, however, that the number of blocks (δ_i) times the block sizes ($block_i$) do not need to sum to one. We can accommodate this by introducing overshoot and undershoot variables ϖ^+, ϖ^- into the budget constraint,

$$\sum_{i=1}^{n} \delta_i block_i + \varpi^+ - \varpi^- = 1. \tag{3.13}$$

Overshoots and undershoots need to be penalized in the objective function with an "appropriate" cost factor. Our portfolio optimization model now becomes

$$
\begin{aligned}
&\min \sum_i \sum_j w_i w_j \sigma_{ij} + c\left(\varpi^+ + \varpi^-\right) \text{ subject to} \\
&\sum_i w_i \mu_i = \mu \\
&w_i = \delta_i block_i \\
&\sum_i w_i + \varpi^+ - \varpi^- = 1 \\
&w_i \geq 0 \\
&\delta_i \in \{0,1\} \\
&\varpi^+, \varpi^- \geq 0.
\end{aligned} \tag{3.14}
$$

The cost factor c needs to be carefully chosen.

3.2.4 Tracking Indices with a Small Number of Stocks

Trading desks at investment banks very often face a similar problem. They are asked by clients to construct tracking baskets; that is, a group of stocks that is constrained in size to a maximum number of stocks. Let us denote the holdings of the target portfolio (the benchmark portfolio to be tracked at minimum tracking error) by b_i. The problem of tracking an index with a small number of stocks now becomes

$$
\begin{aligned}
&\min \sum_i \sum_j \left(w_i - b_i\right)\left(w_j - b_j\right)\sigma_{ij} \text{ subject to} \\
&\sum_i w_i = 1 \\
&w_i \leq \delta_i \cdot \text{large number} \\
&\sum_i \delta_i = \#assets \\
&w_i \geq 0 \\
&\delta_i \in \{0,1\}.
\end{aligned} \tag{3.15}
$$

The implementation of (3.15) is a straightforward change of Code 3.5 and is left to the reader as an exercise.

3.3 Transaction Costs

3.3.1 Turnover Constraints

Turnover constraints are implemented by practitioners to heuristically safeguard against transaction costs. The implicit assumption behind this indirect treatment of transaction costs is that if transaction costs are proportional and equal across assets, it is sufficient to control turnover that is directly related to transaction costs. In reality, transaction costs differ across assets and do not change proportionally with trade size. However, we will use turnover constraints as a starting point for the next sections.

So far we have not needed to know the initial holdings ($w_i^{initial}$) when constructing a portfolio, as we assumed no costs to turn our portfolio into cash and vice versa. Here, in addition to the vector of initial holdings, we need

1. two new sets of variables—assets bought w_i^+ (positive weight changes) and assets sold w_i^- (negative weight changes);
2. a budget constraint for each asset that requires that the final asset weight w_i equal the initial weight $w_i^{initial}$ plus transactions: $w_i = w_i^{initial} + w_i^+ - w_i^-$; and
3. the turnover constraint itself, which limits the total turnover τ that is allowed to go on in portfolios to a specified number:

$$\sum_{i=1}^{n}\left(w_i^+ - w_i^-\right) \le \tau,\ w_i^+ \ge 0,\ w_i^- \ge 0 \tag{3.16}$$

The resulting minimization problem is shown in (3.17).

$$\begin{aligned}
&\min \sum_i \sum_j w_i w_j \sigma_{ij} \text{ subject to} \\
&\sum_i w_i \mu_i = \mu \\
&\sum_i w_i = 1 \\
&w_i = w_i^{initial} + w_i^+ - w_i^- \\
&\sum_{i=1}^{n}\left(w_i^+ + w_i^-\right) \le \tau \\
&w_i^+ \ge 0 \\
&w_i^- \ge 0 \\
&w_i \ge 0.
\end{aligned} \tag{3.17}$$

System (3.17) can now be implemented in SIMPLE. Note again that summation allows us to write down the optimization problem in virtually the same way we would on paper; this is illustrated in Code 3.7.

Turnover constraints effectively anchor the optimized portfolio around the initial holdings. Small turnover figures allow only small deviations, while the reverse is true for large turnover figures. Turnover constraints do not allow an optimal treatment of transaction costs, as transaction costs might vary considerably across asset classes. It is also not intuitive that an asset manager stops trading in June if he has already spent his turnover budget of 40% per year, even though investment opportunities are still around. The next section will therefore allow for proportional transaction costs.

```
MV.model.turnover <- function(S, w.initial,
  mu.target, to)
{
  nobs <- nrow(S)
  n <- Set(1:ncol(S))
  m <- Set(1:nrow(S))
  i <- Element(set=n)
  s <- Element(set=m)

  mu.bar <- apply(S, 2, mean)
  S <- Parameter(S, index=dprod(s,i))
  mu.bar <- Parameter(as.array(mu.bar), index=i)
  mu.target <- Parameter(as.numeric(mu.target),
     changeable=T)

  w <- Variable(index=i)
  to <- Parameter(as.numeric(to), changeable=T)
  w.initial <- Parameter(as.array(w.initial),
     index=i)
  w <- Variable(index=i)
  w.plus <- Variable(index=i)
  w.minus <- Variable(index=i)
  dev <- Variable(index=s)

  Sum((S[s,i]-mu.bar[i])*w[i],i) == dev[s]
  risk <- Objective(type="minimize")
  risk ~ 1/nobs*Sum(dev[s]^2,s)
  Sum(mu.bar[i]*w[i],i) == mu.target
  w.initial[i]+w.plus[i]-w.minus[i] == w[i]

  Sum(w.plus[i]+w.minus[i],i) <= to
  Sum(w[i],i) == 1
```

```
    w.plus[i] >= 0
    w.minus[i] >= 0
    w[i] >= 0
}
```

Code 3.7 Optimization with Turnover Constraints

3.3.2 Proportional Costs

So far, it has been cost-free to shift portfolio allocations.[9] However, in the real world, there are transaction costs associated with buying and selling securities. Let us assume that transaction costs are paid at the beginning of the period. In order to model transaction costs, we have to modify the budget constraint, as the costs associated with our transactions have to be paid out of the existing budget, that is, they have to be financed from asset sales,

$$\sum_{i=1}^{n}\left(w_i^- - w_i^+\right) - \sum_{i=1}^{n}\left(tc_i^+ w_i^+ + tc_i^- w_i^-\right) \geq 0, w_i^+ \geq 0,\ w_i^- \geq 0, \quad (3.18)$$

where $tc_i^{\pm}$ are the proportional transaction costs for buying and selling. The first summation denotes the proceeds from net selling, while the second summation denotes the associated costs. Transaction costs lead to an indirect return reduction, as the amount on which asset returns can be earned is reduced from the start of the investment period, i.e., $\Sigma_i w_i < \Sigma_i w_i^{initial} = 1$. We can incorporate (3.18) into a new budget constraint by adding transaction costs to the summation of holdings that are left after transactions have been paid:

$$\begin{aligned}
&\min \sum_i \sum_j w_i w_j \sigma_{ij} \quad \text{subject to} \\
&\sum_i w_i\left(1+\mu_i\right) = 1+\mu \\
&\sum_{i=1}^{n} w_i + \sum_{i=1}^{n}\left(tc_i^+ w_i^+ + tc_i^- w_i^-\right) = 1 \\
&w_i = w_i^{initial} + w_i^+ - w_i^- \\
&w_i \geq 0 \\
&w_i^+ \geq 0 \\
&w_i^- \geq 0.
\end{aligned} \quad (3.19)$$

Note that as now $\Sigma_i w_i < 1$, we need to write $\Sigma_i w_i (1+\mu_i) = 1+\mu$ instead of $\Sigma_i w_i \mu_i = \mu$. Since (3.19) is very close to (3.17), we only need some minor changes to accommodate transaction costs. For simplicity, we assumed transaction costs to be equal across assets. We can easily change this by providing vectors of transaction costs instead, together with the relevant indexing; this is illustrated in Code 3.8.

```
MV.model.tc <- function(S, w.initial, mu.target,
     tc)
{
  nobs <- nrow(S)
  n <- Set(1:ncol(S))
  m <- Set(1:nrow(S))
  i <- Element(set=n)
  s <- Element(set=m)

  mu.bar <- apply(S, 2, mean)
  S <- Parameter(S, index=dprod(s,i))
  mu.bar <- Parameter(as.array(mu.bar), index=i)
  mu.target <- Parameter(as.numeric(mu.target),
     changeable=T)

  w <- Variable(index=i)
  tc <- Parameter(as.numeric(tc), changeable=T)
  w.initial <- Parameter(as.array(w.initial),
     index=i)
  w <- Variable(index=i)
  w.plus <- Variable(index=i)
  w.minus <- Variable(index=i)
  dev <- Variable(index=s)

  Sum((S[s,i]-mu.bar[i])*w[i],i) == dev[s]
  risk <- Objective(type="minimize")
  risk ~ 1/nobs*Sum(dev[s]^2,s)
  Sum((1+mu.bar[i])*w[i],i) >= 1+mu.target
  w.initial[i] + w.plus[i] - w.minus[i] == w[i]
  Sum(w[i], i) + Sum(w.plus[i]*tc +
     w.minus[i]*tc,i) == 1
  w[i] >= 0
  w.plus[i] >= 0
  w.minus[i] >= 0
}
```

Code 3.8 Optimization with Transaction Constraints

We can illustrate the code with a numerical example. First we generate a set of scenarios for a universe of four assets, assuming equally weighted initial holdings:

```
S <- matrix(rmvnorm(100,
     mean=c(0.02,0.04,0.05,0.08),
     cov=diag(rep(0.2,4))),ncol=4)
w.initial <- rep(1/4,4)
```

Now we can run a single optimization (shown in Code 3.9) and investigate the optimal solution.

```
MV.system.tc <- System(MV.model.tc, S, w.initial,
  mu.target = 0.06, tc = 0.01)
solution <- solve(MV.system.tc, trace = T)
solution$variable$w$current
         1          2         3         4
 0.2056925 0.09749321 0.4485728 0.2442299

attr(, "indexes"):
[1] "i"
sum(solution$variable$w$current)

[1] 0.9959884

round(solution$variable$w.plus$current,digits = 3)

 1 2     3 4
 0 0 0.199 0
attr(, "indexes"):
[1] "i"

round(solution$variable$w.minus$current,digits = 3)

     1     2 3     4
 0.044 0.153 0 0.006
attr(, "indexes"):
[1] "i"
```

Code 3.9 Set of Scenarios for a Universe of Four Assets

Two interesting points are worth mentioning. First, asset weights no longer add up to 100% (99.6% instead), as transaction costs are dragging down initial wealth. Second, adjustments are made either via buying or selling, but no asset is bought and sold at the same time. Although this would have no direct effect on the net change in holdings, it would induce overly high transaction costs.

3.3.3 Fixed Costs

Fixed transaction costs are another go/no-go situation; that is, fixed costs arise as soon as we trade in a particular asset (independent of trade size), while they are zero if no trade takes place. We can hence model the budget constraint for a combination of fixed and proportional costs according to

$$\underbrace{\sum_{i=1}^{n} w_i}_{holdings} + \overbrace{\sum_{i=1}^{n} \left(\delta_i^+ + \delta_i^-\right) \cdot f_i}^{fixed} + \underbrace{\sum_{i=1}^{n} \left(tc_i^+ w_i^+ + tc_i^- w_i^-\right)}_{proportional} = 1. \quad (3.20)$$

Final holdings plus fixed and proportional transaction costs have to add up to the initial budget. Fixed costs are summed with the use of integer variables that take on a value of one if trading takes place and zero otherwise. We assumed fixed costs to be the same for all assets. Extensions are trivial. Variables in (3.20) are defined as follows:

$$\begin{aligned}
& w_i^+ \leq \delta_i^+ \cdot \text{large number} \\
& w_i^- \leq \delta_i^- \cdot \text{large number} \\
& \delta_i^\pm \in \{0,1\} \\
& w_i^+ \geq 0 \\
& w_i^- \geq 0.
\end{aligned} \quad (3.21)$$

For ease of notation, we used constant fixed costs f_i across assets. An extension of notation is trivial. The portfolio optimization problem now becomes

$$\begin{aligned}
& \min \sum_i \sum_j w_i w_j \sigma_{ij} \quad \text{subject to} \\
& \sum_i w_i \left(1+\mu_i\right) = 1+\mu \\
& \sum_{i=1}^{n} w_i + \sum_{i=1}^{n} \left(\delta_i^+ + \delta_i^-\right) \cdot f_i + \sum_{i=1}^{n} \left(tc_i^+ w_i^+ + tc_i^- w_i^-\right) = 1 \\
& w_i = w_i^{initial} + w_i^+ - w_i^- \\
& w_i \geq 0, w_i^+ \geq 0, w_i^- \geq 0 \\
& w_i^+ \leq \delta_i^+ \cdot \text{large number} \\
& w_i^- \leq \delta_i^- \cdot \text{large number} \\
& \delta_i^\pm \in \{0,1\}.
\end{aligned} \quad (3.22)$$

The S-PLUS optimization code for (3.22) is given in Code 3.10 and is a straightforward extension of previous code on portfolio optimization with proportional transaction costs.

```
MV.model.tc <- function(S, w.initial, mu.target,
     tc, fc)
{
  nobs <- nrow(S)
  n <- Set(1:ncol(S))
  m <- Set(1:nrow(S))
  i <- Element(set=n)
  s <- Element(set=m)

  mu.bar <- apply(S, 2, mean)
  S <- Parameter(S, index=dprod(s,i))
  mu.bar <- Parameter(as.array(mu.bar), index=i)
  mu.target <- Parameter(as.numeric(mu.target),
     changeable=T)
  w <- Variable(index=i)
  tc <- Parameter(as.numeric(tc), changeable=T)
  fc <- Parameter(as.numeric(fc), changeable=T)
  w.initial <- Parameter(as.array(w.initial),
     index=i)

  w <- Variable(index=i)
  w.plus <- Variable(index=i)
  w.minus <- Variable(index=i)
  dev <- Variable(index=s)

  Sum((S[s,i]-mu.bar[i])*w[i],i) == dev[s]

  risk <- Objective(type="minimize")
  risk ~ 1/nobs*Sum(dev[s]^2,s)
  Sum((1+mu.bar[i])*w[i],i) >= 1+mu.target
  w.initial[i] + w.plus[i] - w.minus[i] == w[i]
  dummy.plus  <-
     IntegerVariable(index=i,type=binary)
  dummy.minus <-
     IntegerVariable(index=i,type=binary)
  w.plus[i] <= 5*dummy.plus[i]
  w.minus[i] <= 5*dummy.minus[i]
  Sum(w[i], i) +
     Sum(w.plus[i]*tc+w.minus[i]*tc,i) +
     Sum(dummy.plus[i]*fc,i) +
     Sum(dummy.minus[i]*fc,i) == 1
```

```
    w[i] >= 0
    w.plus[i] >= 0
    w.minus[i] >= 0
}
```

Code 3.10 Optimization with Fixed Transaction Constraints

Note that the existence of fixed transaction costs will, ceteris paribus, lead to a larger focus on a small number of assets with which transactions will be performed for any moment in time. Across time, a small number of large adjustments will be preferable to a large number of infinitesimal rebalancing trades.

Exercises

1. Check (3.1). Reformulate it in terms of individual asset betas. What can we say about (3.2) for the minimum variance portfolio?

2. Equal-weight benchmarks try to achieve diversification by investing equal amounts into all assets. However, equal amounts do not lead to equal risk contributions. Construct an equal risk benchmark; (i.e., a portfolio where the percentage contribution for each asset equals the inverse of the number of assets). How would you evaluate such a concept theoretically and practically?

3. Suppose you are given a covariance matrix of returns, a vector of expected returns, and a vector of benchmark weights. Run
   (a) a benchmark-relative optimization without constraints,
   (b) a benchmark-relative optimization with beta neutrality constraint (beta with respect to the benchmark equals one), and
   (c) a total return optimization.
   (d) Plot all three results in active return and active risk space. Explain.

4. Suppose you have ten bonds, each with different yields, durations, and convexities. Write a short program that calculates the maximum yield portfolio for a given duration and convexity target under the constraint that the optimizer needs to pick exactly four bonds, with each bond having at least 20% weight.

5. Suppose you run a standard mean-variance optimization. Suppose further that you include six group constraints (a group constraint limits the sum of a group of assets to fall within boundaries). How would you implement the additional constraint that at least four of these constraints need to be satisfied? Hint: Transform the set of constraints

$$\sum_{i \in group_1} w_i \geq c_1, \cdots, \sum_{i \in group_6} w_i \geq c_6$$

   into

$$\sum_{i \in group_1} w_i \geq \delta_1 c_1, \ldots, \sum_{i \in group_6} w_i \geq \delta_6 c_6, \sum_k \delta_k \geq 4.$$

6. Assume an investor faces two group constraints, but regards them as mutually exclusive. (He wants either constraint one or two to hold, but not

at the same time.) How would you implement this? Hint: If we introduce a large number E, the constraints

$$\sum_{i \in group_1} w_i \geq c_1, \quad \sum_{i \in group_2} w_i \geq c_2$$

can easily be transformed into

$$\sum_{i \in group_1} w_i \geq c_1 + E\delta_1, \quad \sum_{i \in group_2} w_i \geq c_2 + E\left(1-\delta_s\right).$$

7. Transform (3.14) into NUOPT for S-PLUS code.

8. Solve the transaction costs problems described in Section 3.3 with the use of `solveQP()`. Hint: The optimization problem of (3.17) can be rewritten in matrix form:

$$\begin{aligned} &\min_{\mathbf{w}} \mathbf{w}^T \mathbf{\Omega} \mathbf{w} \\ &\mathbf{w}^T \boldsymbol{\mu} = \mu \\ &\mathbf{w}^T \mathbf{1} = 1 \\ &\mathbf{w} = \mathbf{w}^{initial} + \mathbf{w}^+ - \mathbf{w}^- \\ &\mathbf{w}^+ - \mathbf{w}^- \leq \tau \\ &\mathbf{w}, \mathbf{w}^+, \mathbf{w}^- \geq 0. \end{aligned}$$

9. Within the main text, we assumed that transaction costs are fixed and/or proportional. In reality, trading costs are to a large extent made up of liquidity costs (market impact) that are nonlinearly increasing with trade size. Grinold and Kahn (2000, p. 452, Equation 16.4) suggest a transaction cost model of the form

$$tc = commission + \%bid/ask - spread + \theta \sqrt{\frac{trade\ volume}{daily\ volume}}.$$

Linearize and implement this cost function into the code of Section 3.3.

Endnotes

[1] See McCarthy (2000, p. 103).
[2] See Chow and Kritzman (2001, p. 58).
[3] See De Bever et al. (2000, p. 283).
[4] See Rustem and Settergren (2002).
[5] See Chow et al. (1999).
[6] See Shectman (2001).
[7] See Wang (1999).
[8] See Mitra et al. (2003) or Brandimarte (2002).
[9] Inclusion of transaction costs can be found in Mitchell and Braun (2002) or Lobo, Fazel, and Boyd (2002).

4 Resampling and Portfolio Choice

Inputs for portfolio optimization problems are notorious for being measured with substantial estimation error. This is particularly troubling because optimization routines are often characterized as error maximization algorithms, leveraging errors in inputs rather than mitigating their effect. Consequently, financial economists and statisticians have relied on resampling techniques in order to understand the impact of estimation error in means and covariances (inputs) on the distribution of portfolio weights (outputs).[1] In statistics, resampling methods are referred to as bootstrap methods, and there are two basic types: the **parametric bootstrap**, where one fits a parametric model and samples from the fitted parametric model, and the **nonparametric bootstrap**, where one samples directly from the data without fitting a parametric model. See, for example, Efron and Tibshirani (1998) and Davison and Hinkley (1999) for details. In this chapter, we concentrate primarily on the parametric bootstrap using a fitted multivariate normal distribution, as is common in applications to finance.

Throughout the first three sections to follow, a simple numerical example will be used to illustrate the pitfalls of using the center of the resampled weight distribution for portfolio construction exercises. We need to rely on numerical examples in combination with Monte Carlo simulation, as no closed-form solutions are available.

4.1 Portfolio Resampling

Suppose we have estimated a mean vector and a covariance matrix of returns (in the following, we always assume returns come in the form of excess returns) from annual historical data with length n_{hist},

$$\hat{\mathbf{\Omega}}_0 = \begin{pmatrix} 400 & & \\ 210 & 255 & \\ 40 & 15 & 25 \end{pmatrix}, \quad \hat{\mu}_0 = \begin{pmatrix} 6.08 \\ 4.56 \\ 0.94 \end{pmatrix}, \quad n_{hist} = 30. \tag{4.1}$$

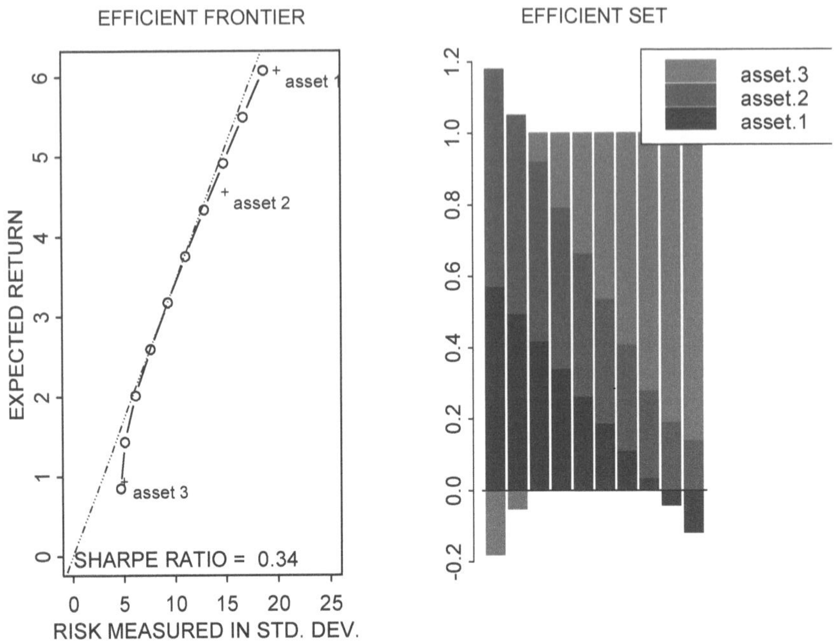

Figure 4.1 Markowitz Portfolios (with Short Selling Allowed)

The unconstrained efficient frontier and corresponding efficient set weights are shown in Figure 4.1. The maximum Sharpe ratio portfolio has weights

$$\mathbf{w}^*_{sharpe} = \frac{\hat{\boldsymbol{\Omega}}_0^{-1}\hat{\boldsymbol{\mu}}_0}{\mathbf{1}'\hat{\boldsymbol{\Omega}}_0^{-1}\hat{\boldsymbol{\mu}}_0} = \left(\tfrac{1}{6} \quad \tfrac{2}{6} \quad \tfrac{3}{6}\right)^T.$$

Investors holding 100% in the maximum Sharpe ratio portfolio exhibit a risk aversion of $\lambda = \mu/\sigma^2 = \mathbf{1}'\hat{\boldsymbol{\Omega}}_0^{-1}\hat{\boldsymbol{\mu}}_0 = 0.038$. As the maximum Sharpe ratio portfolio is the most prominent in finance, we will focus on this portfolio. With the exception of the minimum variance portfolio (which does not require return estimates), everything said in this chapter also applies to all other portfolios on the efficient frontier.

We know that $\hat{\boldsymbol{\Omega}}_0$ and $\hat{\boldsymbol{\mu}}_0$ have been estimated with error. In general, $\hat{\boldsymbol{\Omega}}_0$ is an $n \times n$ matrix, where n denotes the number of assets (here $n = 3$), whereas $\hat{\boldsymbol{\mu}}_0$ is an $n \times 1$ vector. The process of resampling will draw data for a number n_{draw} of returns for each of the n assets from the multivariate normal $N\left(\hat{\boldsymbol{\mu}}_0, \hat{\boldsymbol{\Omega}}_0\right)$. We can use the newly created block of data in the form of an $n_{draw} \times n$ matrix of asset returns (here 30×3) to construct a new mean vector and covariance matrix estimates $\hat{\mathbf{\Omega}}_1$ and $\hat{\boldsymbol{\mu}}_1$. It is often natural to set $n_{draw} = n_{hist}$, but this is not necessary. Obviously, the original and the resampled matrices will differ due to sampling error. The degree of difference will depend on n_{draw}. If we make n_{draw} small, our estimates will fluctuate greatly, while we will find much less difference for a large n_{draw}.[2] Repeating this procedure n_{sim} times, we create a large number of varying input vectors: $(\hat{\mathbf{\Omega}}_1, \hat{\boldsymbol{\mu}}_1, \ldots, \hat{\mathbf{\Omega}}_{n_{sim}}, \hat{\boldsymbol{\mu}}_{n_{sim}})$. We now ask ourselves what choices we would make if we repeatedly constructed optimal portfolios $\mathbf{w}_i$ from these resampled inputs and what insights can be gained from this exercise.

In order to ensure that decisions are indeed comparable across simulations, we assume that investors maximize $U(\mathbf{w}) = \mathbf{w}^T\boldsymbol{\mu} - \frac{\lambda}{2}\mathbf{w}^T\mathbf{\Omega}\mathbf{w}$, where the first-order conditions lead to the familiar formulas $\mathbf{w} = \lambda^{-1}\mathbf{\Omega}^{-1}\boldsymbol{\mu}$ $\boldsymbol{\mu} = \lambda\mathbf{\Omega}\mathbf{w}$ for optimal weights and implied returns. Note that in the current example $\lambda = 0.038$ remains constant through all simulations. As a start, we sample $n_{draw} = n_{hist} = 30$ returns from $N(\hat{\boldsymbol{\mu}}_0, \hat{\mathbf{\Omega}}_0)$ and compute the sample mean vector (for simplicity, covariances are assumed to be measured without error) and the corresponding optimal portfolio with a full investment constraint (i.e., weights need to add up to 100%). This is then repeated n_{sim} times for $n_{sim} = 1, 2, \cdots, 500)$.

We measure the distance between the center of the weight distribution and the original maximum Sharpe ratio portfolio that was constructed without taking estimation error into account (i.e., the maximum Sharpe ratio for the portfolio based on $\hat{\boldsymbol{\mu}}_0, \hat{\mathbf{\Omega}}_0$) as the squared Euclidean distance

$$\left(\bar{\mathbf{w}} - \mathbf{w}^*_{Sharpe}\right)^T \left(\bar{\mathbf{w}} - \mathbf{w}^*_{Sharpe}\right), \text{ where } \bar{\mathbf{w}} = \frac{1}{n_{sim}} \sum_{i=1}^{n_{sim}} \mathbf{w}_i,$$

where $\mathbf{w}_i$ is the optimal weight vector for the *i*-th simulation.

It can be seen in Figure 4.2 that the distance between the center of the resampled distribution and the maximum Sharpe ratio portfolio converges to zero fairly rapidly as n_{sim} increases. Effectively this means that the center of the weight distribution recovers the original maximum Sharpe ratio portfolio. Alternatively, we can say that $\bar{\mathbf{w}} = \mathbf{w}^*_{Sharpe} + noise$, where the noise goes to zero fairly rapidly as n_{sim} increases. In this case, the use of resampling in creating new portfolios adds only noise to the portfolio construction.

Figure 4.3 visualizes the distribution of portfolio weights for $n_{sim} = 500$. Large positive or negative weights can occur in single simulation runs but will be averaged out. This is true for every number of draws n_{draw} per resampling.

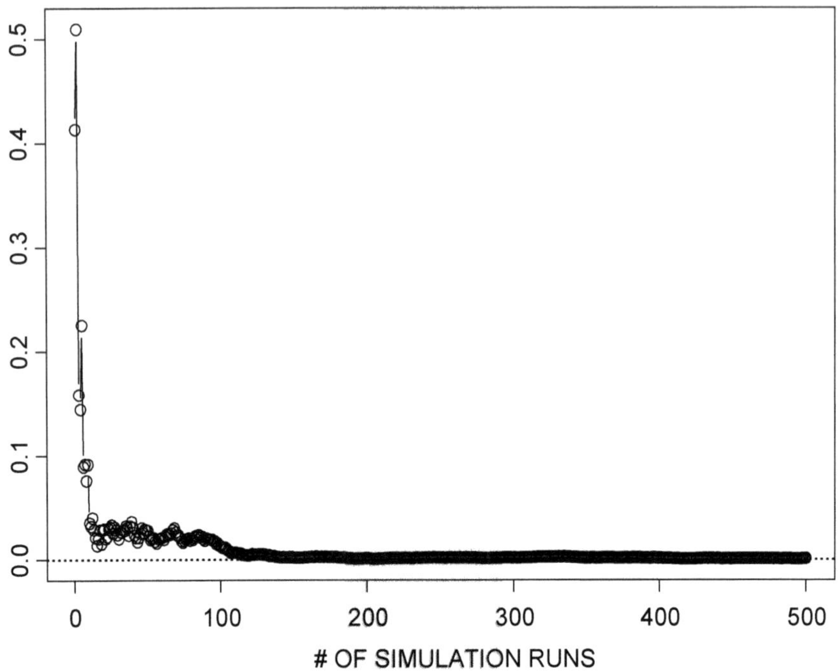

Figure 4.2 Resampling and Convergence (Short-Selling Allowed)

It is apparent from the results that repeatedly drawing average returns and subsequently averaging across optimally constructed portfolio weights, yields the same result as averaging across returns in the first place and then using the averaged returns for portfolio optimization. We could have seen that without having to go through the simulation exercise[3]:

$$\begin{aligned}\bar{\mathbf{w}} &= \frac{1}{n_{sim}}\sum_{i=1}^{n_{sim}} \mathbf{w}_i = \frac{1}{n_{sim}}\sum_{i=1}^{n_{sim}} \lambda^{-1}\mathbf{\Omega}^{-1}\hat{\boldsymbol{\mu}}_i \\ &= \lambda^{-1}\mathbf{\Omega}^{-1}\frac{1}{n_{sim}}\sum_{i=1}^{n_{sim}} \hat{\boldsymbol{\mu}}_i \\ &= \lambda^{-1}\mathbf{\Omega}^{-1}\bar{\boldsymbol{\mu}}.\end{aligned} \tag{4.2}$$

Note that we simulated the effect of estimation error on the distribution of portfolio weights. Neither the average portfolio nor its risk changed. However, we know that if investors are uncertain about their inputs, estimation risk will add to investment risk and the world will become a riskier place. Computing average weights based on resampling is unable to catch this effect, as it is not designed to do so.[4] However, straightforward bootstrap resampling of quantities such as the Sharpe ratio and the return and risk of the tangency portfolio can

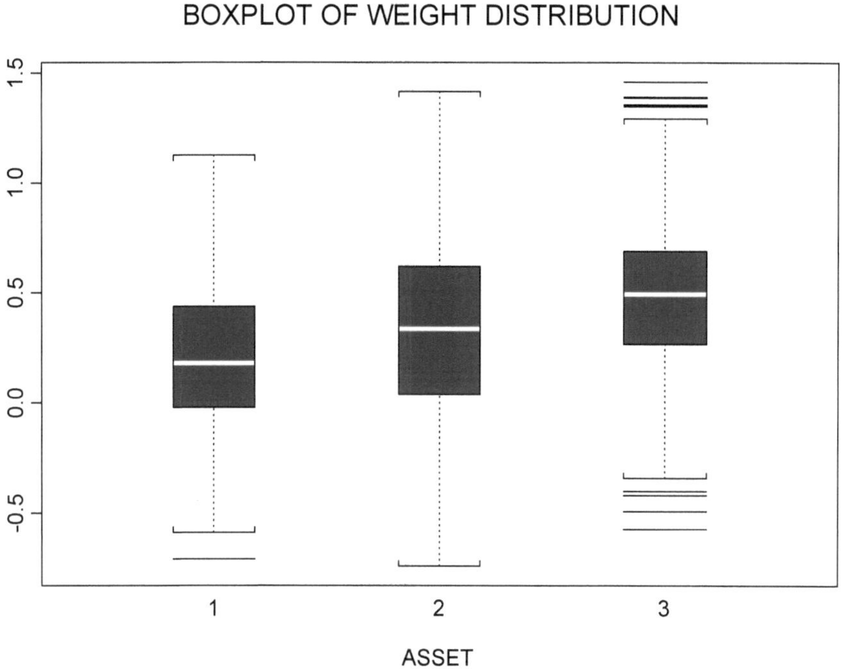

Figure 4.3 Distribution of Resampled Weights (Short-Selling Allowed)

indeed provide measures of uncertainty of operating points (see Sections 4.6 and 6.9.4).

In order to replicate the results above, readers can use Code 4.1, which works for both long/short `(short=T)` and long-only `(short=F)` optimization.

```
portfolio.resampling <- function(cov, fcst, n.sim,
  n.draw, short)
{
  resampled.pf <- matrix(0,ncol=ncol(cov),
     nrow=(n.sim+2))
  frontier.uc <- portfolioFrontier(cov,fcst,
     max.ret=max(fcst),n.ret=1000,
     unconstrained=short)
  iopt <- order(frontier.uc$returns/
     frontier.uc$sd)[1000]
  lambda <- frontier.uc$returns[iopt]/
     frontier.uc$sd[iopt]^2
  resampled.pf[(n.sim+1),] <-
     frontier.uc$weights[,1]
  resampled.pf[(n.sim+2),] <-
     frontier.uc$weights[,iopt]
```

```
   group <- matrix(rep(1, ncol(cov)), nrow=1)
   if(short==T) {
      bUP <- c(rep( Inf, ncol(cov)))
      bLO <- c(rep(-Inf, ncol(cov)))
   }
   if(short==F) {
      bUP <- c(rep(1, ncol(cov)))
      bLO <- c(rep(0, ncol(cov)))
   }
   cUP <- c(1)
   cLO <- c(1)
   for(i in 1:n.sim){
      x <- rmvnorm(n.draw, fcst, cov)
      cov.r <- var(x)
      fcst.r <- apply(x,2,mean)
      resampled.pf[i,] <- solveQP(-lambda*cov,
         fcst.r, group, cLO, cUP, bLO, bUP, ,
         type=maximize, trace=F)$variables$x$current
      cat(" run ", i, "\n")
   }
   list(resampled=resampled.pf)
}
```

Code 4.1 Portfolio Resampling and Weight Convergence

The first part of the function calculates the mean-variance frontier without estimation error. We can also infer the maximum Sharpe ratio portfolio from this (assuming expected returns and covariances are derived using the risk premium rather than total return).

4.2 Resampling Long-Only Portfolios

So far, we have allowed for short-selling in portfolio construction. We have seen that in this case the average resampled portfolio only adds noise to Markowitz portfolios. In this section, we will drop the possibility of going short in individual assets and return to more conventional portfolio optimization using a long-only constraint. Apart from this, we will perform the same calculations as in the previous section.

The first thing to note about Figure 4.4 is that distance (deviation from the estimation error-free solution) is much smaller when short-selling is not allowed, as the long-only constraint reduces the opportunities to leverage on information. We can also see that the distance measure in our simulations does not converge to zero. This means that repeatedly sampling with $n_{draw} = 30$ does

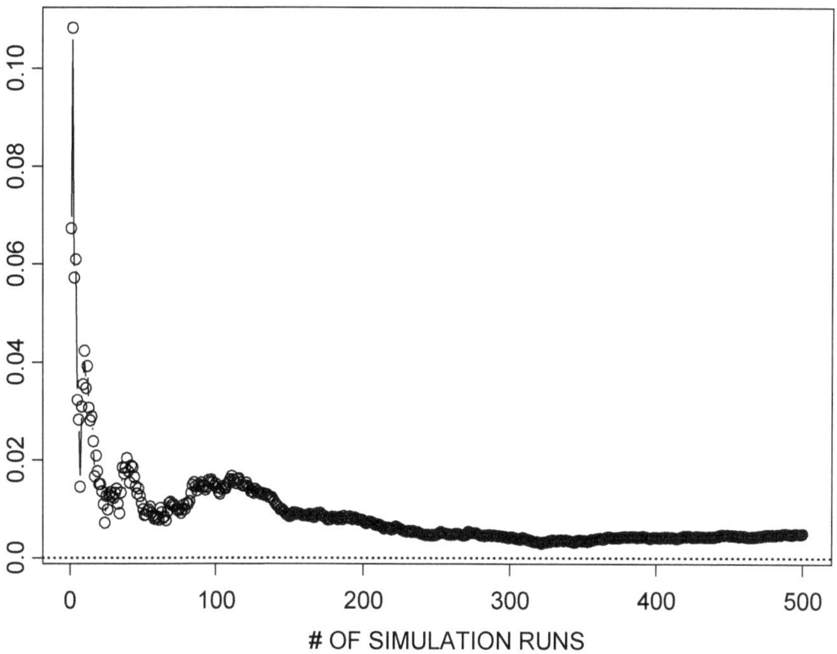

Figure 4.4 Resampling and Convergence (no Short-Selling Allowed)

not recover the Markowitz solution. Hence we get $\bar{\mathbf{w}} = \mathbf{w}^*_{Sharpe} + bias + noise$, where the noise goes to zero as the number of simulations increases but the bias does not. A look at Figure 4.5 provides the reason for this bias.

Weights that are less than zero due to a downward bias in some simulations can no longer be implemented. Hence, averaging will not lead back to the Markowitz solution, as individual assets are now either in or out but never short. The higher the volatility of an asset and/or the smaller n_{draw}, the more pronounced this effect will be. The next section will elaborate on this in more detail.

4.3 Introduction of a Special Lottery Ticket

In order to magnify the effect we just learned in the previous section and to show its relevance for asset allocation decisions, we will introduce a special lottery ticket with zero risk premium into our analysis. Lottery tickets are investments that offer diversification, as they are by definition uncorrelated with all other assets. Since our lottery ticket has zero expected excess return, it exposes investors to high volatility with no expected reward. Broadening the investment universe with lottery tickets should not improve the efficient frontier by pushing it up and to the left. Any asset allocation mechanism that systematically invests in lottery tickets should be treated with utmost caution. The following calculations are based on a lottery ticket with 60% volatility, 0% expected return, and zero covariance with existing assets.

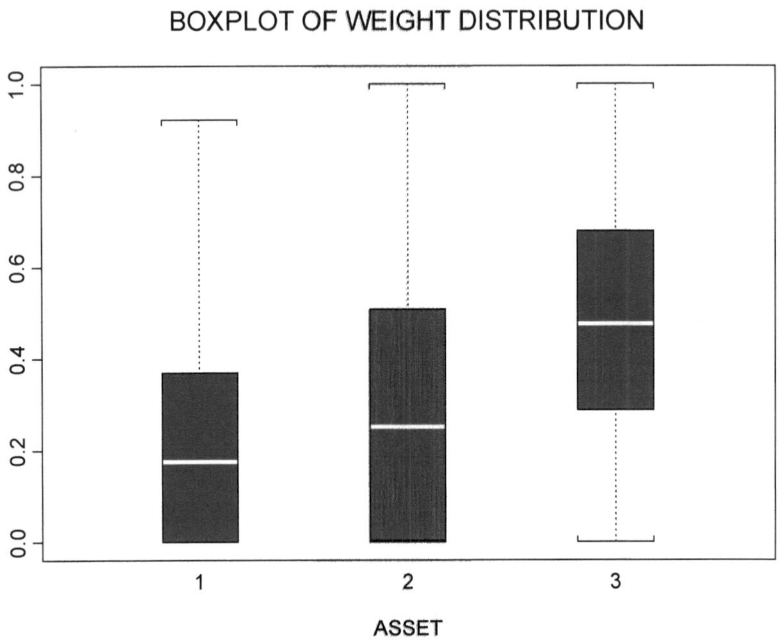

Figure 4.5 Distribution of Resampled Weights (no Short-Selling Allowed)

We repeat the previous calculations, where asset 4 represents the lottery ticket. Figure 4.6 and Figure 4.7 summarize the results. Note that the maximum Sharpe ratio portfolio derived from traditional mean-variance analysis does not allocate to the lottery ticket. Introducing a lottery ticket increases our distance measure in Figure 4.6 for a sufficiently large number of simulations. This should come as no surprise, as allocations to the lottery ticket amount to as high as 22% for some allocation runs, while we can never short the lottery ticket, even for those runs with large negative average returns. It is the long-only constraint that essentially transforms asset volatility into portfolio allocations. However, this does not necessarily mean that the higher the volatility of our lottery ticket the larger the allocation will become, as there are two separate effects at work. Higher volatility induces an upward bias into the average resampled weight, but at the same time higher volatility makes the lottery ticket less attractive, as it worsens the risk-return trade-off for any given risk aversion. While the first effect becomes obviously predominant for the maximum return portfolio, its exact trade-off depends on the risk aversion.

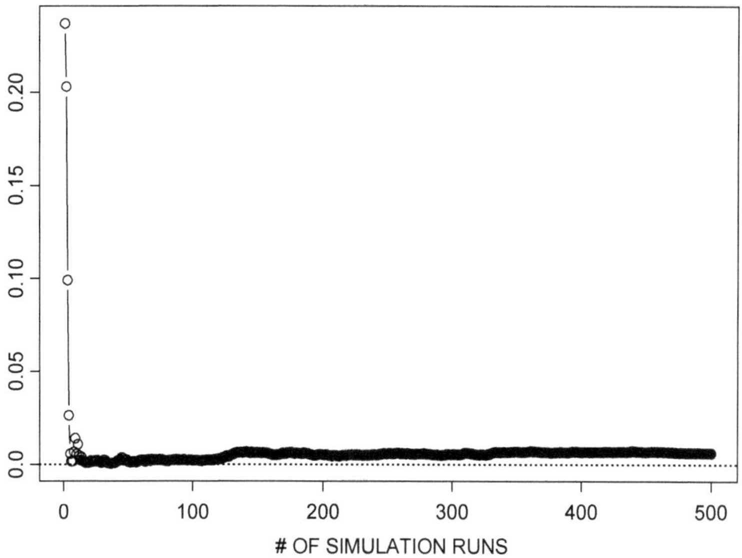

Figure 4.6 Resampling and Convergence (Lottery Ticket and Long-only Constraint)

Figure 4.8 shows that for a reasonably high risk aversion of $\lambda = 0.038$, increasing the volatility of the lottery asset will reduce the average allocation due to the higher risk. The volatility bias is still present, but the direct risk effect more than compensates for the upwards bias induced by high average returns for some simulation runs. For a low risk aversion of $\lambda = 0.01$, this effect also exists, but it starts at higher volatility levels. Up to a volatility level of 30%, the resampling bias dominates. From then on, the direct risk effect leads to smaller allocations even though the long-only constraint leads to more and more serious artifacts.

At this point, it is interesting to see what happens in a world that is affected by the same uncertainty about the correct inputs but that differs in institutional constraints. In short: does the introduction of a lottery ticket also have biased allocations if we are allowed to engage in short-selling? Note that allowing short-selling will not decrease the amount of estimation risk in the world. If anything, the opportunity to go short will increase the estimation error, as the optimizer can now establish long and short positions between similar highly correlated assets that look almost risk-free but yield large returns. Obviously, those almost arbitrage situations are most likely to be created by estimation error.

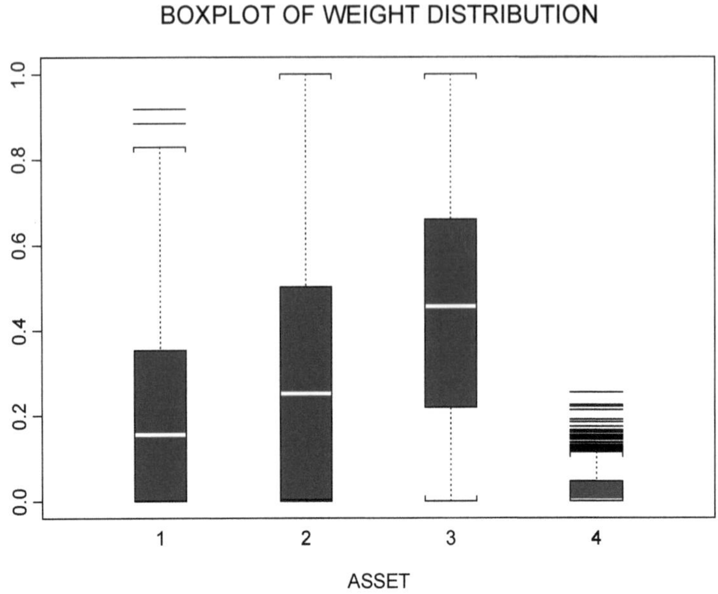

Figure 4.7 Distribution of Free Sampled Weights (Lottery Ticket and Long-Only Constraint)

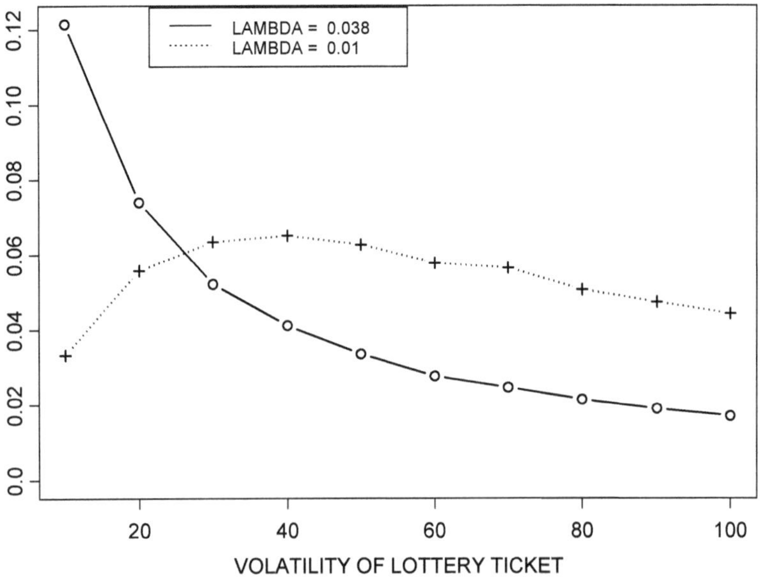

Figure 4.8 Risk Aversion and Volatility Bias

We see in the simulation results of Figure 4.9 that with short-selling allowed, the weights for the lottery ticket allocation scatter symmetrically around an average weight of 0%. Large positive allocations are counterbalanced on average by large negative allocations. Resampling without short-selling constraints helps us appreciate the dispersion in outcomes, while at the same time the average resampled weight is the same as the Markowitz weight.

Another way to look at portfolio resampling is to back out the implied returns of the average resampled portfolio. For $n_{draw} = 30$ and $n_{sim} = 500$, we arrive at average resampled weights $\overline{\mathbf{w}} = (0.21 \quad 0.32 \quad 0.44 \quad 0.03)^T$. In this case, one can check that the implied returns $\boldsymbol{\mu}_{implied} = \lambda\hat{\boldsymbol{\Omega}}_0\overline{\mathbf{w}}$ differ substantially from our original forecasts $\boldsymbol{\mu}_{implied} = (6.45 \quad 4.67 \quad 0.92 \quad 4.15)^T$. The risk premium for the lottery ticket in the latter case is more than 4%, compared with 0% for the portfolio with short-selling allowed. However, this is not plausible, as estimation error should not affect expected returns. By definition there is no uncertainty about expected returns. Estimation error without additional information should instead be reflected in the inflation of risk estimates, which now contain investment risk as well as estimation risk.[5]

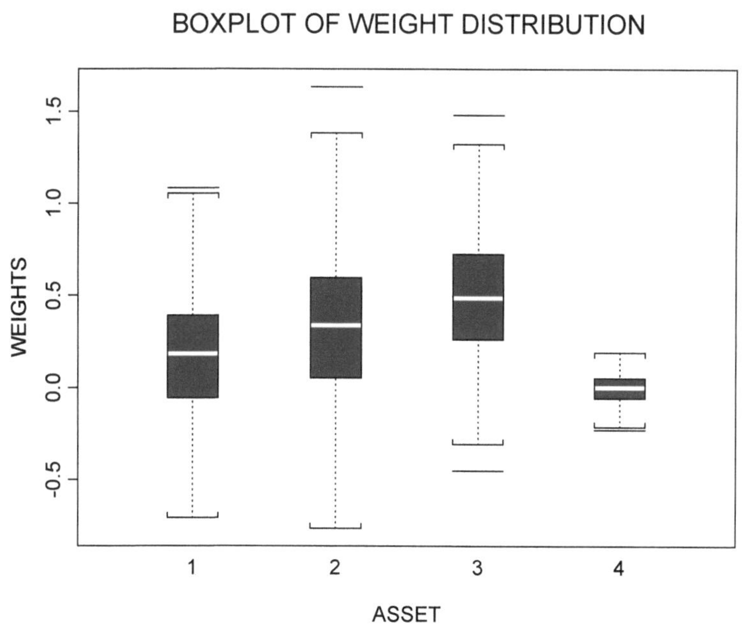

Figure 4.9 Distribution of Free Sampled Weights (Lottery Ticket without Long-only Constraint)

In order to appreciate the impact that the number of draws per resampling n_{draw} has on the allocation of our lottery ticket for a long-only portfolio, we repeat a large number of resamplings $n_{sim} = 100{,}000$ with antithetic variance reduction for various levels of n_{draw}. The results of these simulations are plotted in Figure

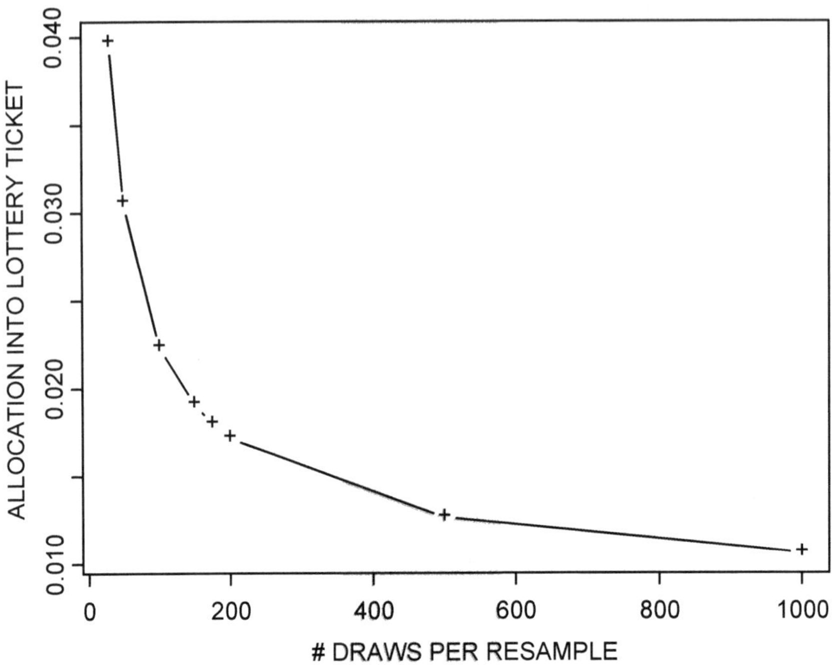

Figure 4.10 Allocation into Lottery Ticket versus n_{draw}

4.10. As the number of resamplings increases, and consequently the variance of estimated parameters decreases, the allocation into our lottery ticket decreases. At first sight this seems to be a confirmation of the concept of resampling. After all, a large number of draws per resampling means confidence in our inputs, in which case we would expect to recover the Markowitz solution. However, it is important to understand that no such effect exists if we allow short-selling. The average allocation into the lottery ticket would be independent of the number of draws even though the estimation error is the same. It is the long-only constraint that transforms asset volatility into asset allocation, implicitly raising the expected return for highly volatile assets.

4.4 Distribution of Portfolio Weights

The resampling procedure that results in a sequence of new covariance matrix and mean vector estimates allows us to generate a resampled set of optimal portfolio weights, thereby giving us an estimate of the distribution of portfolio weights. This in turn allows us to test whether two portfolios are statistically different using an appropriate distance in n-dimensional vector space. It may be tempting to use the simple Euclidean distance measure for the distance of a vector $\mathbf{w}_i$ of portfolio weights from the vector $\mathbf{w}_p$ of portfolio weights of another portfolio given by

$$\left(\mathbf{w}_p - \mathbf{w}_i\right)^T \left(\mathbf{w}_p - \mathbf{w}_i\right). \tag{4.3}$$

However, this is not the appropriate distance for correlated returns, and instead, under appropriate conditions, the proper statistical distance is given by

$$\left(\mathbf{w}_p - \mathbf{w}_i\right)^T \mathbf{\Omega}_{\mathbf{w}}^{-1} \left(\mathbf{w}_p - \mathbf{w}_i\right), \tag{4.4}$$

where $\bar{\mathbf{\Omega}}_{\mathbf{w}}$ is the variance-covariance matrix of portfolio weights $\mathbf{w}_i$ and $\mathbf{w}_p$ is the mean value of $\mathbf{w}_i$. When the $\mathbf{w}_i$ are normally distributed, this test statistic is distributed as a χ^2 with degrees of freedom equal to the number of assets. In the statistical literature this distance is known as the **Mahalanobis distance**, and an intuitive explanation of the distance is provided in Section 6.6.[6]

Suppose for simplicity that we have two assets with 10% mean and 20% volatility each. Suppose further that the correlation between the two assets is zero and the risk aversion coefficient is $\lambda = 5$. The optimal solution without estimation error is given below:

$$\mathbf{w}^* = \begin{bmatrix} w_1^* \\ w_2^* \end{bmatrix} = \lambda^{-1}\mathbf{\Omega}^{-1}\boldsymbol{\mu} = 0.2\begin{bmatrix} \frac{1}{(0.2)^2} & 0 \\ 0 & \frac{1}{(0.2)^2} \end{bmatrix}\begin{bmatrix} 0.1 \\ 0.1 \end{bmatrix} = \begin{bmatrix} 0.5 \\ 0.5 \end{bmatrix}. \tag{4.5}$$

A resampled version of (4.5) is now easily obtained with a few lines of S-PLUS code (see Code 4.2), assuming that the returns are normally distributed.

```
# inputs
Cov <- diag(rep(0.2^2,2))
mu.bar <- c(rep(0.1,2))
n.sim <- 1000
n.draw <- 60
lambda <- 0.2
# simple resampling function
resampling <- function(Cov, mu.bar, n.sim, n.draw,
  lambda)
{
  resampled.weights <- matrix(0, n.sim, ncol(Cov))


  for(i in 1:n.sim) {
     resampled.returns <- rmvnorm(n.draw, mu.bar,
```

```
        Cov)
      VarCov <- var(resampled.returns)
      Mean <- apply(resampled.returns,2,mean)
      w <- lambda*solve(VarCov)%*%Mean
      resampled.weights[i,] <- t(w)
  }
  list("resampled.weights"=resampled.weights)
}
# plot results
x <- resampling(Cov, mu.bar, n.sim, n.draw,
  lambda)$resampled.weights
plot(x[,1],x[,2],xlab="weight asset 1",
  ylab="weight asset 2")
```

Code 4.2 Portfolio Resampling and Weight Distribution

Note that for illustrative purposes we have calculated optimal portfolios without full investment constraints in the code above. Because these portfolios do not require holdings to add up to one, one might be tempted to conclude that these are not portfolios. But one could think of cash as a third (filling) asset, as cash would leave the marginal risks of the portfolio, as well as the total risk of the risky portion of the portfolio, unchanged. While the optimal solution weight is 50% for both assets, Figure 4.11 shows that the estimated weights are scattered around this solution. Comparing the vector difference with an appropriate percentage point (e.g., the upper 95% point) of a chi-squared distribution with two degrees of freedom yields a measure of how statistically different a portfolio is from the optimum.

From the definition of optimal weights, one sees that for our simple example the covariance matrix of the resampled weights is given by

$$\begin{aligned} \mathbf{\Omega_w} &= \operatorname{cov}(\lambda - 1\mathbf{\Omega}^{-1}\hat{\boldsymbol{\mu}}) \\ &= \lambda^{-2}\mathbf{\Omega}^{-1}\operatorname{cov}(\hat{\boldsymbol{\mu}})\mathbf{\Omega}^{-1} \\ &= \lambda^{-2}\mathbf{\Omega}^{-1}\frac{\mathbf{\Omega}}{n_{draw}}\mathbf{\Omega}^{-1} \\ &= \frac{(.2)^2}{n_{draw}}\mathbf{\Omega}^{-1} \\ &= \frac{1}{n_{draw}}\begin{bmatrix} 1 & 0 \\ 0 & 1 \end{bmatrix}. \end{aligned} \tag{4.6}$$

Since $n_{draw} = 60$ in our example, this gives $1/\sqrt{60} = .13$ as the standard errors of the weights. This is consistent with the display in Figure 4.11.

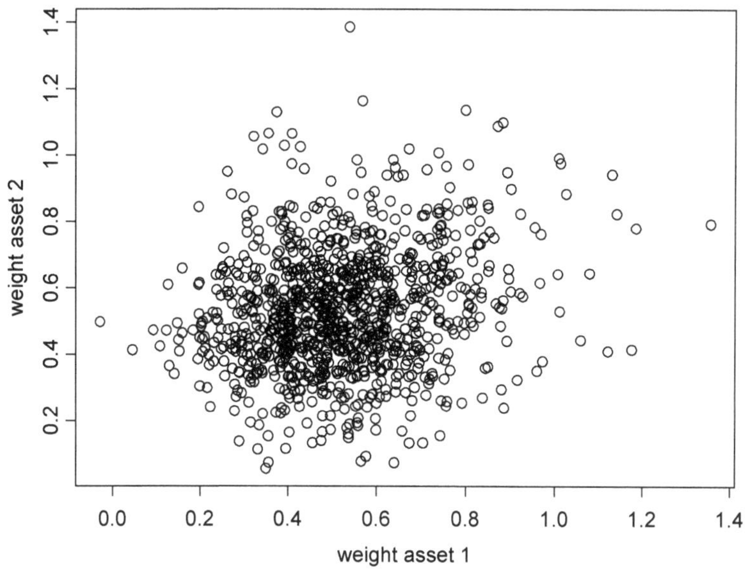

Figure 4.11 Estimation Error and Portfolio Weights

We remark that, for fully invested portfolios, the n-dimensional vector of weights will lie on an $n-1$ dimensional hyperplane that intersects the coordinate axes at the value one. (In this simple two-dimensional case, the weights lie along a line through the points (0,1) and (1,0).) In such cases, we can simply look at the distribution of $n-1$ of the weights in the $n-1$ dimensional subspace. In our simple example above, this would amount to looking at just one weight, which is not very interesting.

Now, to be ever so slightly realistic, let's consider the estimated bivariate distribution of the weights based on observed data that are assumed to be normally distributed according to an estimated mean and covariance for the weights, $\hat{\mathbf{w}}^*$ and $\widehat{\mathbf{\Omega}}_W$. For simplicity, we will use the true optimal weights $\mathbf{w}^* = (.5,.5)'$, leaving it to the reader to repeat the experiment with $\hat{\mathbf{w}}^*$, and compute the estimate $\widehat{\mathbf{\Omega}}_W$ directly from the resampled weights (rather than using the previous formula). The resulting bivariate density is

$$p(w_1, w_2) = \frac{1}{2\pi \det\left(\hat{\mathbf{\Omega}}_w\right)^{\frac{1}{2}}} e^{-\frac{1}{2}\begin{bmatrix} w_1 - w_1^* \\ w_2 - w_2^* \end{bmatrix}' \hat{\mathbf{\Omega}}_W^{-1} \begin{bmatrix} w_1 - w_1^* \\ w_2 - w_2^* \end{bmatrix}}.$$

Code 4.3 generates the perspective plot of Figure 4.12 and the contour plot of Figure 4.13. For the case depicted in these figures, the estimated inverse covariance matrix of the weights was

$$\hat{\mathbf{\Omega}}_W^{-1} = \begin{bmatrix} 27.93 & 0.005 \\ 0.005 & 27.76 \end{bmatrix}.$$

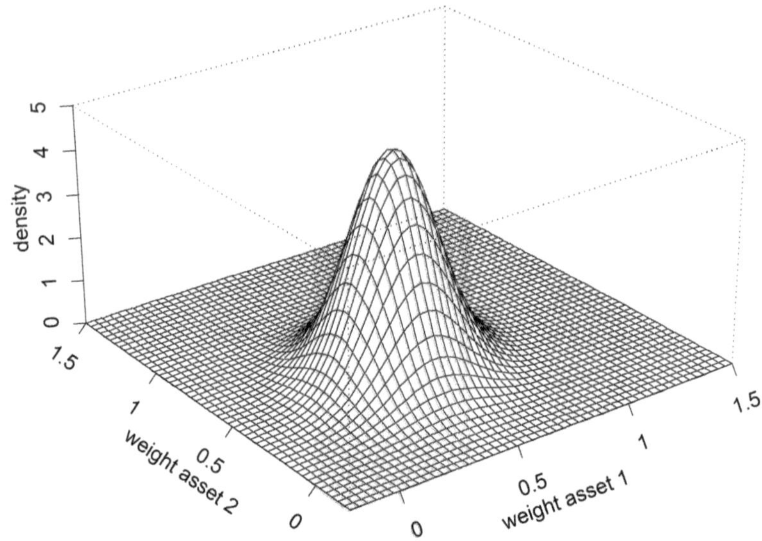

Figure 4.12 Bivariate Normal Weight Distribution for Resampled Portfolio Weights

```
Cov.w <- var(x)
w1 <- seq(-0.2, 1.5,length=100)
w2 <- seq(-0.2, 1.5,length=100)

f1 <- function(w1,w2)
{
  S <- solve(Cov.w)
  d <- (w1-0.5)^2*S[1,1]+(w2-0.5)^2*S[2,2]+
      2*(w1-0.5)*(w2-0.5)*S[1,2]
  1/(2*pi*sqrt(det(Cov.w)))*exp(-1/2*d)
}

z <- outer(w1,w2,f1)
graphsheet()
persp(w1, w2, z, xlab="weight asset 1",
```

```
  ylab="weight asset 2", zlab="density")
graphsheet()
contour(w1, w2, z, nlevels=10,
  xlab="weight asset 1", ylab="weight asset 2")
points(x[,1], x[,2])
```

Code 4.3 Portfolio Resampling and Weight Distribution

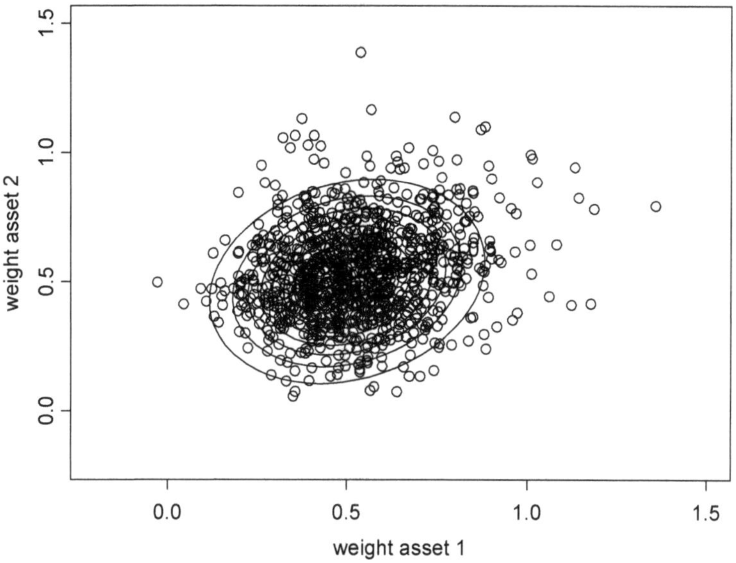

Figure 4.13 Weight Distribution and Lines of Constant Density

Michaud (1998) uses a different distance measure that is widely applied in asset management. His measure recognizes that two portfolios with the same risk and return might actually exhibit different allocations. The distance between two portfolios is defined as

$$\left(\mathbf{w}_p - \mathbf{w}_i\right)^T \hat{\boldsymbol{\Omega}}_0 \left(\mathbf{w}_p - \mathbf{w}_i\right), \tag{4.7}$$

which is equivalent to the squared tracking error. The procedure runs as follows:

Step 1. Define a portfolio against which to test the difference. Calculate (4.4) for all resampled portfolios.

Step 2. Sort the portfolios by tracking error in descending order (highest on top).

Step 3. Define TE_α as the critical tracking error for the $\alpha\%$ level (i.e., if 1000 portfolios are resampled and the critical level is 5%, then look at the tracking error of a portfolio that is 50th from the top). Hence, all portfolios for which $\left(\mathbf{w}_p - \mathbf{w}_i\right)^T \hat{\mathbf{\Omega}}_0 \left(\mathbf{w}_p - \mathbf{w}_i\right) \geq \mathrm{TE}_\alpha^2$ are labeled statistically different.

Step 4. Calculate the minimum and maximum allocations for each asset within the acceptance region.

For a three-asset example, the uncertainty about the optimal weights can be visualized, but it becomes "quite hard" for higher dimensions.

It should be noted that similarity is defined with regard to the optimal weight vector rather than in terms of risk and return. Two portfolios could be very similar in terms of risk and return but very different in allocation. This is well-known, as risk/return points below the frontier are not necessarily unique. Even so, this test procedure is intuitive. It should be noted that the dispersion in weights is large, so it will be difficult to reject the hypothesis that both portfolios are statistically equivalent even if they are not. The power of the suggested test is expected to be low.

4.5 Theoretical Deficiencies of Portfolio Construction via Resampling

4.5.1 Aggregation Problems

Constructing "optimal" portfolios using portfolio resampling requires that we average portfolios in some way (e.g., we average portfolios that carry either the same rank or the same risk-return trade-off).[7] In the case of no long-only constraints, the concept of resampled efficiency will coincide with Markowitz efficiency in the large sample limit (i.e., resampled efficiency in finite sample sizes equals Markowitz efficiency plus noise). Note that even though all inputs are measured with error, resampled efficiency will not pick this up. Asset risk remains unchanged even though the world becomes much riskier in the presence of estimation error.

In the case of long-only constraints, the situation changes considerably. As assets can never be short, we will see that for some resamplings the maximum return portfolio will be 100% cash. This leads to a sampling of cash into the maximum return portfolio. Another consequence is that we cannot engineer portfolios that exhibit low λ's (without long-only constraints, we could have always shorted assets with a negative risk premium), which makes the similarity of rank- and lambda-based approaches questionable. Note also that the inclusion of cash in the maximum return portfolio contrasts both with intuition and portfolio theory. In the case of estimation error, investors will still hold a combination of cash and a market portfolio with the same composition as in the case of no estimation error, with more weight being put on cash, as cash carries no investment risk and is free of estimation error.

Finally, we note that the average is a poor indicator for the center of a distribution that is asymmetric due to heavy truncation at both ends (between 0% and 100%).

4.5.2 Overdiversification

A portfolio construction methodology that allocates to *every* single asset in the universe across *all* portfolios along the efficient frontier creates overdiversification. The combination of the long-only constraint and portfolio resampling will allocate even to dominated assets as long as a lucky draw makes them attractive, while the worst that can happen in all other allocations is a zero weight. Hence, the increase in risk per unit of expected return is not due to estimation error but rather due to overdiversification.

4.5.3 Optionality Problem

Suppose two assets possess the same expected return but one of them has a significantly higher volatility. One could think of this as an international fixed income allocation on a hedged and unhedged basis. Most practitioners (and the mean-variance optimizer) would exclude the higher-volatility asset from the solution unless it has some desirable correlations. How would resampled efficiency deal with these assets? Repeatedly drawing from the original distribution will result in draws for the volatile asset with highly negative returns as well as highly positive returns. Quadratic programming will heavily invest in this asset in the latter case and short the asset in the former case. However, as shorting is not allowed for portfolios with long-only constraints, this will result in positive allocation for draws with high positive average return and zero allocations for draws with high negative average return. This is different from an unconstrained optimization, where large long positions would be offset on average by large negative positions. Consequently, an increase in volatility will yield an increase in the average allocation, and a worsening Sharpe ratio would be accompanied by an increase in weight. This is not a plausible result. It arises directly from the averaging rule in combination with a long-only constraint that results in assets being either in or out but never negative. This behavior is a kind of optionality in which the holder is hurt in terms of bias in the weights whenever a long-only constraint forces otherwise negative coefficients to be positive and less than one in value.

4.5.4 Statistical Foundation Issues

Estimation Error Heritage. All resamplings are derived from the same initial estimates $\hat{\boldsymbol{\Omega}}_0$, $\hat{\boldsymbol{\mu}}_0$ of the covariance matrix and mean returns. However, the true distribution is unknown. Hence, all resampled portfolios will suffer from the deviation of the estimates $\hat{\boldsymbol{\Omega}}_0$, $\hat{\boldsymbol{\mu}}_0$ from their true values $\boldsymbol{\Omega}_{true}$, $\boldsymbol{\mu}_{true}$, in the same way. Averaging will not help very much in this case, as the averaged weights are the result of an input vector, which itself is very uncertain. Hence, it is fair to say that all resampled portfolios inherit the same fundamental estimation error. The utility of normal distribution parametric resampling relies on the assumption that $\hat{\boldsymbol{\Omega}}_0$, $\hat{\boldsymbol{\mu}}_0$, is reasonably close to $\boldsymbol{\Omega}_{true}$, $\boldsymbol{\mu}_{true}$. If this is not the case, the estimation error in $\hat{\boldsymbol{\Omega}}_0$, $\hat{\boldsymbol{\mu}}_0$ is passed on to $\hat{\boldsymbol{\mu}}_1, \hat{\boldsymbol{\Omega}}_1$, $\hat{\boldsymbol{\mu}}_2, \hat{\boldsymbol{\Omega}}_2$, ..., which one might call "estimation error heritage" (see Figure 4.14).[8]

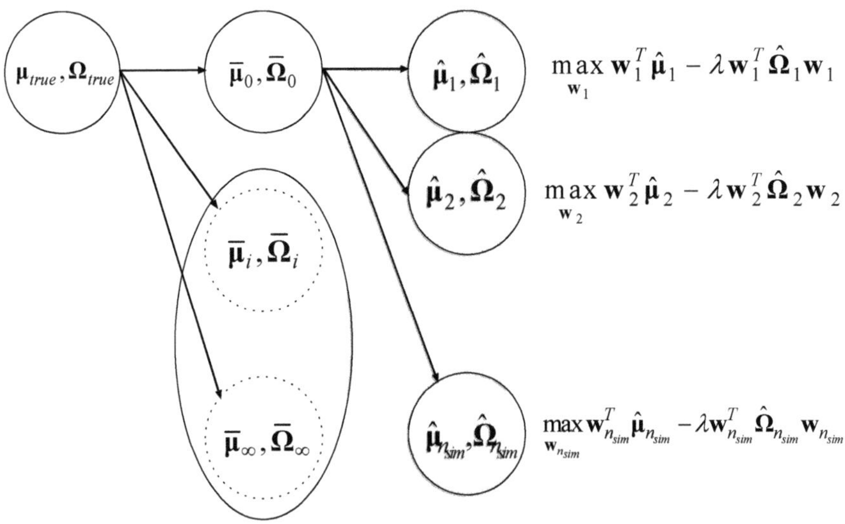

Figure 4.14 Resampling and Estimation Error Inheritance

Parametric Bootstrap Limitations. We know that asset returns are not normally distributed and in some cases are quite non-normal. That makes use of a normal distribution parametric bootstrap highly suspect. One might well turn to a multivariate non-normal distribution such as a multivariate *t* distribution. This requires careful estimation of the degrees of freedom and robust estimation of the mean vector and covariance matrix. Another approach is to use a nonparametric bootstrap that makes no assumptions about the distribution of the returns, as discussed in Sections 4.6 and 6.9.4.

Resampling Bayes. Sometimes it is argued that $\hat{\boldsymbol{\Omega}}_0$, $\hat{\boldsymbol{\mu}}_0$ does not need to be estimated from historical data but can also be the result of Bayesian calculations. However this is entirely against the spirit of Bayesian statistics. Once we calculate the predictive distribution, we have already put in all our subjectivity, and the Bayesian has to accept the result. Resampling from predictive distributions in order to construct better portfolios is pointless.

4.5.5 Lack of Decision-Theoretic Foundation

Resampled efficiency has no decision-theoretic foundation and as such it is questionable whether its use is fiduciary. What resampling actually achieves is some sort of return shrinkage in the presence of long-only constraints. Backing out implied returns from average resampled portfolios already revealed to us that low returns of relatively high-risk assets tend to be adjusted upward and vice versa. The advantage of this form of shrinkage over classical shrinkage methods is that portfolios constructed from it add up to 100%. This is not the case for the statistical shrinkage model, which in addition may still lead to concentrated corner portfolios. However, while we have perfect control over the latter, this cannot be said about the implied returns from resampling.

4.6 Bootstrap Estimation of Error in Risk-Return Ratios

4.6.1 The Problem

Reported risk-return ratios relate average returns to alternative measures of risk and hence involve the ratio of a random numerator and denominator (due to sampling error). As such, point estimates of these ratios are easy to calculate, but confidence intervals are much more difficult to obtain. However, we need confidence intervals for any kind of statistical inference (and hence for decision making). While asymptotic normal distributions have been obtained for the Sharpe ratio under idealized conditions,[9] the idealized conditions do not always hold, and furthermore asymptotic distributions may be poor approximations in finite sample sizes. There is little guidance on the small-sample behavior of risk-adjusted performance measures or on the number of data points needed to justify the use of asymptotic results. Moreover, these analytical solutions are either extremely difficult to work out or simply do not exist for modifications of the popular Sharpe ratio that focus more on downside risk. As an example, we look at the well-known Sortino ratio which relates average return to the standard deviation of downside returns. What we need is a general method that provides us with standard errors and confidence intervals for arbitrary risk-return ratios, sample sizes, and distributions.

4.6.2 Bootstrapping Theory as an Alternative

Suppose we observe a series of excess returns $r_1, r_2, \cdots, r_m$.[10] Ex-post-risk-return ratios $\hat{\varsigma}$ are calculated as the ratio of the average return per unit of risk. For illustrative purposes, we focus on the Sharpe and Sortino ratios given below. Both ratios differ with respect to the risk measure used. The sample calculations for these two ratios are

$$\begin{aligned} \textit{Sharpe ratio} &= \frac{\frac{1}{m}\sum_{i=1}^{m} r_i}{\sqrt{\frac{1}{m-1}\sum_{i=1}^{m}(r_i - \bar{r})^2}}, \\ \textit{Sortino ratio} &= \frac{\frac{1}{m}\sum_{i=1}^{m} r_i}{\sqrt{\frac{1}{m-1}\sum_{i=1}^{m} I(r_i < 0)(r_i)^2}}, \end{aligned} \tag{4.8}$$

where $I(r_i < 0)$ denotes the indicator function. The Sharpe ratio[11] employs the symmetric standard deviation of returns risk measure in the denominator, equally penalizing downside and upside deviations from the sample mean return. The denominator asymmetric risk measure in the Sortino ratio[12] includes only negative returns in its calculation of squared returns. High ratios are preferable, everything else being equal, as they indicate a better return per unit of risk taken.

We include the Sortino ratio for three reasons. First, it better captures the risk if returns are non-normally distributed, as is the case for hedge fund returns for example, and is particularly relevant when the distribution has a negative skew. Second, it is well-known that the Sortino ratio suffers more from estimation error, as it uses roughly half as many data points in the denominator risk measure relative to the Sharpe ratio. Third, no large sample approximations exist. If the small sample distribution of $\hat{\varsigma}$ is far from normal, classical methods are biased and unreliable.

In any case, the analytic formulas for the large sample distributions of the ratios above are extremely hard to come by. In order to overcome this problem we rely on nonparametric bootstrapping techniques. Nonparametric bootstrap resampling treats the empirical distribution function of the current sample as *the* nonparametric approximation of the true distribution—in the absence of further information, it is the best we have. It then repeatedly draws from the empirical distribution and recalculates the statistic of interest many times to arrive at the bootstrap sampling distribution. The bootstrap sampling distribution can then be used to construct standard errors, confidence intervals, and hypothesis tests; see, for example, Efron and Tibshirani (1998) and Davison and Hinkley (1999).

As an example, suppose that we are given 160 monthly returns on the HFR fund-of-funds index ranging from January 1990 to April 2003. We use the JPM one month cash rate from DataStream to calculate a risk-free rate. The nonparametric bootstrapping procedure is as follows.

1. Randomly draw 160 (original sample size) returns with replacement from the original sample.
2. Calculate a new risk-return ratio $\hat{\varsigma}_b^*$ based on the resampled returns.
3. Repeat this procedure for $b = 1,...,B$ times, arriving at $\hat{\varsigma}_1^*, \hat{\varsigma}_1^*, \cdots, \hat{\varsigma}_b^*, \cdots \hat{\varsigma}_B^*$ resampled ratios.

The bootstrap sampling distribution of $\hat{\varsigma}_b^*$ can now be used to judge whether the sampling distribution of $\hat{\varsigma}$ for small samples is normal and hence whether traditional sampling theory approximations might not be so bad after all. Setting $B = 10{,}000$ and using Code 4.4 along with the S-PLUS functions `qqplot` and `histogram`, we get the results in Figure 4.15 and Figure 4.16 . In Figure 4.15, we see that for the Sharpe ratio all resampled realizations plot very close to a straight line, and so we conclude that the Sharpe ratio is quite normally distributed. The same cannot be said about the Sortino ratio, for which the normal Q-Q plot has substantial deviations from linearity at both ends, being heavy-tailed to the right and short-tailed to the left. As we suspected, the histograms in Figure 4.16 show that the Sortino ratio has a much larger dispersion in resampled outcomes than the Sharpe ratio and hence a much larger estimation error. While a small-sample normal approximation looks reasonable for the traditional Sharpe ratio, such an approximation is likely to be largely misleading for the Sortino ratio.

```
B <- 10000
sharpe.ratio <- function(x){
  mean(x,na.rm=T)/stdev(x,na.rm=T)
}
sortino.ratio <- function(x){
  mean(x,na.rm=T)/sqrt(mean(pmin(x,0)^2,na.rm=T))
}
simple.bs <- bootstrap(x,sortino.ratio,B)
```

Code 4.4 Simple Bootstrap

We now use the 2.5% and 97.5% percentiles of the bootstrap distribution to obtain a symmetric 95% confidence interval $CI\left(\hat{\varsigma}_{2.5\%}^*, \hat{\varsigma}_{97.5\%}^*\right)$. The Sortino ratio confidence interval is (.11, .92), and the Sharpe ratio confidence interval is *CI*(.08, .41).

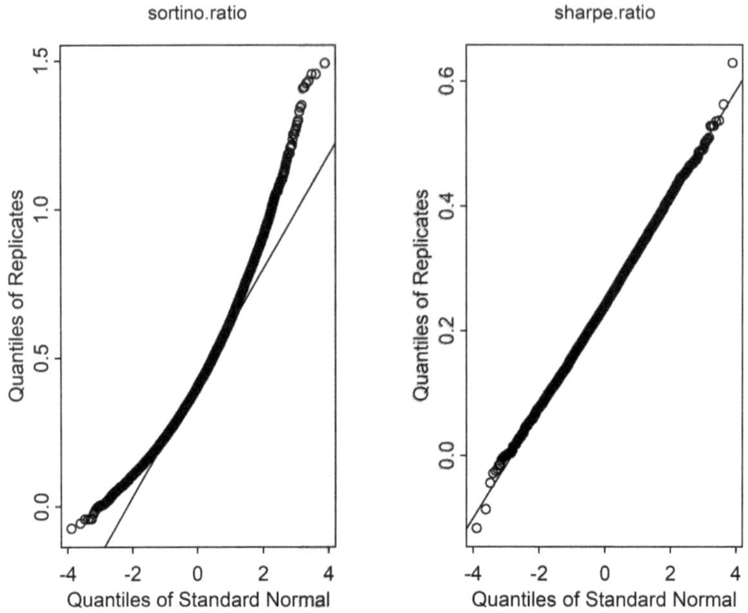

Figure 4.15 Q-Q Plots for Bootstrapped Sampling Distribution

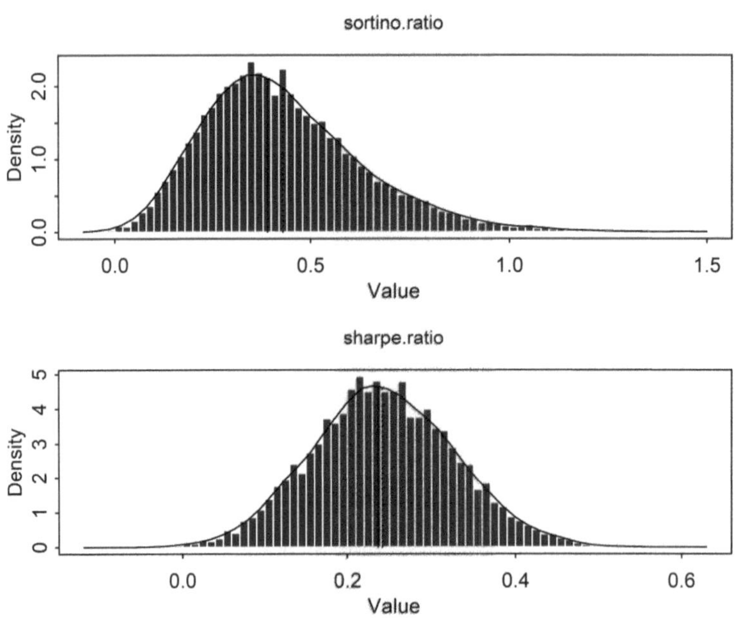

Figure 4.16 Bootstrapped Sampling Distribution

4.6.3 Increasing the Confidence Interval Coverage Probability Accuracy with the Double Bootstrap

So far we have relied on the 95% interval from a simple bootstrap procedure. However, the bootstrap is an approximate method, and it suffers to a greater or lesser extent from finite sample bias. Consequently, our 95% interval covers the true ratio with a probability that is at least somewhat different than .95. One way to increase the accuracy of the coverage probability of our confidence interval for the ratios is to use the **double bootstrap**, which can be thought of as "bootstrapping the bootstrap." It is known that under reasonable conditions the double bootstrap reduces the bias in the coverage probability.[13] The double bootstrap (Code 4.5) involves the following calculation.[14]

1. Perform the simple bootstrap as described above. Save all $b = 1,\ldots,B$ resampled data sets as well as the resampled ratios $\hat{\varsigma}_b^*$. This is called *first stage-resampling.*
2. For each of the B resampled data sets, start a second round of $z = 1,\cdots,Z$ resamples, leading to a total of $B \cdot Z$ resamples denoted as $\hat{\varsigma}_{bz}^{**}$. For each $\hat{\varsigma}_b^*$, there exists a new set of Z resampled ratios $\hat{\varsigma}_{b1}^{**},\cdots,\hat{\varsigma}_{bZ}^{**}$. These are the *second-stage resamples.*
3. For each $\hat{\varsigma}_b^*$, calculate the percentage of second-stage resamples $\hat{\varsigma}_{b1}^{**},\cdots,\hat{\varsigma}_{bZ}^{**}$ that fall below the original sample estimate of the risk-return ratio $\hat{\varsigma}$, namely, calculate $u_b = \frac{1}{Z}\sum_{z=1}^{Z} I\left(\hat{\varsigma}_{bz}^{**} < \hat{\varsigma}\right)$. We choose $B = 1000$ and $Z = 200$.

```
double.bs <- function(data, statistic, B, Z)
{
  call(statistic)
  outer.sample <- matrix(
     sample(data, size=length(data)*B, replace=T),
     nrow=B, ncol=length(data))
  outer.bs <- apply(outer.sample, 1, statistic)
  inner.bs <- matrix(0, nrow=B, ncol=Z)
  prob <- matrix(0, ncol=1, nrow=B)
  estimate <- statistic(data)
  for(i in 1:B) {
     inner.bs[i,] <- bootstrap(outer.sample[i,],
        statistic, Z, trace=F)$replicates
     cat("run #", i)
```

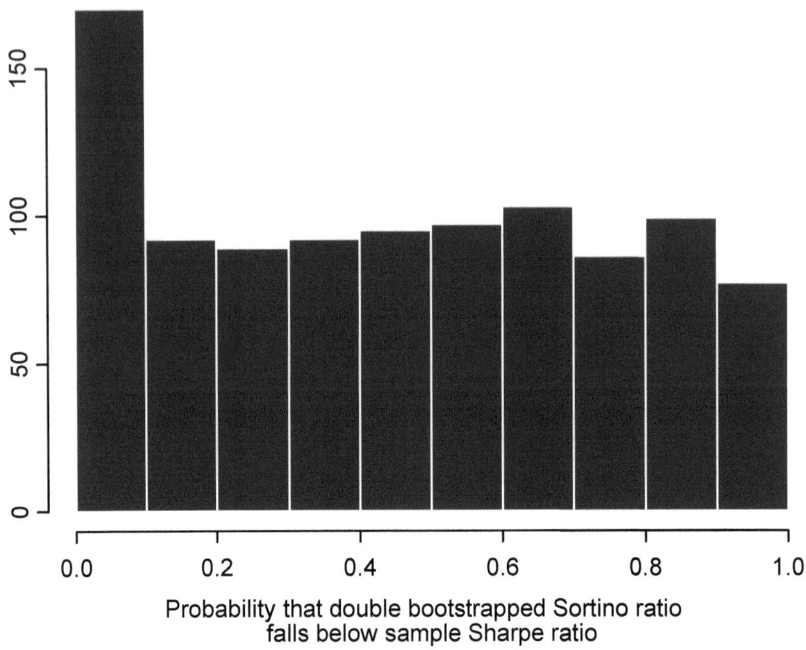

Figure 4.17 Sample Histogram of $u_b = \frac{1}{Z}\sum_{z=1}^{Z} I\left(\hat{\varsigma}_{bz}^{**} < \hat{\varsigma}\right)$

```
    }
    for(i in 1:B) {
       prob[i] <- sum(inner.bs[i,]<estimate)/Z
    }
    prob
}
```

Code 4.5 Double-Bootstrapping Code

Under ideal conditions, u_b follows a uniform distribution. Figure 4.17 shows that this assumption is clearly violated for the double-bootstrapped Sortino ratios $\hat{\varsigma}_{b1}^{**},\cdots,\hat{\varsigma}_{bZ}^{**}$.[15] Finally, we calculate the 2.5% and 97.5% percentiles of u_b and use these values to adjust the first-stage resample confidence band to $CI\left(\hat{\varsigma}_{u_{2.5\%}}^{*}, \hat{\varsigma}_{u_{97.5\%}}^{*}\right)$. Our resulting double-bootstrap confidence interval $CI\left(\hat{\varsigma}_{7.9\%}^{*} = 0.18, \hat{\varsigma}_{96.\%}^{*} = 0.96\right)$ is moved to the right, with a higher lower bound of 0.18 instead of 0.11 (representing the 8% quantile rather than the 2.5% quantile), and a higher upper bound of .96 instead of .92.

Renewed interest in the significance of risk-return ratios has been focused on closed-form solutions for the well-known Sharpe ratio, and there is increasing interest in downside risk measures such as that in the Sortino ratio. This section provided a nonparametric methodology to evaluate the properties of the Sharpe and Sortino ratios' sampling distributions, as well as a method to compute confidence intervals without having to rely on asymptotic approximations. We have seen that while the distribution of the Sharpe ratio is well-approximated by a normal distribution, the Sortino ratio has a quite non-normal distribution, and the double-bootstrap methodology leads to a significantly refined confidence interval.

Exercises

1. This exercise points out an important linear regression model formulation of the Markowitz portfolio optimization without a long-only constraint, a context in which one can obtain standard errors of portfolio weights without resampling. Suppose we have n time series of excess returns (total return minus cash rate) with m observations each. We can combine these excess returns in a matrix **X** (each column contains one return series). Regressing these excess returns against a constant $\mathbf{1}_{mx1} = \mathbf{X}_{mxn}\mathbf{w}_{nx1} + \mathbf{u}_{mx1}$ yields $\mathbf{w} = \left(\mathbf{X}^T\mathbf{X}\right)^{-1}\mathbf{X}^T\mathbf{1}$. These weights correspond to a portfolio that can be interpreted as the closest to a portfolio with zero risk (a vector of ones shows no volatility) and unit return. This would be an arbitrage opportunity. Rescaling the optimal weight vector (so that all weights sum to one) will yield $\mathbf{w}^*_{Sharpe} = \left(\bar{\mathbf{\Omega}}_0^{-1}\bar{\boldsymbol{\mu}}_0\right)/\left(\mathbf{I}^T\bar{\mathbf{\Omega}}_0^{-1}\bar{\boldsymbol{\mu}}_0\right)$, the maximum Sharpe ratio portfolio. This framework can also be used to test restrictions on individual regression coefficients (estimated portfolio weights), as well as restrictions on groups of assets, and test whether they are significantly different from zero.[16]
   (a) Generate a hypothetical data set and use the linear regression command `lm()` in S-PLUS to calculate optimal portfolios.
   (b) Test for the significance of individual weights using alternative correlations and sample length.
   (c) Repeat (a), but add the constraints to the regression. Implement individual constraints, group constraints, and the full investment constraint.

2. Try to replicate Figures 4.4 to 4.9.

3. Make an equal-weighted portfolio of six to ten stocks of your choice from the CRSP returns data sets provided with this book, and apply the bootstrap and double bootstrap analysis of Section 4.6 to the Sharpe ratio and Sortino ratio for these data.

4. Repeat Exercise 3 for a new ratio obtained by modifying the Sortino ratio as follows: replace the denominator with the average of the losses below zero. How does the behavior of this ratio compare with that of the Sortino ratio?

5. Take the data from Michaud (1998, p.17, 19, given below in Table 4.1) and generate a graph similar to the graph in Figure 4.18.

Table 4.1 Data from Michaud (1998) for Exercise 5

$\bar{\Omega}_0 =$

Canada	30.25	15.85	10.26	9.68	19.17	16.79	2.87	2.83
France	15.85	49.42	27.11	20.79	22.82	13.30	3.11	2.85
Germany	10.26	27.11	38.69	15.33	17.94	9.098	3.38	2.72
Japan	9.68	20.79	15.33	49.56	16.92	6.66	1.98	1.76
U.K.	19.17	22.82	17.94	16.92	36.12	14.47	3.02	2.72
U.S.	16.79	13.30	9.098	6.66	14.47	18.49	3.11	2.82
U.S. bonds	2.87	3.11	3.38	1.98	3.02	3.11	4.04	2.88
Euro bonds	2.83	2.85	2.72	1.76	2.72	2.82	2.88	2.43

,

$\bar{\mu}_0 =$

Canada	0.39
France	0.88
Germany	0.53
Japan	0.88
U.K.	0.79
U.S.	0.71
U.S. bonds	0.25
Euro bonds	0.27

6. Redo the calculation in Section 4.6 with simulated data. What do you observe?

7. Select eight mid-cap stocks from `midcap.ts`, and compute the following resampled efficient frontiers: (a) resampling with the basic Michaud efficient frontier resampling described in Exercise 5; (b) resampling with a proper parametric bootstrap (i.e., evaluate each resampled portfolio mean and standard deviation by using the sample mean and covariance that generated the portfolio weights for that resampling, not the original sample mean and covariance as proposed by Jorion (1992) and Michaud (1998)); (c) the nonparametric bootstrap as described in Section 6.9.4, with simplified versions of the code provided in that section. What do you conclude about your results in (a) versus (b)? What about (b) versus (c)?

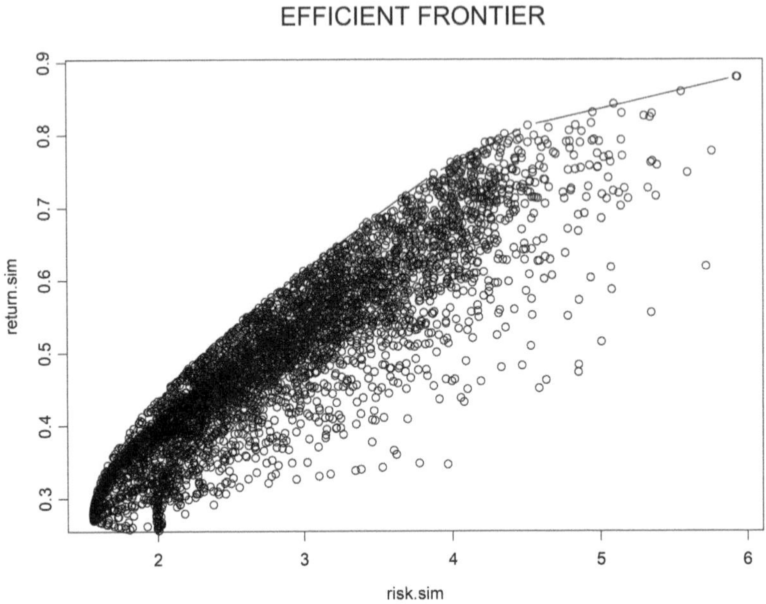

Figure 4.18 Efficient Frontier and Resampled Portfolio

Endnotes

[1] Robert Michaud patented the use of the average resampled portfolio. Readers are referred to U. S. patent # 6003018 or to Michaud (1998). However, the basic idea of portfolio resampling was introduced into the finance literature by Jorion (1992).
[2] See Efron and Tibshirani (1998) for a further discussion on this question.
[3] The assumption of a known fixed risk aversion coefficient is not always realistic, and if instead we use the weight vector $\mathbf{w}_i = \frac{\boldsymbol{\Omega}^{-1}\hat{\boldsymbol{\mu}}_i}{\mathbf{1}'\boldsymbol{\Omega}^{-1}\hat{\boldsymbol{\mu}}_i}$ with estimated risk aversion for the maximum Sharpe ratio for the i-th resample, we do not get this result.
[4] The central motivation of bootstrap resampling as introduced by statisticians is to estimate the distribution, or aspects of the distribution of an estimate such as the mean, standard deviation, or confidence intervals, of complicated statistics for which the standard sampling distribution theory does not apply.
[5] See Chapter 7 on Bayesian methods.
[6] The idea of this test statistic is that it is obviously not enough to look at weight differences only. Small weight differences for highly correlated assets might be of higher significance than large weight differences for negatively correlated assets.
[7] The Michaud approach referenced in Endnote 1 uses the rank-based approach.
[8] In spite of this apparent limitation, bootstrap resampling methods are able to do a quite decent job of estimating the distribution (or a summary such as standard error) of a statistic for which one does not have a decent sampling-distribution approximation; see, for example, Efron and Tibshirani (1998).
[9] See Lo (2002) and Memmel (2003).
[10] We assume here that returns are independently drawn from a single distribution. This is unlikely to be true for hedge fund data, as they exhibit serial correlation. One way to deal with this would be to fit an autoregressive model to the data and use this parametric specification of the return-generating process for resampling.
[11] See Sharpe (1994) for a review.
[12] See Sortino and Price (1994).
[13] See Section 3.9 and related material in Davison and Hinkley (1999) for details.
[14] See Nankervis (2002).
[15] Formal tests such as the Kolmogorov-Smirnov test as well as the χ^2 adjustment test, provide p-values close to 0%. Hence the null hypothesis that Figure 4.17 comes from a uniform distribution can be safely rejected.
[16] The regression framework puts a central problem of portfolio construction into a different, well-known perspective. Highly correlated asset returns mean highly correlated regressors with the obvious consequences arising from multicollinearity: high standard deviations on portfolio weights (regression coefficients) and identification problems (difficulty of distinguishing between two similar assets). Simply downtesting and excluding insignificant assets will result in an outcome that is highly dependent on the order of exclusion, with no guidance where to start. This is a familiar problem for both the asset allocator and the econometrician.

5 Scenario Optimization: Addressing Non-normality

5.1 Scenario Optimization

5.1.1 Foundations of Scenario Optimization

In the case of portfolio optimization, the uncertainty in the optimization process stems from the uncertainty of returns. One way to solve this problem (in the sense of addressing uncertainty, not estimation error) is to solve a very large-scale deterministic program instead, where a large number of scenarios try to capture randomness. For example, we can simulate 100,000 scenarios for four assets from the predictive distribution of portfolio returns. After the draws have been made, the uncertainty is removed and we are left with solving a deterministic problem. This procedure is called **scenario optimization**. We will see later in this chapter that for many objectives scenario optimization can be solved as a linear program. Key to successful scenario optimization is the quality of the sampled scenarios. In particular, scenarios must be

- **Representative** – Scenarios must offer a realistic description of the relevant problem and not induce estimation error.
- **Parsimonious** – Scenarios should use a relatively small number of samples to save computing time.
- **Arbitrage-free** – Scenarios should not allow the optimization algorithm to find highly attractive solutions that make no economic sense.

Scenario optimization that is based on only a few unrepresentative data might over-adjust and lead to an overly optimistic assessment of what could be achieved, while scenario optimization with a very large set of scenarios and assets might become computationally infeasible.

It is well-known that under normally distributed returns there is no need for scenario optimization, as the efficient set of solutions under arbitrary objective functions would still coincide with the efficient set in a traditional mean-variance optimization. Otherwise complicated calculations become quite simple. As a first step we just find the mean-variance solutions. We then calculate the risk measure on the efficient set of portfolio solutions in a second step. For a given return expectation, the portfolio with the smallest value-at-risk (VaR) or lower partial moment will still be the portfolio with the minimum variance.

In order to deviate from the normality assumption, we have to ask ourselves a series of questions.

- Are returns non-normal?
- Are deviations from normality statistically significant?
- Are deviations stable (i.e., can we forecast them over time)?
- Will the non-normality vanish over time?

There is no general dogmatic answer to the questions above. At the asset class level, this is an empirical problem. However, modelers should also be aware of what will be lost if we discard the normality assumption. We lose portfolio aggregation as well as time aggregation (risk measures often do not have closed forms under non-elliptical distributions). Additionally, we need a new equilibrium model where skewness and kurtosis are also priced. On the instrument level, it is clear that nonlinear derivatives (options, collateralized debt obligations (CDOs), etc.) require the explicit modeling of non-normalities that have been deliberately engineered. We discuss a relevant problem within the set of exercises.

Let us now start with a visual inspection of two return series to illustrate the problem of non-normality. Figure 5.1 shows histogram (empirical frequency distribution), empirical cumulative frequency distribution (versus assumed normal distribution), and Q-Q plots (plots of empirical quantile versus hypothetical quantile of assumed distribution) for monthly returns on emerging market bonds (JPM.EMBI) and U. S. dollar returns versus those for the Japanese yen (USD.YEN).[1]

```
graphsheet()
par(mfrow=c(1,3))
hist(Dollar.Yen)
Normal <- rnorm(10000, mean(Dollar.Yen),
  sqrt(var(Dollar.Yen)))
cdf.compare(Dollar.Yen, Normal, cex=0.7)
ks.gof(Dollar.Yen, Normal)
qqnorm(Dollar.Yen)
qqline(Dollar.Yen)
```

It is straightforward to see that returns on emerging market bonds show negative skewness (too many large negative returns), while currency returns are approximately normal. We can also use the Kolmogorov-Smirnov test (calculating how distant both cumulative distributions are) for a more formal assessment.

```
ks.gof(Dollar.Yen, Normal)

  Two-Sample Kolmogorov-Smirnov Test

data:  Dollar.Yen and Normal

ks = 0.0748, p-value = 0.2116
alternative hypothesis:
  cdf of Dollar.Yen does not equal the cdf of
  Normal for at least one sample point.
```

The high p-value (0.21) confirms that the empirical distribution is not significantly different from the normal distribution. We can also use the Kolmogorov-Smirnov test to check for multivariate normality. Note that individual marginal distributions could all be normally distributed, while the corresponding multivariate distribution still might not be normal. Under multivariate normality, we know that $\mathbf{d}_m^T \bar{\mathbf{\Omega}}^{-1} \mathbf{d}_m$ is distributed as $\chi^2(n)$, where $\mathbf{d}_m$ reflects the distance vector at time m (period returns minus mean return). All we need is to compare the cumulative distribution of $\mathbf{d}_m^T \bar{\mathbf{\Omega}}^{-1} \mathbf{d}_m$ with $\chi^2(n)$. This is a straightforward test for multivariate normality.

Even if period-by-period returns are non-normally distributed, it is most likely that multiperiod returns are (log) normally distributed: the Central Limit Theorem states that the product of independent and identical distributed variables (with finite variance) will approach log-normality after approximately 30 random drawings. We can check this using the built-in `bootstrap()` function to generate 36 month returns from the series of one month returns.

```
JPM.EMBI.36month <- bootstrap(JPM.EMBI,
  prod(1+sample(JPM.EMBI, 36)), 10000)$replicates
hist(JPM.EMBI.36month)
```

Figure 5.2 agrees with our intuition. It looks very much like a log-normal distribution, confirming our previous considerations that non-normality will tend to vanish as we move away from the very short time horizon.

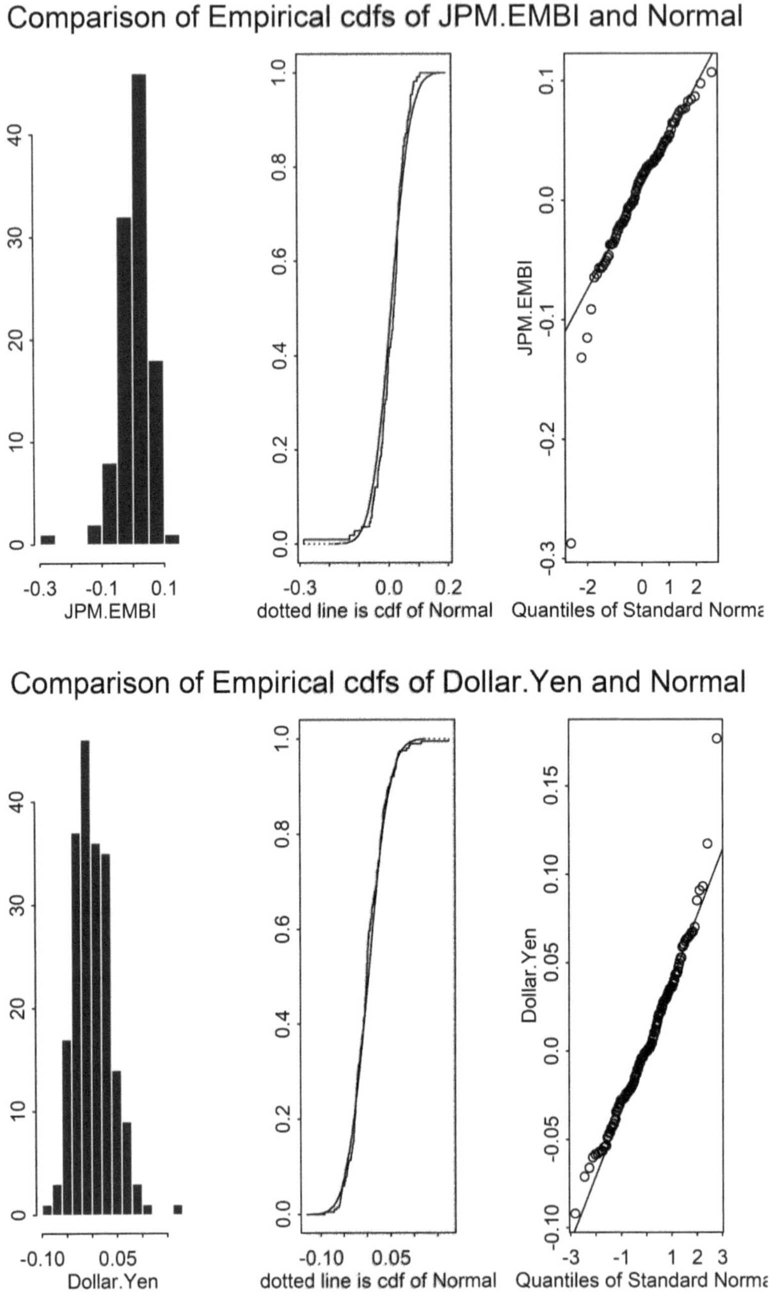

Figure 5.1 Visual Inspection of Asset Returns in S-PLUS

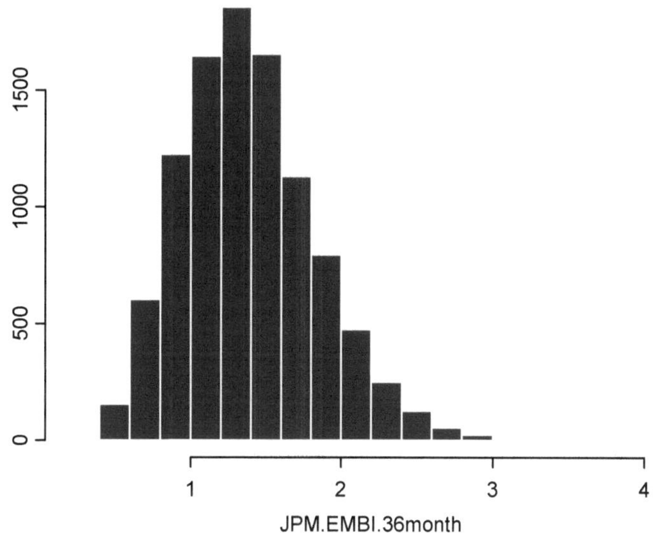

Figure 5.2 Multiperiod Returns for JPM.EMBI

5.1.2 Implied Returns and Arbitrary Preferences (Utilities) and Distributions

One problem with non-normal returns is that we lose the ability to back out implied returns using reversed optimization as seen in Chapter 1. What can we do to back out the implied returns for investors with different preferences under arbitrary return distributions?[2] Suppose our investor maximizes expected utility

$$E(U) = \sum_{s=1}^{m} \pi_s U\left(1 + \sum_{i=1}^{n} w_i r_{is}\right), \tag{5.1}$$

where we use the same notation as throughout the previous chapters. Expected utility is calculated as the average utility over m simulated scenarios. Each scenario is drawn with probability $\pi_s = \frac{1}{m}$. In our example, utility itself is defined as

$$U(1+r) = \begin{cases} \dfrac{(1+r)^{1-\gamma}}{1-\gamma}, & \gamma \geq 0 \\ \ln(1+r), & \gamma = 1, \end{cases} \tag{5.2}$$

where γ denotes the risk aversion coefficient. Note that any series of historic returns can be written as

$$r_{is} = c + \mu_i + \sigma_i z_{is}, \tag{5.3}$$

where we can isolate the degree of non-normality captured in the empirical distribution of z_{is} from our forward-looking assumptions on risk premiums (μ) as well as volatilities (σ). Applying (5.3) to the definition of benchmark returns $r_{bs} = \sum_{i=1}^{n} w_i r_{is}$, we get

$$r_{bs} = c + \mu_b + \sum_{i=1}^{n} w_i \sigma_i z_{is}. \tag{5.4}$$

We know from standard valuation theory that[3]

$$\pi_s^* = \pi_s \frac{U'(W_s)}{\sum_{s=1}^{m} \pi_s U'(W_s)}, \tag{5.5}$$

where π_s^* denotes the risk-neutral probabilities. The risk-neutral probability will be high in states where marginal utility is high (wealth is low). Hence, large weight is given to those states where wealth levels are depressed. Under the assumed utility function (5.3), we get

$$\pi_s^* = \pi_s \frac{(1 + R_{bs})^{-\gamma}}{\sum_{s=1}^{m} \pi_s (1 + R_{bs})^{-\gamma}}. \tag{5.6}$$

Risk-neutral probabilities equalize all expected returns, as they correct for risk via (5.5). Assuming our investor finds the current (benchmark) portfolio optimal, we hence know that he prices all assets according to

$$\sum_{s=1}^{m} \pi_s^* r_{is} = \sum_{s=1}^{m} \pi_s^* r_{bs}. \tag{5.7}$$

Inserting (5.3) and (5.4) into (5.7), we arrive at the implied return for the *i*-th asset,

$$\mu_i = \mu_b + \sum_{s=1}^{m} \pi_s^* \left(\sum_{i=1}^{n} w_i \sigma_i z_{is} - \sigma_i z_{is} \right). \tag{5.8}$$

Changing the risk aversion parameter will change the implied returns. The very risk-averse investor will require large compensations for assets that show considerable "tail risk."

5.1.3 Generation of Non-normally Distributed Scenarios

Suppose we want to generate a set of returns for future scenarios. Also assume we want to dismiss normality and generalize our simulation methodology in two main aspects. First, we aim to allow marginal distributions that could take any arbitrary form (i.e., they could follow a blend of a normal distribution plus an extreme value distribution for large losses, a mixture of normals, etc.). Second, we want to relax the modeling of dependence beyond the concept of correlation, as it is well-known that empirical distributions show tail dependence that is not explained by correlation alone. How can we glue arbitrary return distributions together and still maintain their correlation structure? How can we model tail dependence and still keep the same marginal distributions?

The answer to these questions is the concept of **copula functions**. Recall that a typical Monte Carlo simulation of random returns requires us to draw a uniform random number $u_s \sim uniform(0,1)$ and then invert the cumulative distribution function to arrive at a simulated return observation $r_{i,s} = F_{r_i}^{-1}(u_s)$. What do we do in a multivariate context? An n-dimensional copula is a multivariate cumulative distribution function with uniformly distributed marginals. Alternatively, we can think of it as a random vector of uniformly distributed variables that share a specified dependence structure,

$$C(u_1,\ldots,u_n) = prob\left(\tilde{u}_1 \le u_1,\ldots,\tilde{u}_n \le u_n\right). \tag{5.9}$$

As soon as we know the realization of the uniform random numbers $(u_1,\ldots,u_n)$, we can calculate the marginals from $F_{r_1}^{-1}(u_1),\cdots,F_{r_n}^{-1}(u_n)$. In general, we can say that if $F(r_1,\cdots,r_n)$ denotes a multivariate distribution function with continuous marginals, it will have a unique copula representation,

$$F(r_1,\cdots,r_n) = C\left(F_{r_1},\cdots,F_{r_n}\right). \tag{5.10}$$

We can hence separate the univariate margins and the multivariate dependence structure.[4] This proves to be very convenient in scenario generation. In what follows, we will not elaborate on how best to estimate the copula function (and the marginals) in (5.10). We rather work on the assumption that the marginals and copula are given.

In order to appreciate how scenario-based solutions (which will be presented later in this chapter) differ from a simple mean-variance approach, we use the copula approach to glue four mixtures of normals together, maintaining a specified correlation structure and modeling (symmetric) tail dependence according to a t copula. To simulate a t copula with υ degrees of freedom, we have to proceed according to the following steps.

1. Find the Cholesky decomposition $\mathbf{C}_{Cholesky}$ of the correlation matrix $\mathbf{C}$ (dimension: $n \times n$).
2. Draw a vector of n standard normals $\mathbf{u}$ and calculate $\mathbf{C}_{Cholesky}\mathbf{u}$. Alternatively, you might want to combine both steps and draw directly from a multivariate normal.
3. Draw from $s \sim \chi^2_\upsilon$ and multiply the result of the second step by $\sqrt{\upsilon}/\sqrt{s}$, i.e., calculate $\mathbf{x} = \mathbf{C}_{Cholesky}\mathbf{u}\frac{\sqrt{v}}{\sqrt{s}}$.
4. Each element of $\mathbf{x}$ $(x_1,\cdots,x_n)$ is inserted into the cumulative distribution function to arrive at uniformly distributed variables $u_i \sim t_v(x_i)$.
5. Repeat steps 2 to 4 many (m) times.

What looks like a complicated procedure can be performed in S-PLUS using a single line of code:

```
Corr <- matrix(c(1.0,0.8,0.2,0.2,
                 0.8,1.0,0.6,0.2,
                 0.2,0.6,1.0,0.2,
                 0.2,0.2,0.2,1.0), ncol=4, nrow=4)
m <- 100000
v <- 2
copula <- pt(rmvnorm(m, mean=rep(0,ncol(Corr)),
  cov=Corr)*sqrt(v)/sqrt(rchisq(m,v)),v)
```

Figure 5.3 and Figure 5.4 can be replicated with the following code:

```
x <- matrix(qnorm(copula), ncol=4)
graphsheet()
pairs(x, label=c("asset 1", "asset 2", "asset 3",
  "asset 4"))
xx <- rmvnorm(m, mean=rep(0,ncol(Corr)), cov=Corr)
graphsheet()
pairs(xx,label=c("asset 1", "asset 2", "asset 3",
  "asset 4"))
```

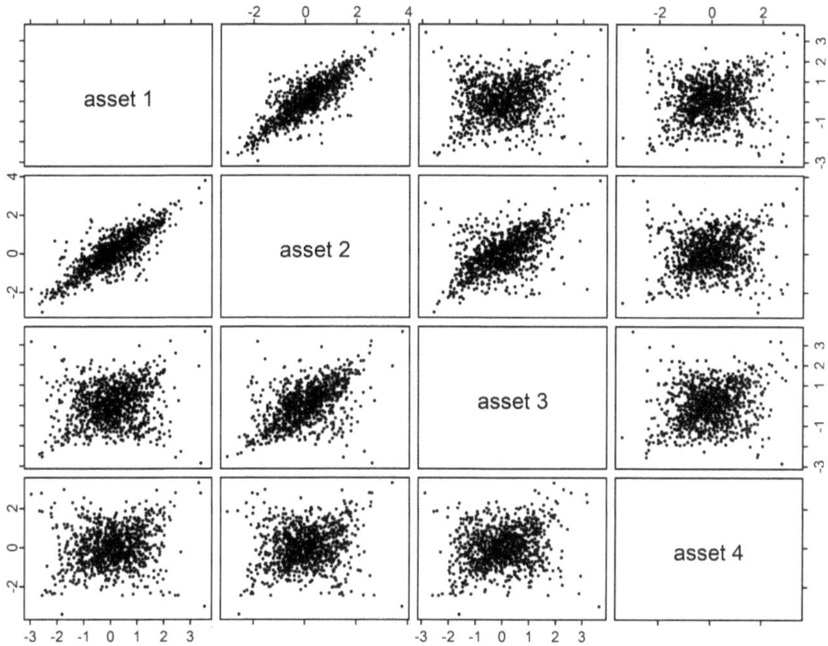

Figure 5.3 *t* copula with 2 Degrees of Freedom and Standard Normal Margins

Figure 5.5 is the result of the commands

```
graphsheet()
plot(x[,1],x[,2], xlab="asset 1", ylab="asset 2",
  pch=1)
points(xx[,1],xx[,2], pch=3)
```

Marginal distributions for each of the four assets are assumed to be drawn from a mixture of normals and are plotted in Figure 5.6.

```
asset.1 <- exp(c(rnorm(500,  0.05, 0.05),
  rnorm(9500, 0.05, 0.05)))-1
asset.2 <- exp(c(rnorm(500, -0.3, 0.01),
  rnorm(9500, 0.08, 0.05)))-1
asset.3 <- exp(c(rnorm(200, +0.4, 0.01),
  rnorm(9800, 0.1, 0.17)))-1
asset.4 <- exp(c(rnorm(500, -0.6, 0.1),
  rnorm(9500, 0.12, 0.25)))-1
```

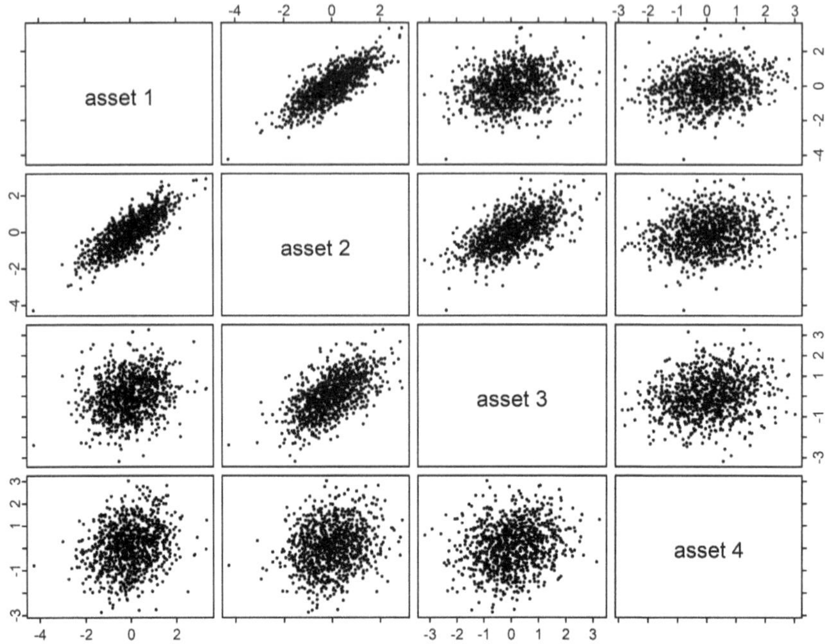

Figure 5.4 Multivariate Standard Normal

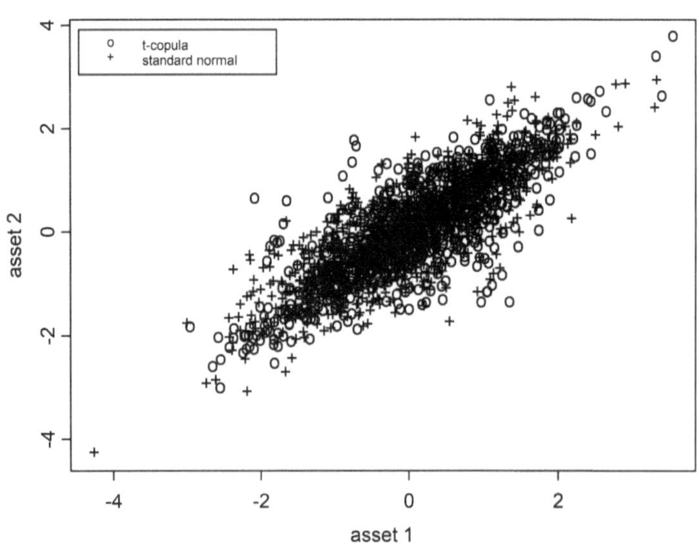

Figure 5.5 Scatterplot for Normal Distribution versus *t* copula

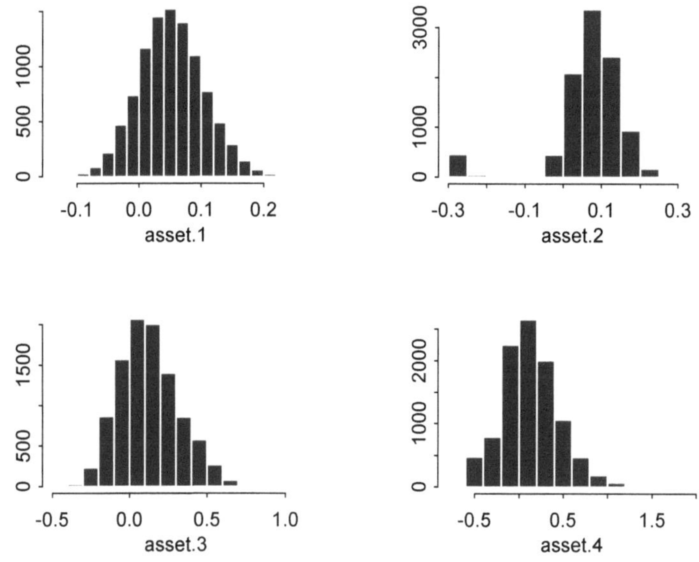

Figure 5.6 Marginal Distributions in Four-Asset Test Case

```
graphsheet()
par(mfrow=c(2,2))
hist(asset.1)
hist(asset.2)
hist(asset.3)
hist(asset.4)
```

Marginal distributions are glued together with the use of a *t* copula and stored in the scenario matrix $\mathbf{S}$. The necessary operations are

```
asset.1 <- matrix(quantile(asset.1, copula[,1]))
asset.2 <- matrix(quantile(asset.2, copula[,2]))
asset.3 <- matrix(quantile(asset.3, copula[,3]))
asset.4 <- matrix(quantile(asset.4, copula[,4]))
S <- cbind(asset.1, asset.2, asset.3, asset.4)
```

We can now code a scenario-based Markowitz optimization (Code 5.1), where portfolio variance is calculated using all scenarios for each weight allocation rather than by supplying a single covariance matrix.

```
MV.model <- function(S, mu.target)
{
  m <- nrow(S)
```

```
  n <- ncol(S)
  mu.bar <- apply(S, 2, mean)
  asset <- Set()
  period <- Set()
  i <- Element(set=asset)
  s <- Element(set=period)
  S <- Parameter(S, index=dprod(s,i))
  mu.bar <- Parameter(as.array(mu.bar), index=i)
  mu.target <- Parameter(mu.target, changeable=T)
  w <- Variable(index=i)
  r <- Variable(index=s)
  r[s] == Sum((S[s,i]-mu.bar[i])*w[i],i)
  risk <- Objective(type="minimize")
  risk ~ Sum(r[s]^2,s)/(m-1)
  Sum(mu.bar[i]*w[i],i) >= mu.target
  Sum(w[i],i) == 1
  w[i] >= 0
}

MV.portfolio <- function(S, mu.target)
{
  call(MV.model)
  MV.system <- System(MV.model, S, mu.target)
  solution <- solve(MV.system, trace=T)
  weight <-
     matrix(round(solution$variable$w$current,
     digit=5)*100, ncol=1)
  risk <- solution$objective
  return(weight,risk)
}

MV.frontier <- function(S, n.pf)
{
  call(MV.portfolio)
  Risk <- matrix(0, ncol=1, nrow=n.pf)
  Return <- matrix(0, ncol=1, nrow=n.pf)
  m <- nrow(S)
  mu.min <- min(apply(S,2,mean))
  mu.max <- max(apply(S, 2, mean))
  mu.range <- seq(mu.min, mu.max,
     (mu.max-mu.min)/(n.pf-1))
  x <- MV.portfolio(S, mu.target=mu.min)
  weight <- x$weight
  Risk[1,1] <- x$risk
  Return[1,1] <- mu.min
```

```
  for(i in 2:n.pf){
    x <- MV.portfolio(S, mu.target=mu.range[i])
    Risk[i,1] <- x$risk
    Return[i,1] <- mu.range[i]
    weight <- cbind(weight,x$weight)
  }
  graphsheet()
  par(mfrow=c(1,2))
  plot(Risk, Return, type="b")
  title("Mean - Variance Frontier")
  barplot(weight)
  title("Frontier Portfolios")
  list("optimal.weights" = weight)
}
```

Code 5.1 Mean-Variance Scenario Optimization

Typing `x <- MV.frontier(S, n.pf=10)` will trace out an efficient frontier with ten portfolios.

```
> x$optimal.weights
numeric matrix: 4 rows, 10 columns.
       [,1]   [,2]   [,3]   [,4]   [,5]   [,6]
[1,] 97.058 88.816 71.458 52.225 32.993 13.760
[2,]  0.000  0.000  8.032 18.572 29.112 39.652
[3,]  2.942 10.403 17.268 23.561 29.854 36.148
[4,]  0.000  0.781  3.243  5.642  8.041 10.440

       [,7]   [,8]   [,9] [,10]
[1,]  0.000  0.000  0.000     0
[2,] 42.869 27.675 12.481     0
[3,] 44.131 56.363 68.596   100
[4,] 13.000 15.962 18.923     0
```

Figure 5.7 shows the solutions for ten return points along the efficient frontier. All portfolios look reasonably diversified. We can use these solutions as a reference point for the following scenario optimizations.

5.2 Mean Absolute Deviation

The first scenario-based alternative to Markowitz optimization is the **Mean Absolute Deviation** model (MAD).[5] It involves the minimization of the probability-weighted (where p_s denotes the probability of scenario s) sum of

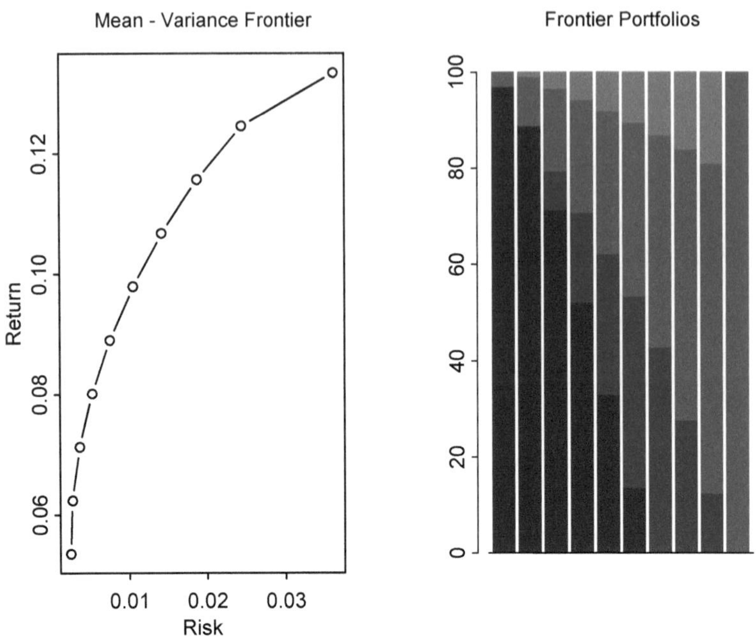

Figure 5.7 Mean-Variance Solutions

absolute deviations subject to the usual constraints. Risk is measured in the context of MAD as an absolute deviation from the mean rather than the squared deviation as in the case of variance.

$$
\begin{aligned}
&MAD = \sum_{s=1}^{m} p_s \left| \sum_{i=1}^{n} w_i \left(r_{i,s} - \bar{\mu}_i \right) \right| \\
&\sum_{i=1}^{n} w_i \bar{\mu}_i = \bar{\mu} \\
&\sum_{i=1}^{n} w_i = 1 \\
&w_i \geq 0.
\end{aligned}
\tag{5.11}
$$

While the square function $(\cdot)^2$ penalizes larger deviations at an increasing rate, this is not the case with MAD. In fact, MAD implies that an additional unit of underperformance relative to the mean creates the same disutility no matter how big the loss already is. However, one advantage of MAD is that we can specify the costs of deviations above and below the mean differently, putting greater weight (costs) on underperformance rather than outperformance. If we work on simulated data ($p_s = \frac{1}{m}$) and denote the absolute deviation of the scenario portfolio return from the average portfolio return by ad_s, we can transform (5.11) into a linear program.

$$
\begin{aligned}
&\min_{ad_s, w_i} \frac{1}{m}\sum_{s=1}^{m} ad_s \\
&\sum_{i=1}^{n} w_i \left(r_{i,s} - \overline{\mu}_i \right) \leq ad_s \\
&\sum_{i=1}^{n} w_i \left(r_{i,s} - \overline{\mu}_i \right) \geq -ad_s \\
&\sum_{i=1}^{n} w_i \mu_i = \overline{\mu} \\
&\sum_{i=1}^{n} w_i = 1 \\
&ad_s \geq 0 \\
&w_i \geq 0.
\end{aligned}
\tag{5.12}
$$

The MAD-based portfolio selection shown here offers a range of appealing properties versus variance-based models:

1. There is no need to calculate a covariance matrix because we use the scenario matrix, which can be constructed from time series of asset returns. However, this is only true if we rely on historical data for scenario generation; simulated scenarios from a parametric distribution have to be drawn using a covariance matrix.
2. Solving a linear program is much easier than mean–variance optimization. The number of constraints ($2m+2$ in the case of MAD) depends on the number of scenarios, not the number of assets.
3. The upper bound on the number of assets in the optimal solution is related to the number of scenarios ($2m+2$ in the case of MAD).

We leave the solution of (5.12) in NuOPT for S-Plus as an exercise and use SIMPLE instead. Let us add one more layer of complexity by attaching different costs to upside and downside deviations, $\sum_{i=1}^{n} w_i \left(r_{i,s} - \overline{\mu}_i \right) \geq 0$ and $\sum_{i=1}^{n} w_i \left(r_{i,s} - \overline{\mu}_i \right) \geq 0$, respectively. The objective now becomes

$$
\begin{aligned}
&\frac{1}{m}\sum_{s=1}^{m} \left(c^+ ad_s^+ + c^- ad_s^- \right) \\
&ad_s^+ = \begin{cases} ad_s & \text{if } ad_s > 0 \\ 0 & \text{else} \end{cases} \\
&ad_s^- = \begin{cases} ad_s & \text{if } ad_s < 0 \\ 0 & \text{else.} \end{cases}
\end{aligned}
\tag{5.13}
$$

The corresponding code is given in Code 5.2.

```
MAD.model <- function(S, cost.up, cost.dn,
     mu.target)
{
  m <- nrow(S)
  n <- ncol(S)
  mu.bar <- apply(S, 2, mean)
  asset <- Set()
  period <- Set()
  i <- Element(set=asset)
  s <- Element(set=period)
  S <- Parameter(S, index=dprod(s,i))
  mu.bar <- Parameter(as.array(mu.bar), index=i)
  mu.target <- Parameter(mu.target, changeable=T)
  cost.up <- Parameter(cost.up, changeable=T)
  cost.dn <- Parameter(cost.dn, changeable=T)
  w <- Variable(index=i)
  up <- Variable(index=s)
  dn <- Variable(index=s)
  up[s] >= 0
  dn[s] >= 0
  up[s]-dn[s] == Sum((S[s,i]-mu.bar[i])*w[i],i)
  risk <- Objective(type="minimize")
  risk ~ Sum((cost.up*up[s]+cost.dn*dn[s]),s)/(m-1)
  Sum(mu.bar[i]*w[i],i) >= mu.target
  Sum(w[i],i) == 1
  w[i] >= 0
}

MAD.portfolio <- function(S, cost.up, cost.dn,
  mu.target)
{
  call(MAD.model)
  MAD.system <- System(MAD.model, S, cost.up,
     cost.dn, mu.target)
  solution <- solve(MAD.system, trace=T)
  weight <-
     matrix(round(solution$variable$w$current,
     digit=5)*100, ncol=1)
  risk <- solution$objective
  return(weight,risk)
}
```

```
MAD.frontier <- function(S, cost.up, cost.dn, n.pf)
{
  call(MAD.portfolio)
  Risk <- matrix(0, ncol=1, nrow=n.pf)
  Return <- matrix(0, ncol=1, nrow=n.pf)
  m <- nrow(S)
  mu.min <- min(apply(S,2,mean))
  mu.max <- max(apply(S, 2, mean))
  mu.range <- seq(mu.min, mu.max,
     (mu.max-mu.min)/(n.pf-1))
  x <- MAD.portfolio(S, cost.up, cost.dn,
     mu.target=mu.min)
  weight <- x$weight
  Risk[1,1] <- x$risk
  Return[1,1] <- mu.min
  for(i in 2:n.pf){
     x <- MAD.portfolio(S, cost.up, cost.dn,
        mu.target=mu.range[i])
     Risk[i,1] <- x$risk
     Return[i,1] <- mu.range[i]
     weight <- cbind(weight,x$weight)
  }
  graphsheet()
  par(mfrow=c(1,2))
  plot(Risk, Return, type="b")
  title("Mean - Mean Absolute Deviation Frontier")
  barplot(weight)

  title("Frontier Portfolios")
  list("optimal.weights" = weight)
}
```

Code 5.2 Scenario Optimization Using Mean Absolute Deviation

Hence, typing `MAD.frontier(S,cost.up=1,cost.dn=1,n.pf=10)` will trace out an efficient frontier with ten portfolios, as shown in Figure 5.8.

```
> x$optimal.weights
        [,1]   [,2]   [,3]   [,4]   [,5]   [,6]
[1,] 94.733 88.889 62.241 32.043  0.012  0.000
[2,]  0.005  0.000 20.447 45.543 73.145 57.860
[3,]  5.261 11.109 15.002 17.091 18.986 30.461
[4,]  0.000  0.002  2.311  5.323  7.857 11.679
```

```
       [,7]   [,8]   [,9]  [,10]
[1,]  0.009  0.000  0.000     0
[2,] 42.540 27.174 11.857     0
[3,] 41.901 52.826 64.188   100
[4,] 15.549 20.000 23.955     0
```

Portfolios constructed with a symmetric understanding of Mean Absolute Deviation (up and down costs of deviations equal one) are close to mean-variance solutions. This is not surprising, as variance is also a symmetric measure of investment risk.

5.3 Semi-variance and Generalized Semi-variance Optimization

5.3.1 Properties of Semi-variance

Mean-variance-based portfolio construction has always suffered from the implicit assumption of normality that dictated that risk be measured as the variance of returns. One of the earliest alternative risk measures is **semi-variance**. While variance uses all return realizations, semi-variance utilizes only those returns that either fall below the average return (lower semi-variance, sv^-) or above the average return (upper semi-variance, sv^+):

$$
\begin{aligned}
sv^- &= \frac{1}{m}\sum_{s=1}^{m}(r_s-\mu)^2\,\delta_s \\
sv^+ &= \frac{1}{m}\sum_{s=1}^{m}(r_s-\mu)^2\,(1-\delta_s) \\
\delta_s &= \begin{cases} 0 & r_s > \bar{\mu} \\ 1 & r_s \le \bar{\mu}. \end{cases}
\end{aligned}
\tag{5.14}
$$

We can combine lower and upper semi-variances to arrive at variance again. To see this, note that if $\delta_s = 1$, it must follow that $1-\delta_s = 0$ and vice versa.

$$
\begin{aligned}
sv^- + sv^+ &= \frac{1}{m}\sum_{s=1}^{m}(r_s-\mu)^2\,\delta_s + \frac{1}{m}\sum_{s=1}^{m}(r_s-\mu)^2\,(1-\delta_s) \\
&= \frac{1}{m}\sum_{s=1}^{m}(r_s-\mu)^2 \\
&= \sigma^2.
\end{aligned}
\tag{5.15}
$$

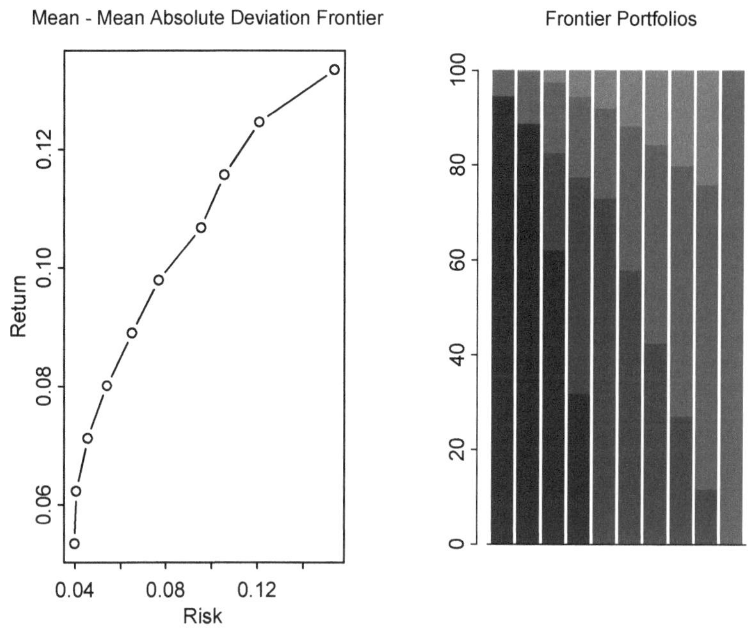

Figure 5.8 Mean Absolute Deviation Frontier

In the case of symmetry (for every return deviation below the mean, there is an equal return deviation above the mean), we get $sv^{-} = sv^{+}$ and

$$2sv = \sigma^2. \tag{5.16}$$

Equation (5.16) often serves as a simple check for symmetry. If the variance is roughly twice the semi-variance, the distribution is close to symmetric. (Symmetry does not imply normality, as a distribution might be symmetric but still exhibit fat tails.) For many assets, we can regress the difference between the variance and twice the semi-variance against a constant to see whether deviations are statistically different from zero:

$$\left(2sv_n - \sigma^2\right) = \alpha + \varepsilon_n. \tag{5.17}$$

Alternatively, we can test whether skewness is persistent across time. Suppose we have observations of returns for a number of time series. We can split the observations into two subperiods and test for persistence in deviations from symmetry,

$$\left(2sv - \sigma^2\right)_{t+1} = a + b\left(2sv - \sigma^2\right)_t + e_{t+1}. \tag{5.18}$$

Persistence is indicated by a significant b and a high R^2.

5.3.2 A General Semi-variance Model

Traditionally, semi-variance optimization has centered around lower semi-variance. However, there is no reason not to merge upper and lower semi-variances into a combined risk measure. This allows us every flexibility in expressing a mixture of risk-averse and risk-seeking behaviors.

The proposed measure uses a weighted linear combination of upper and lower semi-variances and is hence called the **weighted semi-variance model**[6] (see Figure 5.9):

$$sv_{weighted} = \omega sv^{-} + (1-\omega) sv^{+}. \tag{5.19}$$

Note that the range of ω is restricted to lie between zero and one. We can plot the new penalty function (5.19) for various weights.

```
deviation.from.mean.return <- seq(-60, 60, 1)
l.sv <- ifelse(deviation.from.mean.return <= 0,
  deviation.from.mean.return^2, 0)
u.sv <- ifelse(deviation.from.mean.return > 0,
  deviation.from.mean.return^2, 0)
V <- deviation.from.mean.return^2
DB <- 2*ifelse(deviation.from.mean.return <= 0,
  0.75*deviation.from.mean.return^2,
  0.25*deviation.from.mean.return^2)
UB <- 2*ifelse(deviation.from.mean.return <= 0,
  0.25*deviation.from.mean.return^2,
  0.75*deviation.from.mean.return^2)
graphsheet()
par(mfrow=c(2,2))
plot(deviation.from.mean.return, l.sv, type="l",
  ylab="penalty")
title("Lower semi-variance")

plot(deviation.from.mean.return, DB, type="l",
  ylab="penalty")
title("75% weight on lower semi-variance")
plot(deviation.from.mean.return, V, type="l",
  ylab="penalty")
title("Variance")
plot(deviation.from.mean.return, UB, type="l",
```

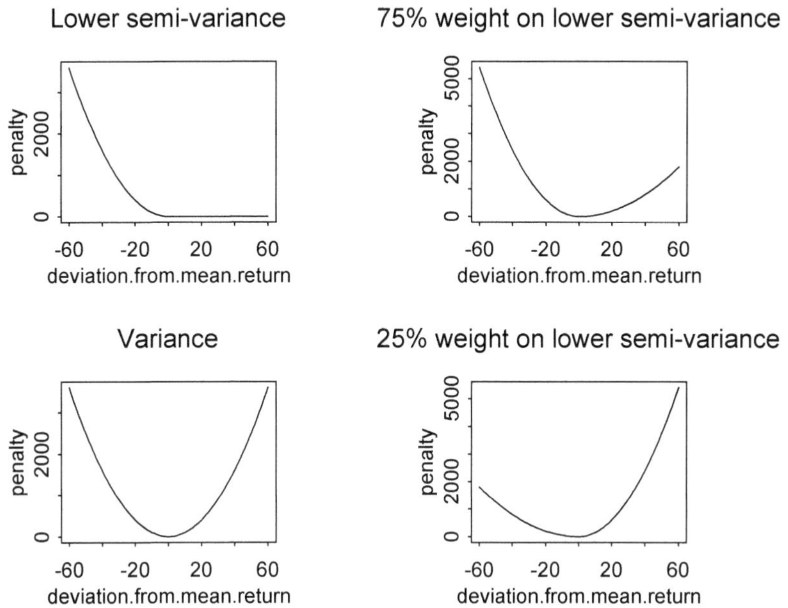

Figure 5.9 Weighted Semi-variance Measure

```
  ylab="penalty")
title("25% weight on lower semi-variance")
```

The weighted semi-variance portfolio optimization model becomes

$$
\begin{aligned}
&\text{minimize} \\
&\omega\left[\frac{1}{m}\sum_{s=1}^{m}\left(r_{pf,s}-\bar{\mu}\right)^2\delta_s\right]+(1-\omega)\left[\frac{1}{m}\sum_{s=1}^{m}\left(r_{pf,s}-\bar{\mu}\right)^2(1-\delta_s)\right] \\
&\sum_{i=1}^{n} w_i r_{i,s} = r_{pf,s} \\
&\sum_{i=1}^{n} w_i \mu_i = \bar{\mu} \\
&\sum_{i=1}^{n} w_i = 1 \\
&\delta_s = \begin{cases} 0 & r_{pf,s} > \bar{\mu} \\ 1 & r_{pf,s} \le \bar{\mu} \end{cases} \\
&w_i \ge 0,
\end{aligned}
\tag{5.20}
$$

where $r_{pf,s}$ denotes the portfolio return in scenario s. Code 5.3 illustrates how the model given in (5.20) can be translated into NuOPT for S-PLUS.

```
WSV.model <- function(S, mu.target,
     downside.weight)
{
  if(downside.weight < 0 | downside.weight > 1)
     stop("downside weight must range between
        0 and 1")
  m <- nrow(S)
  n <- ncol(S)
  mu.bar <- apply(S, 2, mean)
  asset <- Set()
  period <- Set()
  i <- Element(set=asset)
  s <- Element(set=period)
  S <- Parameter(S, index=dprod(s,i))
  mu.bar <- Parameter(as.array(mu.bar), index=i)
  mu.target <- Parameter(as.numeric(mu.target),
     changeable=T)
  dw <- Parameter(downside.weight, changeable=T)
  w <- Variable(index=i)
  up <- Variable(index=s)
  dn <- Variable(index=s)
  up[s] >= 0
  dn[s] >= 0
  up[s]-dn[s] == Sum((S[s,i]-mu.bar[i])*w[i],i)
  risk <- Objective(type="minimize")
  risk ~ Sum(dw*dn[s]^2+(1-dw)*up[s]^2,s)/(m-1)
  Sum(mu.bar[i]*w[i],i) == mu.target
  Sum(w[i],i) == 1
  w[i] >= 0
}

WSV.portfolio <- function(S, mu.target,
  downside.weight)
{
  call(WSV.model)
  WSV.system <- System(WSV.model, S, mu.target,
     downside.weight)
  solution <- solve(WSV.system, trace=T)
  weight <-
     matrix(round(solution$variable$w$current,
     digit=5)*100, ncol=1)
```

```
  risk <- solution$objective
  return(weight,risk)
}

WSV.frontier <- function(S, n.pf, downside.weight)
{
  call(WSV.portfolio)
  Risk <- matrix(0, ncol=1, nrow=n.pf)
  Return <- matrix(0, ncol=1, nrow=n.pf)
  m <- nrow(S)
  mu.min <- min(apply(S, 2, mean))
  x <- WSV.portfolio(S, mu.target=mu.min,
     downside.weight)
  mu.max <- max(apply(S, 2, mean))
  mu.range <- seq(mu.min, mu.max,
     (mu.max-mu.min)/(n.pf-1))
  weight <- x$weight
  Risk[1,1] <- x$risk
  Return[1,1] <- mu.min

  for(i in 2:n.pf)
  {
     x <- WSV.portfolio(S, mu.target=mu.range[i],
        downside.weight)
     Risk[i,1] <- x$risk
     Return[i,1] <- mu.range[i]
     weight <- cbind(weight,x$weight)
  }

  graphsheet()
  par(mfrow=c(1,2))
  plot(Risk, Return, type="b")
  title("Mean - Weighted Semi-variance Frontier")
  barplot(weight)
  title("Frontier Portfolios")
  litle("optimal.weights"=weight)
}
```

Code 5.3 Weighted Semi-variance Model

As usual, we run an example optimization with a simulated data set:

```
> x <- WSV.frontier(S, n.pf=10,
     downside.weight=0.8)
```

```
> x$optimal.weights
      [,1]   [,2]   [,3]   [,4]   [,5]   [,6]
[1,]   100 88.825 77.449 66.080 50.272 34.251
[2,]     0  0.000  0.000  0.019  5.985 12.241
[3,]     0 10.483 19.010 27.736 35.148 42.527
[4,]     0  0.693  3.542  6.166  8.595 10.981

       [,7]   [,8]   [,9] [,10]
[1,] 18.080  1.977  0.000     0
[2,] 18.696 25.061 12.531     0
[3,] 49.850 57.192 68.953   100
[4,] 13.374 15.771 18.516     0
```

Weighted semi-variance solutions invest more cautiously in asset 2 due to the obvious non-normality in its returns. Although the flexibility of the weighted semi-variance model is appealing, little guidance can be given on how to weight upper and lower semi-variances. For most practical applications, investors will hence stick with the lower semi-variance model. Figure 5.10 illustrates the weighted semi-variance frontier.

5.4 Probability-Based Risk/Return Measures

5.4.1 *Shortfall Probability, Lower Partial Moment, and Value-at-Risk*

Regulatory pressures (bankruptcy/default occurs if wealth falls below a liability threshold) as well as investment intuition (probability statements seem to be easier to understand than volatility numbers) often guide investors towards shortfall probability as their preferred measure of risk.[7] We start with the general observation that uncertain investment returns can be decomposed into a threshold return (γ) plus an upside measure, expressed as $\max[r-\gamma,0]$ which is either positive or zero, minus a downside risk measure, denoted by $\max[\gamma-r,0]$ which is also either positive or zero. In combination, we get

$$r=\gamma+\max[r-\gamma,0]-\max[\gamma-r,0]. \tag{5.21}$$

Measures that focus on the downside of a return distribution are called **lower partial moments**, while measures that focus on the upside are called **upper partial moments**. If return distributions become non-normal, risk measures that capture non-normality become attractive. We have already discussed the special case of $\gamma=\bar{\mu}$, in which we distinguished between upper and lower

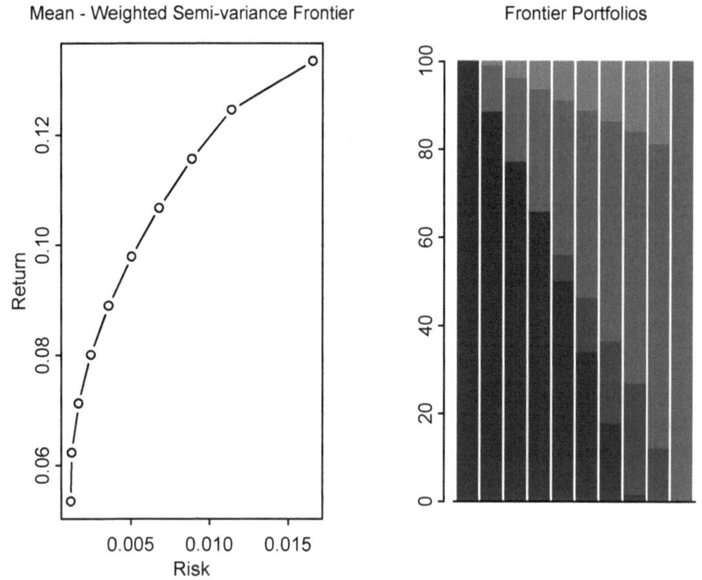

Figure 5.10 Weighted Semi-variance Frontier

semi-variances. Lower partial moments characterize the moments of a return distribution below the specified threshold return. In general, we define the lower partial moment of degree k in its discrete form (working on realized return scenarios rather than on a continuous distribution) as

$$
\begin{aligned}
lpm_{\gamma}^{k} &= \tfrac{1}{m}\sum\nolimits_{s=1}^{m}\left(r_s-\gamma\right)^{k}\delta_s,\\
\delta_s &= \begin{cases}0, & r_s>\gamma\\ 1, & r_s\le\gamma.\end{cases}
\end{aligned}
\tag{5.22}
$$

Again we use the same notation as in the previous chapter, where δ_s denotes an integer variable that assumes either one or zero. Effectively, δ_s will decide which observations enter the calculation on a go/no-go basis and hence can be modeled using integer variables. Apart from the threshold level, we also control the choice of the moment parameter k. For $k=0$, we get the shortfall probability, for $k=1$ we get the average shortfall, and for $k=2,3,4,\cdots$, we find shortfall variance, skewness, kurtosis, etc. In this section, we will focus on $k=0$,

$$lpm_{\gamma}^{0} = \tfrac{1}{m}\sum\nolimits_{s=1}^{m}\left(r_s-\gamma\right)^{0}\delta_s = \tfrac{1}{m}\sum\nolimits_{s=1}^{m}\delta_s,$$

because shortfall probability is closely related to value-at-risk. Note that as value-at-risk denotes the maximum loss (to be calculated) likely to occur in normal circumstances (i.e., 95% of all times, with a significance level of $\alpha = 5\%$),

$$VaR = F^{-1}(\alpha),$$

while the shortfall probability denotes the probability (to be calculated) that the loss will fall below a prespecified loss of amount γ :

$$lpm_{\gamma}^{0} = prob(r \leq \gamma).$$

While we fix probability in value-at-risk calculations, we fix our loss threshold in the calculation of shortfall probability. Hence both measures coincide if we set $\gamma = VaR$:

$$VaR = F^{-1}\left(lpm_{\gamma}^{0}\right).$$

In short, the value-at-risk at a significance level of α denotes a loss with shortfall probability α.[8] After this short digression on value-at-risk, shortfall probability, and lower partial moments, we proceed with the implementation of shortfall probability in NUOPT for S-PLUS. We focus on the calculation of shortfall probability with the use of scenarios, in which case the minimization of shortfall probabilities requires the use of integer variables.

5.4.2 Portfolio Construction and Shortfall Probability

We focus on an investor who aims to minimize shortfall risk (relative to a return threshold) subject to a specified return target. Equations (5.23) and (5.24) provide the appropriate switches,

$$\sum_{i=1}^{n} w_i r_{i,s} - \gamma \geq e - \delta_s E, \tag{5.23}$$

$$\sum_{i=1}^{n} w_i r_{i,s} - \gamma \leq (1 - \delta_s) E, \tag{5.24}$$

where e denotes a very small number and E represents a very large number. Each time the portfolio return is higher than the threshold return, $\delta_s = 0$ will simultaneously satisfy (5.23) and (5.24). Note that $\delta_s = 1$ will only satisfy (5.23). The portfolio construction problem becomes

$$
\begin{aligned}
&\text{minimize } \tfrac{1}{m}\sum\nolimits_{s=1}^{m} \delta_s \\
&\sum\nolimits_{i=1}^{n} w_i \mu_i = \bar{\mu} \\
&\sum\nolimits_{i=1}^{n} w_i = 1 \\
&\sum\nolimits_{i=1}^{n} w_i r_{i,s} - \gamma \geq e - \delta_s E \\
&\sum\nolimits_{i=1}^{n} w_i r_{i,s} - \gamma \leq (1 - \delta_s) E \\
&w_i \geq 0 \\
&\delta_s \in \{0,1\}.
\end{aligned}
\tag{5.25}
$$

The system (5.25) can also be brought into S-PLUS code (Code 5.4):

```
SF.model <- function(S, mu.target, mu.threshold)
{
  m <- nrow(S)
  n <- ncol(S)
  mu.bar <- apply(S, 2, mean)
  names(mu.bar) <- NULL
  asset <- Set()
  period <- Set()
  i <- Element(set=asset)
  s <- Element(set=period)
  S <- Parameter(S, index=dprod(s,i))
  mu.bar <- Parameter(as.array(mu.bar), index=i)
  mu.target <- Parameter(mu.target, changeable=T)
  mu.threshold <- Parameter(mu.threshold,
      changeable=T)
  w <- Variable(index=i)
  dummy <- IntegerVariable(index=s, type=binary)
  Sum(S[s,i]*w[i],i)-mu.threshold <=
      (1-dummy[s])*10
  Sum(S[s,i]*w[i],i)-mu.threshold >=
      -1*dummy[s]*10+0.000001
  risk <- Objective(type="minimize")
  risk ~ 1/m*Sum(dummy[s],s)
  Sum(mu.bar[i]*w[i],i) >= mu.target
  Sum(w[i],i) == 1
  w[i] >= 0
}
```

```
SF.portfolio <- function(S, mu.target,
     mu.threshold)
{
  SF.system <- System(SF.model, S, mu.target,
     mu.threshold)
  nuopt.options(maxitn=1000)
  solution <- solve(SF.system, risk, trace=T)
  weight <-
     matrix(round(solution$variable$w$current,
     digit=5)*100, ncol=1)
  risk <- solution$objective
  return(weight,risk)
}

SF.frontier <- function(S, mu.threshold, n.pf)
{
  call(SF.portfolio)
  Risk <- matrix(0, ncol=1, nrow=n.pf)
  Return <- matrix(0, ncol=1, nrow=n.pf)
  m <- nrow(S)
  n <- ncol(S)
  mu.min <- min(apply(S, 2, mean))
  x <- SF.portfolio(S, mu.target=mu.min,
     mu.threshold)
  mu.min <- t(x$weight) %*% apply(S, 2, mean)/100
  weight <- x$weight
  Risk[1,1] <- c(x$risk[2])


  Return[1,1] <- mu.min
  mu.max <- max(apply(S, 2, mean))
  mu.range <- seq(mu.min, mu.max,
     (mu.max-mu.min)/(n.pf-1))
  for(i in 2:n.pf){
     x <- SF.portfolio(S, mu.range[i],
     mu.threshold)
     Risk[i,1] <- x$risk[2]
     Return[i,1] <- mu.range[i]
     weight <- cbind(weight,x$weight)
  }
  graphsheet()
  par(mfrow=c(1,2))
  plot(Risk, Return, type="b")
  title("Mean - Shortfall Frontier")
  barplot(weight)
```

```
  title("Frontier Portfolios")
  list("optimal.weights" = weight)
}
```

Code 5.4 Shortfall Efficient Model

```
> x <- SF.frontier(S, mu.threshold, n.pf=50)
> x$optimal.weights
       [,1]   [,2]   [,3]   [,4]   [,5]   [,6]
[1,]  0.000  0.000  0.000  0.000  0.063  0.000
[2,] 77.436 68.604 58.863 49.981 40.926 27.247
[3,] 16.393 24.086 25.763 32.667 39.386 50.680
[4,]  6.172  7.310 15.375 17.352 19.625 22.073

       [,7]   [,8]   [,9] [,10]
[1,]  1.467  1.710  0.000     0
[2,] 21.324 11.716  4.907     0
[3,] 54.751 61.965 64.596   100
[4,] 22.458 24.609 30.497     0
```

The inspection of the left part of Figure 5.11 is disappointing. However, as VaR is a nonconvex function with respect to portfolio weights (and hence possesses many local minima), it is not surprising that standard optimization techniques will not always find the optimal solution. After all, heuristics are needed if objective functions are nonconvex. This difficulty in finding optimal portfolios when using VaR as a risk measure in scenario optimization is one of the major obstacles to its use. It is not only inconvenient but also directly related to its theoretical deficiencies (i.e., its lack of subadditivity; see Section 5.6).

5.4.3 Probability of Outperformance

Many portfolio managers and plan sponsors are given performance objectives. Hence their interest is to outperform their given investment targets. The problem of maximizing the probability of outperformance can be written as

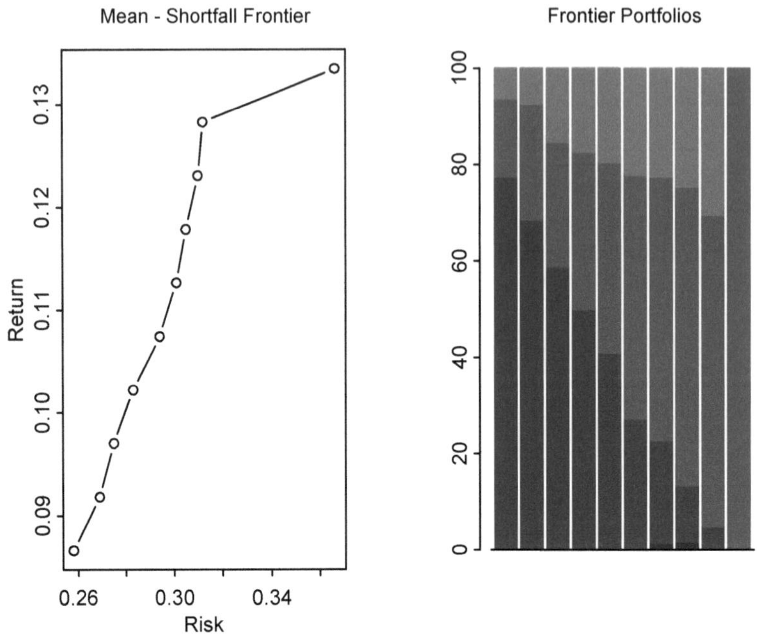

Figure 5.11 Shortfall Efficient Portfolios

$$\begin{aligned}
&\text{maximize } 1-\tfrac{1}{m}\sum\nolimits_{s=1}^{m}\delta_s \\
&\sum\nolimits_{i=1}^{n} w_i = 1 \\
&\sum\nolimits_{i=1}^{n} w_i r_{i,s} - \gamma \ge e - \delta_s E \\
&\sum\nolimits_{i=1}^{n} w_i r_{i,s} - \gamma \le (1-\delta_s)E \\
&w_i \ge 0,\ \delta_s \in \{0,1\}.
\end{aligned} \tag{5.26}$$

We leave the implementation of (5.26) to the reader. Do you think this investment objective makes sense?

5.5 Minimum Regret

Suppose we are again given the scenario matrix $\mathbf{S}$, either from historical returns or from a scenario simulation exercise. As in basic decision theory, we could choose minimax criteria, as illustrated in Figure 5.12 (i.e., we might want to minimize the maximum portfolio loss—minimizing regret).[9] This could be the

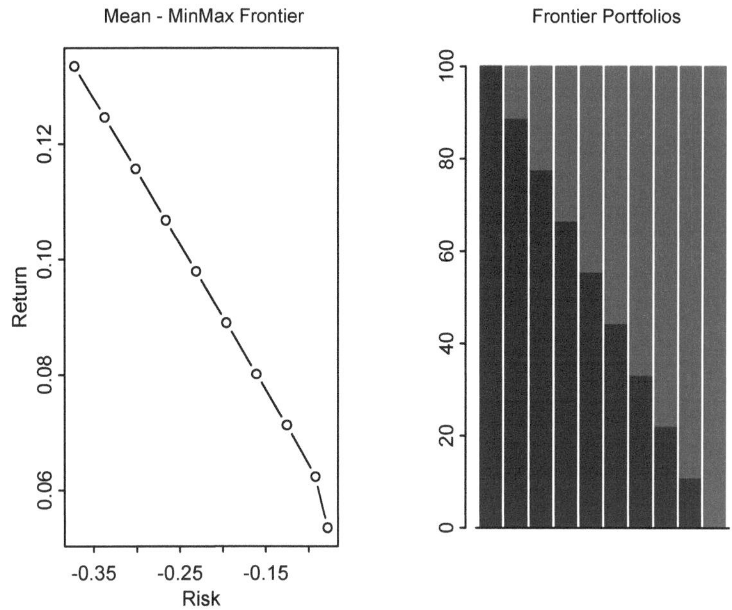

Figure 5.12 MinMax Efficient Frontier

optimal strategy for investors who have to make sure under all means (scenarios) that they never experience a particular size of loss. Focusing on extreme events will have its merits if returns either substantially deviate from normality or if investors are extremely risk-averse. Minimizing the maximum loss can be written as a linear program:

$$\begin{aligned}&\max\ r_{\min}\\&\sum\nolimits_{i=1}^{n} w_i r_{i,s} - r_{\min} \geq 0\\&\sum\nolimits_{i=1}^{n} w_i \mu_i = \bar{\mu}\\&\sum\nolimits_{i=1}^{n} w_i = 1\\&w_i \geq 0.\end{aligned} \tag{5.27}$$

The first constraint $\sum\nolimits_{i=1}^{n} w_i r_{i,s} - r_{\min} \geq 0$ ensures that there is no scenario for which the portfolio return is worse than the minimum return. As $r_{\min}$ is a variable as well as the objective in the system (5.27), it will take on the value of the minimum maximum loss. An alternative (equivalent) formulation to (5.27) is to maximize return subject to the restriction that there is no scenario for which the portfolio return falls below a threshold return $r_{\min}$.

$$\begin{aligned}
&\max \sum_{i=1}^{n} w_i \mu_i \\
&\sum_{i=1}^{n} w_i r_{i,s} \geq r_{\min} \\
&\sum_{i=1}^{n} w_i = 1 \\
&w_i \geq 0.
\end{aligned} \tag{5.28}$$

We choose the program of (5.27) for implementation in NuOPT for S-Plus (shown in Code 5.5).

```
MinMax.model <- function(S, mu.target)
{
  m <- nrow(S)
  n <- ncol(S)
  mu.bar <- apply(S, 2, mean)
  asset <- Set()
  period <- Set()
  i <- Element(set=asset)
  s <- Element(set=period)
  S <- Parameter(S, index=dprod(s,i))
  mu.bar <- Parameter(as.array(mu.bar), index=i)
  mu.target <- Parameter(mu.target, changeable=T)
  w <- Variable(index=i)
  mu.Min <- Variable()
  Sum(S[s,i]*w[i],i)-mu.Min >= 0
  MinMax <- Objective(type="maximize")
  MinMax ~ mu.Min
  Sum(mu.bar[i]*w[i],i) == mu.target
  Sum(w[i],i) == 1
  w[i] >= 0
}

MinMax.portfolio <- function(S, mu.target)
{
  call(MinMax.model)
  MinMax.system <- System(MinMax.model, S,
     mu.target)
  solution <- solve(MinMax.system, trace=T)
  weight <-
     matrix(round(solution$variable$w$current,
```

```
      digit=5)*100, ncol=1)
   risk <- solution$objective
   return(weight,risk)
}

MinMax.frontier <- function(S, n.pf)
{
   call(MinMax.portfolio)
   Risk <- matrix(0, ncol=1, nrow=n.pf)
   Return <- matrix(0, ncol=1, nrow=n.pf)
   m <- nrow(S)
   mu.min <- min(apply(S,2,mean))
   x <- MinMax.portfolio(S, mu.target=mu.min)
   weight <- x$weight
   Risk[1,1] <- x$risk
   Return[1,1] <- mu.min
   mu.max <- max(apply(S, 2, mean))
   mu.range <- seq(mu.min, mu.max,
      (mu.max-mu.min)/(n.pf-1))
   for(i in 2:n.pf){
      x <- MinMax.portfolio(S,
      mu.target=mu.range[i])
      Risk[i,1] <- x$risk
      Return[i,1] <- mu.range[i]
      weight <- cbind(weight,x$weight)
   }
   graphsheet
   par(mfrow=c(1,2))
   plot(Risk, Return, type="b")
   title("Mean - MinMax Frontier")
   barplot(weight)
   title("Frontier Portfolios")
   list("optimal.weights" = weight)
}
```

Code 5.5 Regret Minimization

```
> x <- MinMax.frontier(S, n.pf=50)
> x$optimal.weights
     [,1]   [,2]   [,3]   [,4]   [,5]   [,6]
[1,]  100 88.889 77.778 66.667 55.556 44.444
[2,]    0  0.000  0.000  0.000  0.000  0.000
[3,]    0 11.111 22.222 33.333 44.444 55.556
[4,]    0  0.000  0.000  0.000  0.000  0.000
```

```
       [,7]   [,8]   [,9]  [,10]
[1,] 33.333 22.222 11.111      0
[2,]  0.000  0.000  0.000      0
[3,] 66.667 77.778 88.889    100
[4,]  0.000  0.000  0.000      0
```

Minimizing maximum regret leads to concentrated portfolios. The highly non-normal assets 2 and 4 never enter the optimal solution.

5.6 Conditional Value-at-Risk

5.6.1 CVaR, Tail Conditional Loss, and VaR

Suppose we sampled discrete realizations of portfolio returns from a continuous distribution to arrive at m realizations $\{r_s\}_{s=1,\ldots,m}$ of random returns. To make matters transparent, just think of this as a sequence of returns $\{6,-10,-3,4,5,\cdots,1\}_{m=100}$.[10] We now define the order statistics (simply by ordering the returns starting with the smallest return from the left) $r_{1:m} \le r_{2:m} \le \ldots \le r_{m:m}$ that result in the sorted returns $\{-10,-5,-3,-3,-3,-3,-3,\cdots,16\}_{m=100}$. If we need to estimate the $\alpha\%$-quantile (value-at-risk), we simply look for

$$VaR_\alpha = r_{\alpha m:m}. \tag{5.29}$$

If we set $\alpha = 5\%$, $m = 100$, we arrive at $VaR = r_{5:100} = -3$ (5th out of 100 returns) in the example above.

The estimator for the expected loss in $\alpha\%$ of all cases, also called **conditional value-at-risk** ($CVaR$) or **expected shortfall** (ES), is calculated from

$$ES_\alpha = \frac{1}{m\alpha}\sum_{s=1}^{m\alpha} r_{s:m}. \tag{5.30}$$

For the example above we get

$$ES_{5\%} = \tfrac{1}{5}\sum_{s=1}^{5} r_{s:100} = \tfrac{1}{5}(-10-5-3-3-3) = -4.8.$$

A measure similar to (5.30) is the "tail conditional loss" (TCL) defined as $E(r|r \le VaR_\alpha)$, which looks the same at first sight.

$$TCL_\alpha = \frac{\sum_{s=1}^{m} \delta_s r_s}{\sum_{s=1}^{m} \delta_s} \text{ where } \delta_s = \begin{cases} 1, & r_s \le r_{\alpha m:m} \\ 0, & \text{otherwise.} \end{cases} \quad (5.31)$$

This, however, is only true for continuous distributions, as the discrete example above shows.

$$TCL_\alpha = (-10-5-3-3-3-3-3)/7 = -4.3.$$

Tail conditional loss and expected shortfall differ. In general, we will find that expected shortfall is at least as large as tail conditional loss.

$$ES_\alpha = TCL_\alpha + \phi(TCL_\alpha - VaR_\alpha), \quad (5.32)$$

where

$$\phi = \frac{prob(r \le VaR)}{\alpha} - 1 \ge 0.$$

The reason for this is that for discrete distributions $prob(r \le VaR) \ge \alpha$. In our example, we find that $prob(r \le VaR) = \frac{7}{100} = 7\%$, which is larger than 5%. Substituting the appropriate values into (5.32), we get

$$-4.8 = -4.3 + \left(\frac{7\%}{5\%} - 1\right)(-4.3-3).$$

We now show various ways to calculate the numbers above in S-PLUS. Suppose we simulate a mixture (of two normals) to generate a data set for the sample calculations below.

```
returns <- c(rnorm(50, -0.4, 0.3),
  rnorm(950, 0.07, 0.2))
graphsheet()
par(mfrow=c(1,2))
hist(returns)
qqnorm(returns)
qqline(returns)
```

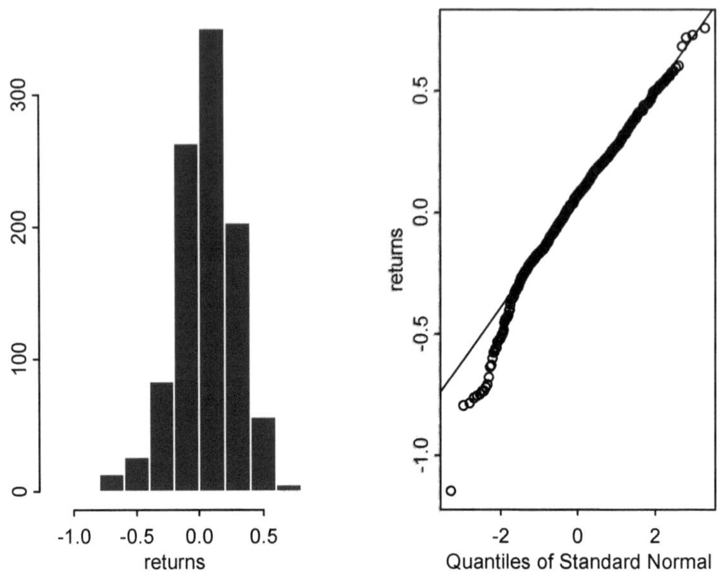

Figure 5.13 Sample Data

The data are plotted in Figure 5.13. It is apparent that the distribution differs significantly from the normal distribution in its left tail (the Q-Q plot deviates substantially from a straight line). S-PLUS functions that generate the required risk measures are given below (in Code 5.6) and can also be used to generate the distribution of our estimated risk measure via repeated resampling (bootstrapping). Bootstrapped results are shown in Figure 5.14.

```
VaR <- function(returns, alpha){
  sort(returns)[trunc(length(returns)*alpha)]
}

CVaR <- function(returns, alpha){
  mean(sort(returns)[1:trunc(length(returns)*alpha)
     ])
}

TCL <- function(VaR, returns){
  mean(returns[returns<=VaR(returns, alpha)])
}

bs.VaR <- bootstrap(returns, VaR(returns, alpha))
bs.CVaR <- bootstrap(returns, CVaR(returns, alpha))
```

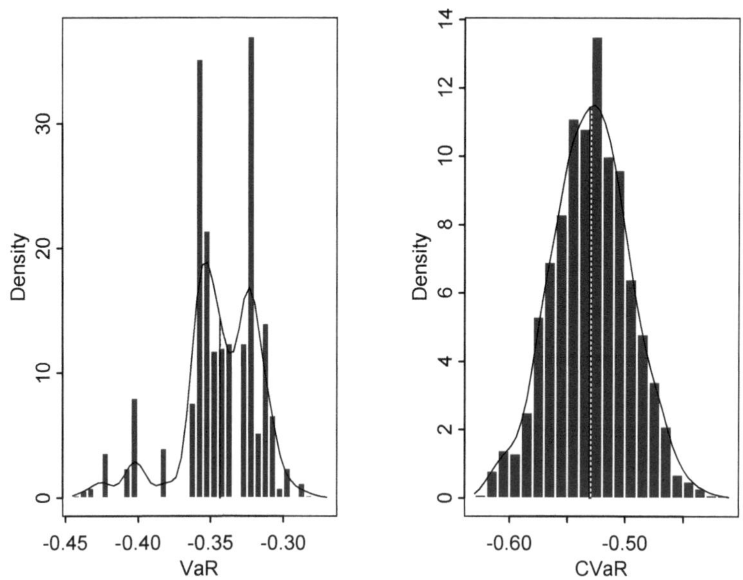

Figure 5.14 Resampled VaR and CVaR Calculations

```
graphsheet()
par(mfrow=c(1,2))
plot(bs.VaR, xlab="VaR", main="")
plot(bs.CVaR, xlab="CVaR", main="")
```

Code 5.6 Bootstrap Distributions of Various Risk Measures

This way, the bootstrap function can be used to investigate which risk concept (VaR, CVaR, or volatility) needs more data to be estimated with the same precision. Note that we can address the resampled risk measures (see Figure 5.15) directly using

```
bs.VaR <- bootstrap(returns,
  VaR(returns, alpha))$replicates
```

for example. We can then use the `boxplot()` command to visualize the estimation error in all three risk measures. For this purpose, we remove the means of the estimates of our risk measures to plot them on the same level in Code 5.7.

```
bs.VaR <- bootstrap(returns,
  VaR(returns, alpha))$replicates

bs.CVaR <- bootstrap(returns,
```

```
  CVaR(returns, alpha))$replicates

bs.Vol <- bootstrap(returns,
  Vol(returns))$replicates

demean <- function(x){x-mean(x)}

boxplot(demean(bs.VaR), demean(bs.CVaR),
  demean(bs.Vol),
  names=c("VaR", "CVaR", "Volatility"))
```

Code 5.7 Estimation Error in Various Risk Measures

We see that the estimation error for CVaR is larger than for volatility (where precision is highest) or VaR. This makes intuitive sense, as CVaR looks deeper into the tail and is hence outlier-dependent, while large outliers do not affect VaR (which is economically a central weakness). Volatility, on the other hand, uses all the data in a sample and arrives at a very precise estimate (that might be completely useless in the case of serious non-normality).

Rather than using sampled data, we could also employ the numerical integration techniques offered by S-PLUS to calculate the required risk measures. Suppose we are given the distribution behind Figure 5.13 in continuous form, $pf(x,\mu_1,\sigma_1)+(1-p)f(x,\mu_2,\sigma_2)$, where $f(x,\mu,\sigma)=\frac{1}{\sigma\sqrt{2\pi}}\exp\left(-\frac{(x-\mu)^2}{2\sigma^2}\right)$. The value-at-risk for the 2.5% level is –45.89%. To check this, just calculate

$$p\int_{-\infty}^{-45.89\%} f(x,\mu_1,\sigma_1)+(1-p)\int_{-\infty}^{-45.89\%}(x,\mu_2,\sigma_2)=2.5\%. \tag{5.33}$$

```
integrand.1 <- function(x){
  (0.05*dnorm(x,-0.4,0.3) + 0.95*dnorm(x,0.07,0.2))
}
integrate(integrand.1,  - Inf, -0.4589)$integral
[1] 0.02499479
```

We can also calculate the expected shortfall, or conditional value-at-risk (see Code 5.8),

$$\frac{p\int_{-\infty}^{-45.89\%} xf(x,\mu_1,\sigma_1)dx+(1-p)\int_{-\infty}^{-45.89\%} xf(x,\mu_2,\sigma_2)dx}{p\int_{-\infty}^{-45.89\%} f(x,\mu_1,\sigma_1)dx+(1-p)\int_{-\infty}^{-45.89\%} f(x,\mu_2,\sigma_2)dx}=-65.366\%\,. \tag{5.34}$$

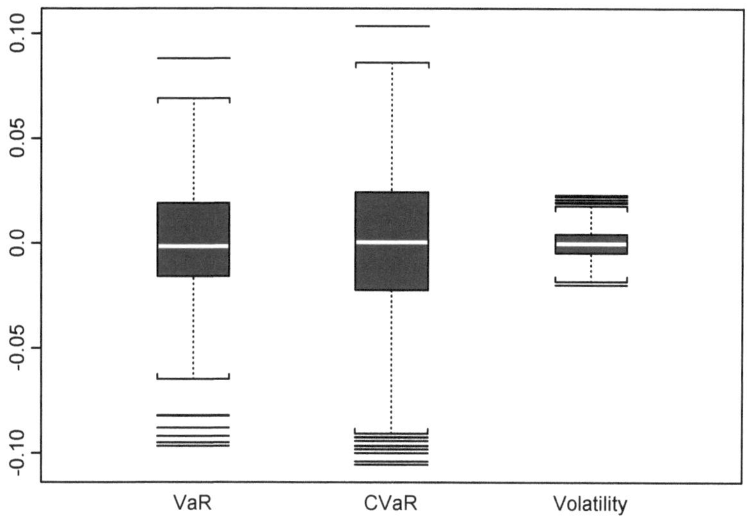

Figure 5.15 Boxplot for Resampled Risk Measures

```
integrand.2 <- function(x){
  (0.05*dnorm(x,-0.4,0.3) +
     0.95*dnorm(x,0.07,0.2))*x
}
integrate(integrand.2,-Inf,
  -0.458915)$integral/integrate(integrand.1,-Inf,
  -0.458915)$integral
[1] -0.6536664
```

Code 5.8 Risk Estimates via Numerical Integration

Numerical integration allows us to calculate arbitrary risk figures as soon as a continuous distribution has been fit to the data.

5.6.2 What Do We Require from a Risk Measure?

For portfolio managers, risk managers, and plan sponsors, there is the vital question: what properties are needed for a statistic of portfolio returns to qualify as a risk measure? The answer to this question has been given through a complete set of axioms. They define what has been called a **coherent risk measure.**[11] A coherent risk measure is a function (that translates returns into a risk figure) that is

1. **Monotonous**. Larger losses translate into higher risks.

2. **Positive homogeneous**. If we multiply holdings (positions, exposures) by a linear factor, risk also rises by this factor.
3. **Invariant to translations**. Adding a constant to our losses does not change risk.
4. **Subadditive**. The risk of a portfolio is at most the combined risks of the single positions.

The last axiom catches diversification. Adding two portfolios together must not create higher risks than both on a stand-alone basis. These axioms define the nature of the concept with a minimum set of precise formulations (requirements). A risk measure that violates one of the axioms above will lead to paradoxical results. Note that return statistics that do not fit into the axiomatic framework cannot be called risk measures (by the very definition).

Let's see how VaR, CVaR, volatility, and shortfall probability do in a simple setting. A plan sponsor budgets risks given to individual managers. His scenarios of two very diversifying managers (i.e., negative returns can only occur in different states) are given in Table 5.1.[12]

Table 5.1 Data for Manager Combination: Active Returns

Scenario	Manager 1	Manager 2	Manager 1+2	Probability
1	–20%	2%	–9%	3%
2	–3%	2%	–0.5%	2%
3	2%	–20%	–9%	3%
4	2%	–3%	–0.5%	2%
5	2%	2%	2%	90%

We can now calculate the risk measures mentioned above. The outcome is summarized in Table 5.2. While volatility and CVaR are decreasing as we move from a stand-alone approach to a combination of managers, this is not the case for shortfall probability and VaR. Hence, both statistics are not suitable risk measures.

Table 5.2 Risk Measures in Multiple-Manager Example

Risk Measure	Manager 1	Manager 2	Manager 1+2
Volatility	3.80%	3.80%	2.63%
VaR	–3%	–3%	–9%
Shortfall Probability	5%	5%	10%
CVaR	–13.20%	–13.20%	–5.60%

The reason for these paradoxical results lies in the concept of value-at-risk. It ignores the large –20% losses that are waiting undetected in the tail of the distribution. However, when we average across portfolios, these returns will be diversified into the portfolio risk measure and will increase risk, as they have been ignored before. No investor would find a risk measure that attaches equal weight to small losses and complete bankruptcy satisfying. Value-at-risk fails in detecting tail risks.

Value-at-risk and shortfall probability are not coherent risk measures (i.e., they should not be called risk measures at all). Value-at-risk will only be subadditive in special circumstances (i.e., for so-called elliptical distributions, such as the normal distribution, Student's *t* distribution, and the Cauchy distribution). In fact, value-at-risk and volatility share the same properties when the underlying distribution is elliptical.[13] However, the distributions of many assets involved in portfolio construction do not belong to this class. They either naturally deviate from asset class characteristics (hedge fund, credit risk, etc.) or are deliberately created to do so using heavily skewed distributions. Interestingly, value-at-risk, which once was regarded as the Holy Grail in risk management, fails when return distributions are not elliptical. Casually stated, value-at-risk cannot deal with non-normality. Table 5.3 gives a concise summary of our discussion.

Table 5.3 VaR versus CVaR

Criterion	VaR	CVaR
Subadditivity?	No	Yes
Tail risk measure?	No	Yes
Handle nonnormality?	Constrained to cases for non-normal elliptical distributions	Yes
Data requirements?	Needs more data than the volatility measure to be measured with the same precision	Needs more data than VaR to be measured with the same precision

It is difficult to understand why value-at-risk is still so popular. We do warn against its use in portfolio optimization, as the problems above are likely to increase further when a portfolio optimizer leverages axiomatic shortcomings of VaR.

5.6.3 The Use of CVaR in Portfolio Construction

Not only does CVaR offer a much sounder theoretical basis for risk management decisions but it is also computationally more efficient. While portfolio optimization using VaR becomes a complicated integer-programming problem, CVaR optimization only requires well-established and widely available linear programming tools. First, we define an auxiliary variable,

$$e_s = \max\left[0, VaR - \sum_{i=1}^{n} w_i r_{i,s}\right]. \tag{5.35}$$

Suppose $VaR = -20$ (percent), while the portfolio return for scenario s turns out to be $\sum_{i=1}^{n} w_i r_{i,s} = -25$. Equation (5.35) would then find an excess of $e_s = \max[0,(-20)-(-25)] = \max[0,5] = 5$. Conditional value-at-risk equals value-at-risk plus the average of all losses in excess of value-at-risk. Hence we can write

$$CVaR = VaR - \left(\tfrac{1}{m}\sum_{s=1}^{m} e_s\right)/\alpha. \tag{5.36}$$

Note that $\frac{1}{m}\sum_{s=1}^{m} e_s$ reflects the average excess loss across all m scenarios. In order to scale this loss up to the average excess loss, if an excess loss occurs, we have to divide it by the probability of an excess loss (α). As VaR is a negative number, $VaR - \left(\frac{1}{m}\sum_{s=1}^{m} e_s\right)/\alpha$ will be even more negative. A risk-averse investor will hence want to maximize $CVaR$ (–5 is larger than –20). The complete portfolio optimization problem can now be written down as

$$\begin{aligned}
&\max_{w_i, e_s, VaR} \; VaR - \left(\tfrac{1}{m}\sum_{s=1}^{m} e_s\right)/\alpha \\
&\sum_{i=1}^{n} w_i \mu_i \geq \bar{\mu} \\
&e_s \geq VaR - \sum_{i=1} w_i r_{i,s} \\
&e_s \geq 0 \\
&w_i \geq 0.
\end{aligned} \tag{5.37}$$

A closer look at (5.37) will reveal some of the mechanics. Note that excesses are forced to be positive. Any excess ($e_s \geq VaR - \sum_{i=1} w_i r_{i,s}, e_s \geq 0$) will hence have a negative impact on the objective function. Excesses can be kept small by choosing the appropriate set of weights in order to prevent $VaR - \sum_{i=1} w_i r_{i,s}$ from becoming a large positive number.

If portfolio returns in all scenarios were positive, how could we prevent e_s from becoming negative? The optimizer can now increase VaR and therefore positively impact the objective function. However, moving VaR up too much will result in increasing excesses that will counterbalance this effect. The CVaR model is given in Code 5.9.

```
CVaR.model <- function(S, alpha, mu.target)
{
  m <- nrow(S)
  n <- ncol(S)
  mu.bar <- apply(S, 2, mean)
  asset <- Set()
  period <- Set()
  mu.VaR <- Set(1)
  i <- Element(set=asset)
  s <- Element(set=period)
  mu.VaR <- Element(set=mu.VaR)
  S <- Parameter(S, index=dprod(s,i))
  mu.bar <- Parameter(as.array(mu.bar), index=i)
  mu.target <- Parameter(mu.target, changeable=T)
  w <- Variable(index=i)
  e <- Variable(index=s)
  mu.VaR <- Variable(index=mu.VaR)
  e[s] >= mu.VaR[1]-Sum(S[s,i]*w[i],i)
  e[s] >= 0
  CVAR <- Objective(type="maximize")
  CVAR ~ mu.VaR[1]-(1/m)*(1/alpha)*Sum(e[s],s)
  Sum(mu.bar[i]*w[i],i) == mu.target
  Sum(w[i],i) == 1
  w[i] >= 0
}
```

Code 5.9 CVaR Optimization

Suppose we run the model on $m = 100$ scenarios with $\alpha = 0.05$ and $\bar{\mu} = 0.04$. What are the values for VaR and CVaR?

```
mu.target <- 0.04
alpha <- 0.05
S.mvnorm <- matrix(rmvnorm(100, mean=c(0.02,
   0.04, 0.05, 0.08), cov=diag(rep(0.2,4))),
   ncol=4)
CVaR.system <- System(CVaR.model, S.mvnorm,
   alpha, mu.target)
solution <- solve(CVaR.system, trace=T)
```

The value for CVaR can be obtained from `solution$objective`, which returns a value of `-0.7469582`. We compare this to VaR, which can be retrieved from `solution$variables$mu.VaR$current` and amounts to `-0.478263`. Finally, we plot the cumulative distribution for portfolio returns as well as for a normal distribution in Figure 5.16 and the CVaR frontier and optimal portfolios in Figure 5.17.

```
normal <- rnorm(100000, mean(returns),
  sqrt(var(returns)))
cdf.compare(returns, normal)

CVaR.portfolio <- function(S, alpha, mu.target)
{
  call(CVaR.model)
  CVaR.system <- System(CVaR.model, S, alpha,
     mu.target)
  solution <- solve(CVaR.system, trace=T)
  weight <-
     matrix(round(solution$variable$w$current,
     digit=5)*100, ncol=1)
  risk <- -solution$objective
  return(weight,risk)
}

CVaR.frontier <- function(S, alpha, n.pf)
{
  call(CVaR.portfolio)
  Risk <- matrix(0, ncol=1, nrow=n.pf)
  Return <- matrix(0, ncol=1, nrow=n.pf)
  mu.min <- min(apply(S,2,mean))
  mu.max <- max(apply(S,2,mean))
  mu.range <- seq(mu.min, mu.max,
     (mu.max-mu.min)/(n.pf-1))
  x <- CVaR.portfolio(S, alpha, mu.target=mu.min)
  weight <- x$weight
  Risk[1,1] <- x$risk
  Return[1,1] <- mu.min
  for(i in 2:n.pf){
     x <- CVaR.portfolio(S, alpha,
        mu.target=mu.range[i])
     Risk[i,1] <- x$risk
     Return[i,1] <- mu.range[i]
     weight <- cbind(weight,x$weight)
```

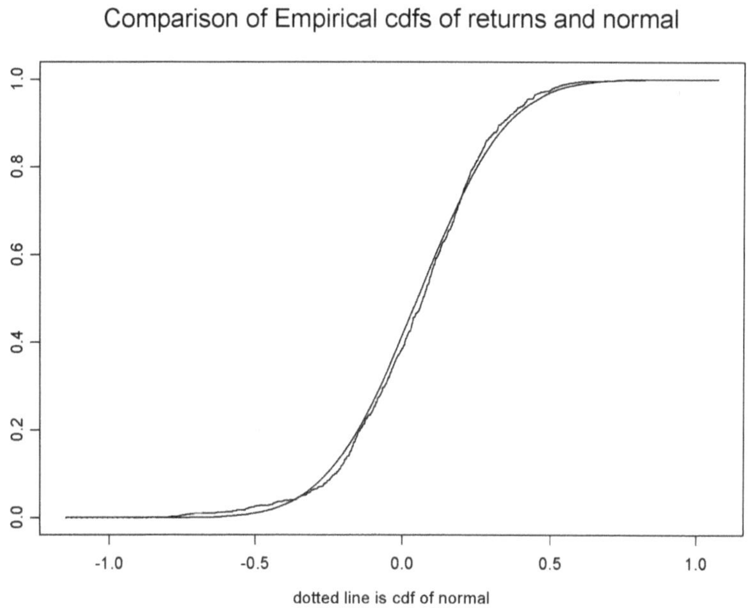

Figure 5.16 Cumulative Distribution of Portfolio Returns

```
  }
  graphsheet()
  par(mfrow=c(1,2))
  plot(Risk, Return, type="b")
  title("Mean - CVaR Frontier")
  barplot(weight)
  title("Frontier Portfolios")
  list("optimal.weights" = weight)
}
```

Code 5.10 CVaR Frontier

```
> x <- CVaR.frontier(S, alpha, n.pf=20)
> x$optimal.weights

        [,7]    [,8]    [,9]   [,10]   [,11]   [,12]
[1,] 67.944 62.631 57.295 52.035 46.684 41.375
[2,]  0.000  0.000  0.000  0.000  0.000  0.000
[3,] 26.918 31.703 36.253 41.548 45.949 50.761
[4,]  5.138  5.666  6.452  6.416  7.368  7.864

       [,13]   [,14]   [,15]   [,16]   [,17]   [,18]
[1,] 36.067 30.777 25.496 20.204 14.912  9.646
```

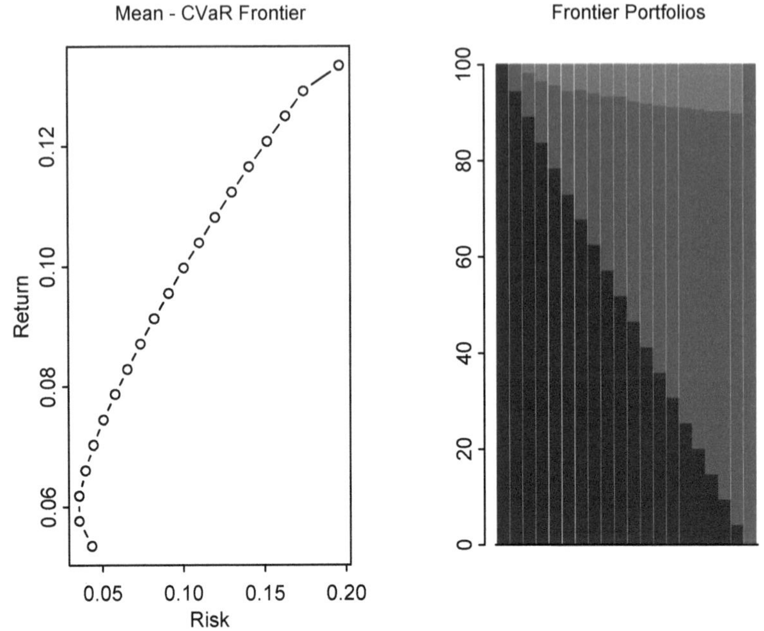

Figure 5.17 CVaR Frontier and Optimal Portfolios

```
[2,]  0.000  0.000  0.000  0.000  0.000  0.000
[3,] 55.592 60.589 65.685 70.666 75.647 80.882
[4,]  8.341  8.634  8.819  9.130  9.441  9.471

      [,19] [,20]
[1,]  4.338     0
[2,]  0.000     0
[3,] 85.705   100
[4,]  9.957     0
```

Conditional value-at-risk looks deeply into the tail of a distribution. In contrast with mean-variance-based solutions, we note that the highly non-normal assets 2 and 4 enter and are given much less weight.

5.6.4 VaR Approximation Using CVaR

It is well-known that VaR for discrete distributions is a nonsmooth, nonconvex, and multiextremum function with respect to w_i. VaR and the related shortfall risk are therefore difficult to optimize, as we have already seen in Section 5.4.2. This section will present a heuristic that attempts to minimize VaR by solving a sequence of CVaR problems.[14] As CVaR presents an upper bound on VaR (CVaR will always be greater than VaR, as it adds the average of the excess losses to VaR), one approach to minimizing VaR is to minimize the upper bound in a sequence of CVaR problems, gradually discarding scenarios that exhibit losses larger than VaR.

Step 0. Let us assume we have generated $m = 1000$ scenarios for $n = 4$ assets. We start with a standard CVaR minimization for a prespecified α and return target $\bar{\mu}$ as described in (5.37) using all m scenarios. The resulting CVaR (call it α-CVaR) represents the first upper bound on VaR for the prespecified α (call this α-VaR). Now we split the set of total scenarios into active scenarios (those used for further CVaR minimizations) and inactive scenarios (those discarded). From all scenarios that show losses larger than α-VaR, we discard a fraction ξ. For example, if 50 portfolio returns fall below VaR, we discard the largest 25 losses for further use.

Step 1. Start a new CVaR optimization on the remaining set of active scenarios (if we discard 25 out of 1000 scenarios we are left with 975 scenarios). However, we have to modify our CVaR optimization in two important respects. First we need to take into account that we discarded a number of scenarios. As we are interested in the α-VaR, we need to ensure that the α_1-CVaR optimization in Step 2 focuses on the same quantile. The new α_1 for the α_1-CVaR optimization needs to satisfy

$$\alpha_i = 1-(1-\alpha)\left(1-\frac{\sum_{step=0,i} discarded\ scenarios\ in\ step_i}{m}\right)^{-1}.$$

An example has been calculated in Table 5.4. As the number of discarded scenarios rises, we need to go further into the tail to maintain the quantile with respect to the original set of scenarios.

Table 5.4 Evolution of α_i for α=0.05 and ξ=0.5

Step i	# Discarded Scenarios	α_i
0	0	5.00%
1	25	2.56%
2	13	1.30%
3	6	0.65%
4	3	0.33%
5	2	0.16%

Second, we need to ensure that the allocation resulting from Step 1 does not create losses for the active scenarios that are larger than those for the inactive scenarios. This is important if we want to gradually reduce the number of active scenarios in a meaningful way. We therefore have to add m constraints and one new free variable (γ) to problem (5.37):

$$\sum_{i=1} w_i r_{i,s} \geq \gamma, \; s \in \textit{active scenario}$$

$$\sum_{i=1} w_i r_{i,s} \leq \gamma, \; s \in \textit{inactive scenario}$$

The constraints above will always be satisfied, as the optimizer could always default to the optimal allocation from the previous optimization (which we used to split scenarios into active and inactive scenarios). As before, we calculate α-VaR for the optimal solution in Step 1 and use it to split up scenarios further (reducing the number of scenarios stepwise) for use in the next step. It is obvious that α_1-CVaR needs to be smaller than α-CVaR, as we discarded the largest losses but adjusted the quantile for the inactive scenarios. Note that the VaR calculation is not affected, as all losses below VaR are equally counted, no matter how large they are. A refined CVaR code that takes account of this is shown in Code 5.11.

```
CVAR.model <- function(S.in, S.out, alpha, mu.bar,
  mu.target, VaR.cutoff)
{
  m.in <- nrow(S.in)
  m.out <- nrow(S.out)
  m <- m.in+m.out
  n <- ncol(S.in)
  asset <- Set()
  period.in <- Set()
  period.out <- Set()
  mu.VaR <- Set(1)
  i <- Element(set=asset)
  s <- Element(set=period.in)
  ss <- Element(set=period.out)
  mu.VaR <- Element(set=mu.VaR)
  S.in <- Parameter(S.in, index=dprod(s,i))
  S.out <- Parameter(S.out, index=dprod(ss,i))
  mu.bar <- Parameter(as.array(mu.bar), index=i)
  mu.target <- Parameter(mu.target, changeable=T)
  w <- Variable(index=i)
  e <- Variable(index=s)
  mu.VaR <- Variable(index=mu.VaR)
```

```
    g <- Variable(VaR.cutoff)
    e[s] >= mu.VaR[1]-Sum(S.in[s,i]*w[i],i)
    e[s] >= 0
    CVAR <- Objective(type="maximize")
    CVAR ~ mu.VaR[1]-(1/m)*(1/alpha)*Sum(e[s],s)
    Sum(mu.bar[i]*w[i],i) == mu.target
    Sum(w[i],i) == 1
    Sum(w[i]*S.in[s,i],i) >= g
    Sum(w[i]*S.out[ss,i],i) <= g
    w[i] >= 0
}
```

Code 5.11 VaR Approximation Using CVaR

Step 2. Repeat Step 1 as long as there are scenarios left to be discarded. The more scenarios we discard, the closer α_i-CVaR and α_i-VaR will become as more and more scenarios with large losses are removed (reducing CVaR), while this removal does not affect VaR as long as α_i is properly set. Implementing the algorithm above leads to the result shown in Figure 5.18 for our sample data set.

5.7 CDO Valuation using Scenario Optimization

Suppose we know the loss distribution of an underlying pool of assets valued today at 100 (i.e., we know $f(l)$). Note that $\tilde{l}$ can assume positive values (losses) as well as negative values (profits). This asset pool is financed via three different tranches. The first tranche is called *equity* (or sometimes a *junior note*). Liabilities for holders of this tranche are limited to l_{α_1}, which denotes the α_1 percentile of the loss distribution. Payoff to equity can be expressed as

$$CF_{equity} = l_{\alpha_1} - \tilde{l} + \max\left(\tilde{l} - l_{\alpha_1}, 0\right). \tag{5.38}$$

If the losses exceed l_{α_1}, the equity is wiped out and losses will start to eat into the second tranche (also called the *mezzanine*). The second tranche promises to pay an amount $l_{\alpha_1} - l_{\alpha_2}$. Losses larger than l_{α_2} lead to a complete loss of the second tranche and eat into the last tranche,

$$CF_{mezzanine} = l_{\alpha_2} - l_{\alpha_1} - \max\left(\tilde{l} - l_{\alpha_1}, 0\right) + \max\left(\tilde{l} - l_{\alpha_2}, 0\right). \tag{5.39}$$

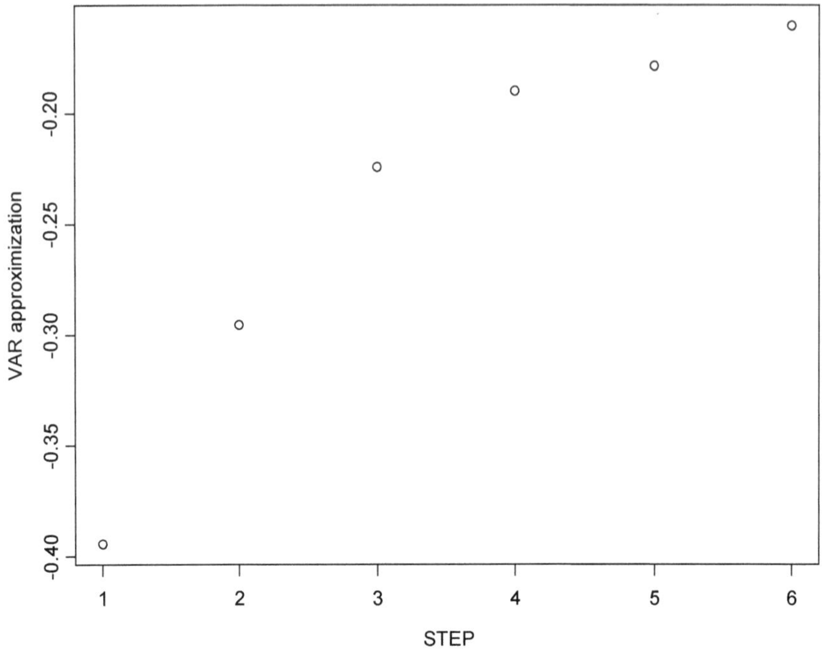

Figure 5.18 VaR Approximation Using CvaR for a Sample Data Set

The last tranche is called a *senior note*. Losses will only eat into the senior note after the first two tranches have been wiped out,

$$CF_{senior} = 100 - l_{\alpha_1} - \left(l_{\alpha_2} - l_{\alpha_1}\right) - \max\left(\tilde{l} - l_{\alpha_2}, 0\right). \tag{5.40}$$

We can see that investors in senior notes write a limited-liability option to holders of mezzanine debt $\max\left(\tilde{l} - l_{\alpha_2}, 0\right)$, while mezzanine debt investors write a limited-liability option $\max\left(\tilde{l} - l_{\alpha_1}, 0\right)$ to equity investors. If we add all positions, we arrive at

$$\begin{aligned} CF_{equity} + CF_{mezzanine} + CF_{senior} &= l_{\alpha_1} - \tilde{l} + \max\left(\tilde{l} - l_{\alpha_1}, 0\right) \\ &\quad + l_{\alpha_2} - l_{\alpha_1} - \max\left(\tilde{l} - l_{\alpha_1}, 0\right) \\ &\quad + \max\left(\tilde{l} - l_{\alpha_2}, 0\right) + 100 \\ &\quad - l_{\alpha_1} - \left(l_{\alpha_2} - l_{\alpha_1}\right) - \max\left(\tilde{l} - l_{\alpha_2}, 0\right) \\ &= 100 - \tilde{l}, \end{aligned} \tag{5.41}$$

which is exactly the payoff to the asset pool. Assuming that the loss distribution is log-normal, we can use standard Monte Carlo simulation techniques to evaluate the attached options. However, how can we value options under arbitrary distributions? One simple way is to use

$$\pi_s^* = \pi_s \frac{U'\left(W_s^{optimal}\right)}{\sum_{s=1}^{m} \pi_s U'\left(W_s^{optimal}\right)}, \tag{5.42}$$

where $W_s^{optimal}$ denotes the wealth in scenario s attached to the optimal solution of $\max_w \sum_{s=1}^{m} \pi_s U(W_s)$ and $W_s = 1 + wr_f + (1-w)r_s$. Cash flows from a particular CDO tranche are valued with

$$value_{tranche_i} = \sum_{s=1}^{m} \pi_s^* \left(\frac{cash - flow_s^{tranche_i}}{1 + r_f} \right) \tag{5.43}$$

assuming a particular form for $U(W_s)$.

We finish this section with an example. Suppose $r_f = 0.03$, $l_{\alpha_1} = 2$, and $l_{\alpha_2} = 10$. The returns of the underlying assets in a hypothetical CDO are assumed to be drawn from a normal distribution with mean 6% and volatility 9%. Applying standard risk-neutral valuation theory, we arrive at the following values for the three tranches:

$$\begin{aligned} value_{junior} &= 6.58 \\ value_{mezzanine} &= 6.42 \\ value_{senior} &= 87.00. \end{aligned}$$

These prices are required to subsequently calculate the returns for the respective tranches, as shown in Figure 5.19 and Figure 5.20.

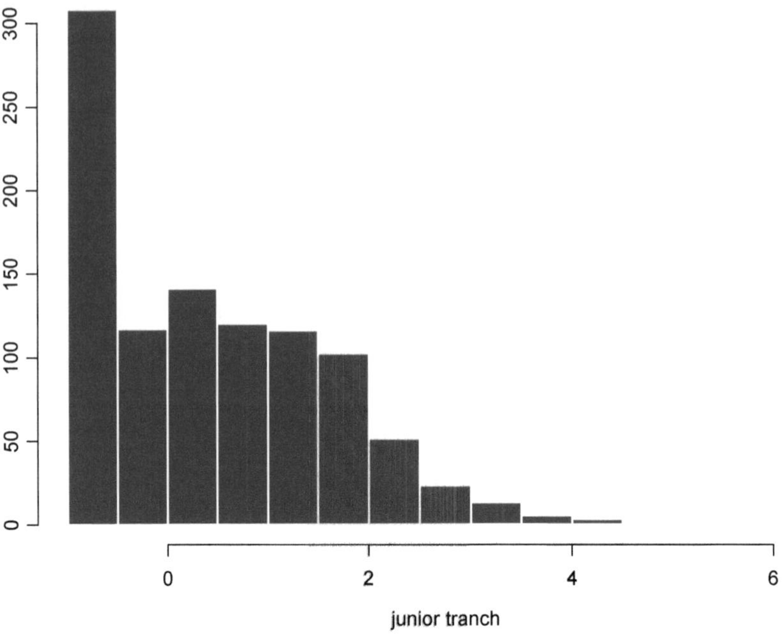

Figure 5.19 Return Distribution for Junior Tranche

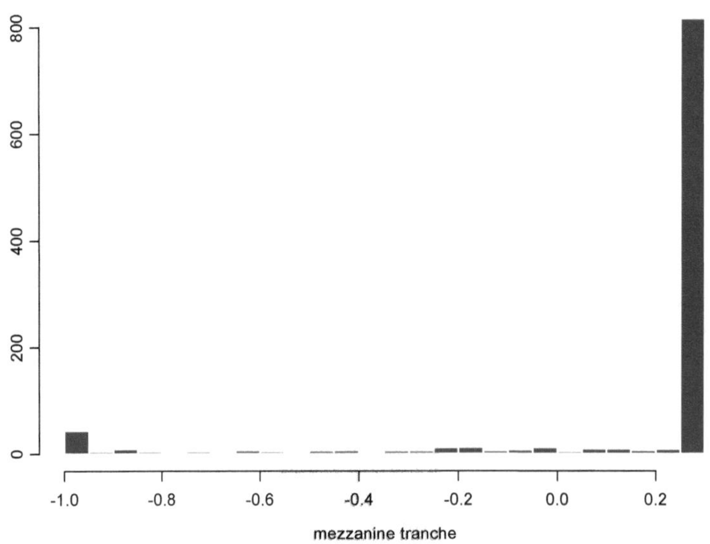

Figure 5.20 Return Distribution for Mezzanine Tranche

Exercises

1. Write a program that uses the methodology in Section 5.1.2.

2. Solve (5.12), first in Excel and then with `solveQP()`. How do we need to rewrite (5.12) in the case of different upper and lower costs of mean deviations?

3. Download data on a high yield corporate bond index.
   (a) Calculate the value of tranches within a standard CDO that is financed via equity (taking the first 5% losses), mezzanine debt (taking the next 10% of losses), and senior debt. Assume a risk-free rate of 3% and use Monte Carlo simulation.
   (b) Use the option prices of (a) to generate scenarios that are consistent with CDO pricing.
   (c) Add at least two more asset classes to the junior note and construct a CVaR efficient frontier.
   (d) How does the CVaR frontier differ from a mean-variance frontier?

4. Can you approximate semi-variance optimization using the MAD model and a piecewise linearization? Hint: See Hamza and Janssen (1995).

5. Extend the failed model (5.25) for shortfall probability to lower partial moments of degrees $k = 1, 2, 3, 4$. What do you observe?

6. Include fixed and proportional transaction costs in the weighted semi-variance model.

7. Assume a mixture of (two) normal distributions and write a program that calculates lower and upper partial moments for arbitrary threshold returns and moments using numerical integration. Check your results using Monte Carlo simulation.

8. Write a program that does the calculations in Section 5.6.4. Experiment with the number of scenarios discarded in each step. What do you observe?

Endnotes

[1] The time series stretches from January 1994 to December 2002 for Emerging Markets (JPM.EMBI) and from February 1986 to December 2002 for the dollar/yen exchange rate (DOLLAR.YEN).
[2] See Grinold (1999).
[3] See Zimmermann (1998, p.67, equation 4.2).
[4] See Sklar (1996).
[5] This model was introduced into the literature by Konno and Yamazaki (1991) and investigated further in Feinstein and Thapa (1993).
[6] See Hamza and Janssen (1995).
[7] Satchell and Sortino (2001) provide an excellent review on downside-based risk measures.
[8] This relationship is only true for continuous distributions, as we show in Section 5.6.1.
[9] See Young (1998).
[10] This section draws heavily on Acerbi and Tasche (2001).
[11] See Artzner et al. (1997).
[12] A similar example can be found in Acerbi et al. (2001).
[13] See Embrechts, McNeil, and Straumann (2002).
[14] The heuristic is described in Larsen, Mausser and Uryasev (2002)

6 Robust Statistical Methods for Portfolio Construction

6.1 Outliers and Non-normal Returns

The value of robust statistical methods in portfolio construction arises because asset returns and other financial quantities often contain **outliers**. Outliers are data values that are well-separated from the bulk of the data values and are not predicted by univariate or multivariate normal distributions. Under normal distribution models, such an outlier sometimes occurs with exceedingly small probability. For example, if we fit a normal distribution to S & P 500 daily returns for various periods of time prior to the stock market crash of 1987, we find that the probability of occurrence of an event of that magnitude is so small that one would have to wait much longer than the history of civilization for another such occurrence.[1] Large outliers of this type are not limited to situations with extreme market movements—one can find many such examples in individual asset returns. For example, the five-year monthly returns for the microcap stock with ticker EVST shown in Figure 6.1 has an extremely large outlier in December 1988 with value 6.88. You can make this plot with the S-PLUS commands given in Code 6.1, in which we first extract EVST monthly stock returns for a five-year span from the `microcap.ts` time series object, and then plot the EVST time series:

```
EVST.returns.ts <- microcap.ts[,"EVST"]
plot(EVST.returns.ts, plot.args = list(type = "b",
  pch = "."), reference.grid = F, ylab = "RETURNS",
  main = "EVST RETURNS")
# Add text "OUTLIER" by left-clicking mouse at
# desired location
text(locator(1),"OUTLIER")
# Add line by left-clicking at each line end-point,
# then right click
lines(locator())
# This command and the next are equivalent
```

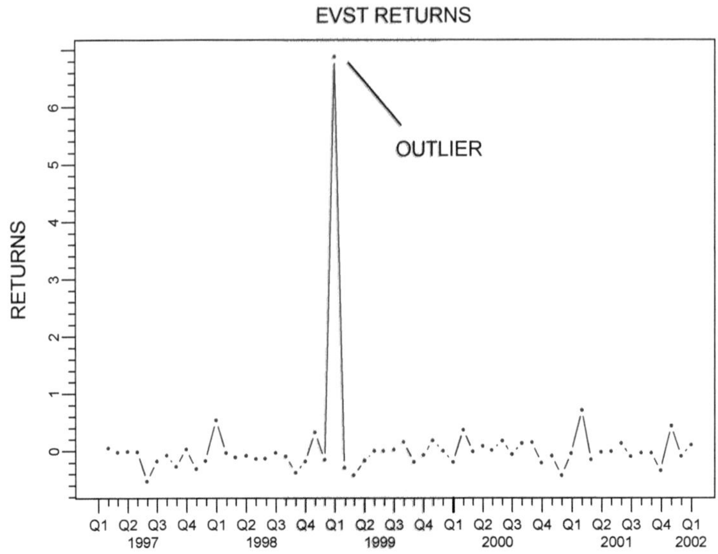

Figure 6.1 Time Series of EVST Returns

```
EVST.returns <- EVST.returns.ts@data[,1]
EVST.returns <- seriesData(EVST.returns.ts)[,1]
```

Code 6.1 Time Series Plot of EVST Returns

Note that `EVST.returns.ts` is an S-PLUS V4 time series object, the first part of which looks like:

```
> EVST.returns.ts
  Positions          EVST
  1/31/1997  0.050847456
  2/28/1997 -0.024193548
  3/31/1997 -0.008264462
        ............
```

At the end of Code 6.1, the data from `EVST.returns.ts` is extracted and converted to a simple S-PLUS vector object.[2] You can now compute the mean and standard deviation of the EVST returns as follows:

```
> mean(EVST.returns)
[1] 0.07568058
> stdev(EVST.returns)
[1] 0.9197129
```

To get the value of the outlier and its time of occurrence, use the command

```
> EVST.returns.ts[EVST.returns.ts@data > 3,]
  Positions      EVST
 12/31/1998 6.878788
```

Now let's compute the probability of getting a return as large or larger than 6.88 for a normal distribution with mean .076 and standard deviation .92:

```
> 1-pnorm(6.88,.076,.92)
[1] 7.038814e-014
```

Under the normal distribution model, you would have to wait an unbelievable amount of time to see the recurrence of such an outlier in the monthly returns of EVST.

We can easily assess the non-normality of these returns using a normal Q-Q plot with a robustly fitted straight line, as shown in Figure 6.2.[3] (See Section 6.5 for a discussion of robust straight-line fitting in the context of estimating the CAPM beta.)

```
> qqnorm(EVST.returns, ylab="EVST.returns")
> qqline(EVST.returns)
```

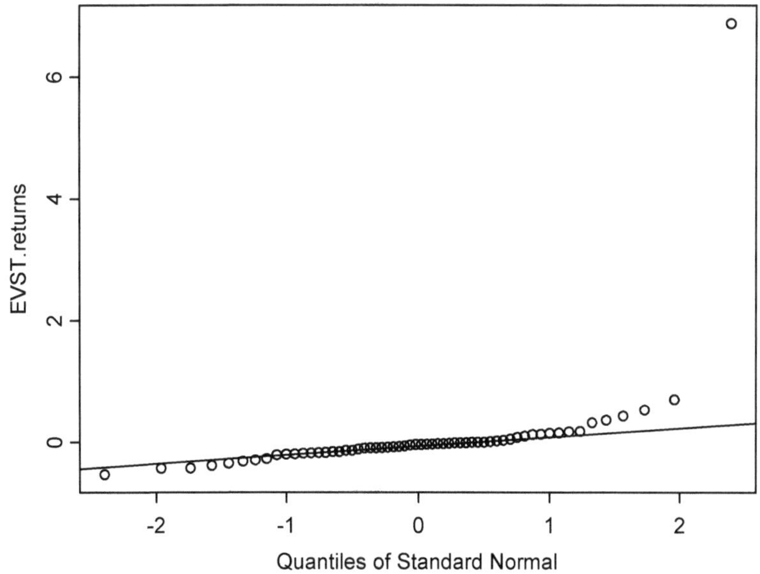

Figure 6.2 Normal Q-Q Plot of EVST Returns with Robust Line Fit

The normal Q-Q plot indicates that the returns are non-normal because of the single outlier and possibly because of the other deviations of the points from a straight line. You can check this easily by making a normal Q-Q plot with the outlier deleted:

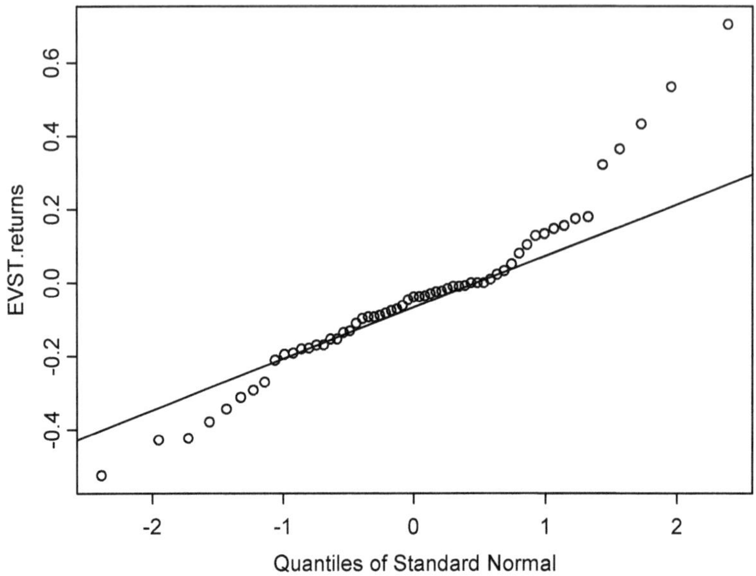

Figure 6.3 Normal QQ-Plot of EVST Returns with Outlier Removed

```
> qqnorm(EVST.returns[EVST.returns<2], ylab = "EVST
    Returns")
> qqline(EVST.returns[EVST.returns<2])
```

The result in Figure 6.3 shows that some outliers and non-normality are still present. How are we to judge non-normality from a Q-Q plot? The answer is to add 95% simulation confidence bands to plots like that of Figure 6.3, which you can do as follows:

```
> EVST.robfit <-
    lmRob(EVST.returns[EVST.returns<2]~1,eff=.95)
> plot(EVST.robfit,which.plots=2)
```

The function `lmRob` is a robust regression-fitting function that can estimate a mean robustly when used with a formula of the form `x ~ 1` as the first argument. The use of `lmRob` for robustly estimating mean returns will be discussed further in Section 6.3. It is used here only because the generic function plot invokes a special plot method for an `lmRob` object that computes and plots 95% simulation envelopes, which are useful for assessing whether residuals (the error term in the model) contain outliers.

Figure 6.4, showing the *standardized* residuals and the 95% simulation envelopes, indicates that there is still some non-normality in the form of incipient outliers (not quite outliers). This is due to a slightly heavier right-hand tail of the returns density.[4]

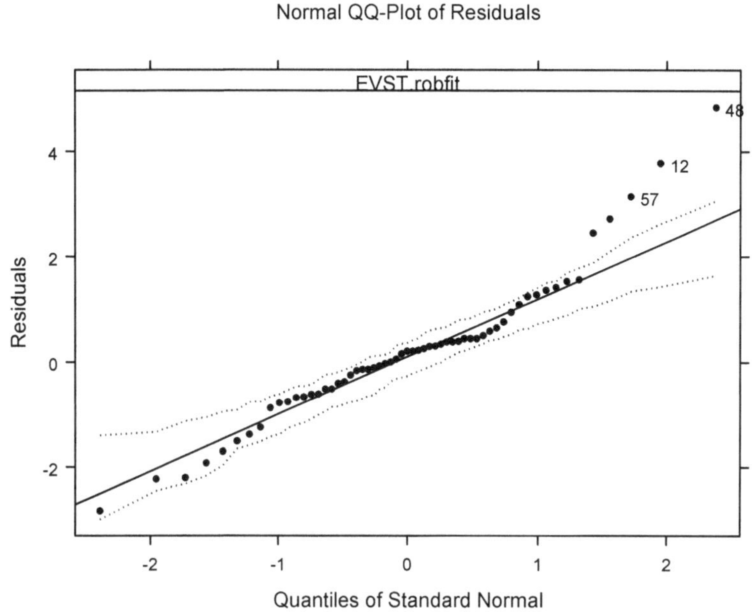

Figure 6.4 Normal Q-Q Plot with 95% Confidence Envelope

The single outlier in the EVST returns is highly **influential** in that it greatly increases the values of the mean and standard deviation relative to the values you would get if December 1988 were not an outlier. You can see the effect by doing the computation with the outlier deleted:

```
> mean(EVST.returns[EVST.returns < 2])
[1] -0.03962633
> stdev(EVST.returns[EVST.returns < 2])
[1] 0.2212708
```

The mean drops to –.040 and the standard deviation drops to .22. The latter is a little over four times smaller than the standard deviation that was calculated using all of the returns (.92). Consider the impact on the sample mean when you add the December 1998 return, assuming it was not originally there. A natural yardstick to measure the change is the standard deviation of the sample mean without the outlier, $.22/\sqrt{60} = .028$, so the influence of adding the outlier is to shift the sample mean by $(.076-(-.04))/.028=4.14$ standard deviations, which is a very large influence indeed.

It is well-known that mean returns are estimated poorly in that the standard deviation of the sample mean is typically a substantial fraction of the (absolute) value of the mean. In this example without the outlier, the standard deviation of the sample mean is .028, which is 70% of the size of the sample mean absolute value of .04. The single outliers increase the value of the sample mean by more than four standard deviations of the sample mean. *This example emphasizes (most painfully) that the problem of accurately estimating asset mean returns is greatly compounded by the presence of outliers*!

6.2 Robust Statistics versus Classical Statistics

One reason that we need to pay attention to outliers is that, as suggested by the example above, virtually all classical statistical parameter estimation and associated model-fitting methods lack **robustness** toward outliers. Even a single outlier can have an arbitrarily large adverse influence on classical model-fitting methods and classical statistical inference. Outliers can adversely influence not only mean and volatility estimates of returns but also covariance and correlation estimates, factor model parameter estimates, and optimal portfolio weights and related quantities such as Sharpe ratios. In data-oriented terms, a **robust statistical model-fitting method** is one that is not much influenced by outliers and provides a good fit to the bulk of the data.

A vivid example of the difference between a classical method and a robust method is shown in Figure 6.5, which displays a time series of annual earnings per share (EPS) from 1984 to 2001 for a company with ticker INVENSYS, along with two straight-line fits, one the classical least squares (LS) fit and one a highly robust fit (ROBUST). We describe the latter in Section 6.5 in the context of estimating the CAPM beta. It is quite clear that there are two outlier values of EPS. The LS line fit is highly influenced by these outliers and consequently does not provide a good fit to the bulk of the data, while the opposite is true of the robust fit, which is influenced very little by the outliers.

This particular example came to one of the authors from an analyst in the corporate finance office of a large, well-known firm. The analyst's task was to compute one-year-ahead forecasts of EPS for hundreds of firms as part of a portfolio stock-selection process. This example indicates clearly that use of LS through years 1998 and 1999 would have produced very poor predictions for years 1999 and 2000, respectively, and can be expected to produce a poor prediction for 2001. It is impossible to predict the time and direction of future outliers with any degree of accuracy, for if this were possible one could make a lot of money with an appropriate investment strategy. Thus, we cannot expect to

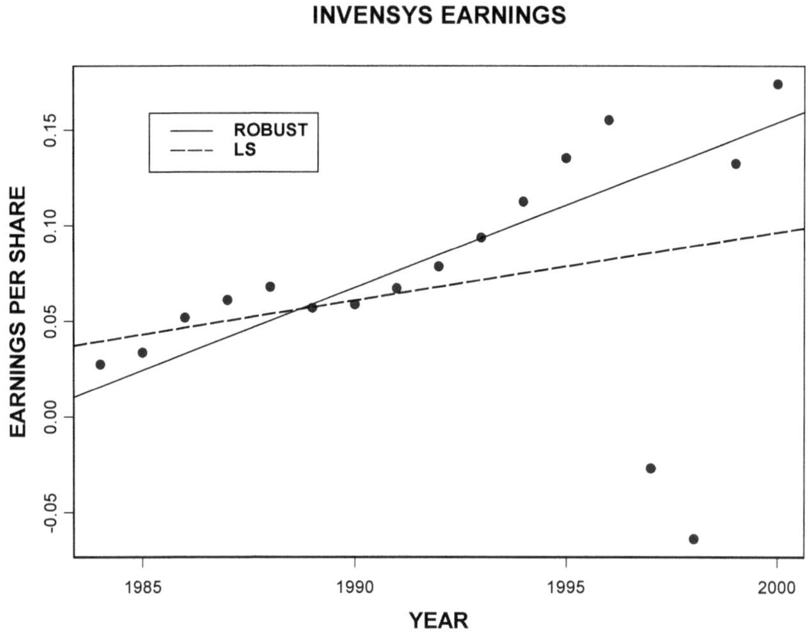

Figure 6.5 EPS versus Time with LS and Robust Line Fits

accurately predict the EPS outliers in 1997 and 1998. However, with a robust fit, we can compute good predictions for future data that are similar to the bulk of the outlier-free historical data.

While the data-oriented description of robust statistical model-fitting methods above has immediate appeal, it is important to know that there are rigorous probability-based statistical modeling foundations for robust statistics. These are probabilistic forms of stability of variance and minimization of bias under outlier-generating heavy-tailed deviations from a nominal (often normal) distribution model. An important approach for model parameter estimation is that of minimizing the maximum bias due to outlier contamination while at the same time achieving high statistical efficiency at the nominal model. For further details, see Martin and Zamar (1993) and Yohai and Zamar (1997).

So far, we have been talking about robustness of model parameter "point" estimates. It is important to note that robustness is also quite important with regard to methods of statistical inference such as hypothesis tests, confidence intervals, and model selection criteria. It turns out that outliers can seriously distort the level and power of a *t* test and the coverage probability and error rate of a confidence interval. A **robust hypothesis test** is one for which neither the nominal error rate nor the power of the test are much affected by the presence of outliers. A **robust confidence interval** is one for which neither the confidence level nor the expected confidence interval length are much affected by outliers.

The importance of robust methods in finance is immediately clear—one does not want an investment decision to be highly influenced by a small number of data points. When the historical asset price or returns data are of limited extent and exhibit at most a small number of outliers, it is typically impossible to predict with any degree of certainty whether there will be outliers in the future investment horizon.[5] In such situations, the bulk of the data is the only predictable part of the data, and an investment decision based on a robust statistical method may be preferred to the use of a procedure that is optimal under normality. On the other hand, robust statistics down-weight the influence of outlying returns, which reduces volatility estimates while reducing mean return estimates that have been influenced only by positive outliers and increasing mean return estimates that have been influenced only by negative outliers. Fund managers may rightly feel uncomfortable about making an investment decision based on a robust estimate. Such fund managers can nonetheless derive value from the use of robust methods as a diagnostic tool by comparing the results of the classical and robust methods. When both results agree, there is little worry about the possibility of outliers influencing the robust method, but when the two methods disagree substantially, the fund manager should be wary and look more closely at the data before making an investment decision.

6.3 Robust Estimates of Mean Returns

Suppose you have a set of identically distributed returns $(r_1, r_2, \cdots, r_n)$ with common mean $\mu = E(r_1)$. You can estimate μ with a variety of robust estimates that are influenced very little by outliers. The simplest and most transparent of these are the sample median and trimmed mean, both of which are based on the set of ordered returns $r_{1:n} \le r_{2:n} \le \cdots \le r_{n:n}$ (the **order statistics**). The **sample median** $\hat{\mu}_{MED}$ is the "middle" order statistic (i.e., the unique middle order statistic when n is odd, and the average of the two middle order statistics when n is even). For example, when $n = 11$, $\hat{\mu}_{MED} = r_{6:11}$, and when $n = 10$, $\hat{\mu}_{MED} = \frac{1}{2}(r_{5:11} + r_{6:11})$. An **α-trimmed mean**, $\hat{\mu}_{trim,\alpha}$, is computed by discarding a fraction α of the largest and smallest order statistic values and computing the sample mean of the remaining data points. For example, when $n = 10$, the 10% trimmed mean is

$$\hat{\mu}_{trim,\,.1} = \frac{1}{8}\sum_{i=2}^{9} r_{i:n}.$$

You can easily compute the median and trimmed mean in S-PLUS as illustrated below for the EVST returns.

```
> median(EVST.returns)
[1] -0.03863926
> mean(EVST.returns,trim=.1)
[1] -0.04756718
```

You can also compute the sample mean with the outlier deleted just as you did in Section 6.1:

```
> mean(EVST.returns[EVST.returns < 2])
[1] -0.0396
```

The result is almost identical to the sample median and not grossly different from the 10% trimmed mean.

While the simplicity of the median and trimmed means make them attractive robust estimators of location, they have some deficiencies that limit their general use. For example, it is not easy to construct confidence intervals for the sample median by any means other than bootstrapping, and the trimmed mean does not generalize nicely to other estimation problems such as fitting factor models. For this reason we introduce so-called **M-estimators** of location, a class of estimators that does generalize to many other model-fitting problems, including (as we shall see in Section 6.4) linear regression models.[6]

A location M-estimator $\hat{\mu}$ is a solution of the minimization problem

$$\min_{\mu} \sum_{t=1}^{n} \rho\left(\frac{r_t - \mu}{\hat{s}}\right), \tag{6.1}$$

where $\hat{s}$ is a robust scale estimate (see Section 6.4) and ρ is a "robustifying" loss function.

We obtain an **estimating equation** for $\hat{\mu}$ by differentiating the objective function (6.1) with respect to μ; this gives the M-estimator estimating equation

$$\sum_{t=1}^{n} \psi\left(\frac{r_t - \hat{\mu}}{\hat{s}}\right) = 0, \tag{6.2}$$

where $\psi = \rho'$. There is an intuitively appealing weighted least squares (WLS) interpretation of the M-estimator: if we set

$$w_t = w\left(\frac{r_t - \hat{\mu}}{\hat{s}}\right) \qquad w(u) = \frac{\psi(u)}{u}, \tag{6.3}$$

then we can rewrite the estimating equation (6.2) as a weighted least squares equation[7]

$$\sum_{i=1}^{n} w_t \cdot (r_t - \hat{\mu}) = 0. \tag{6.4}$$

When $\rho(t) = t^2$, the psi function is the identity function $\psi(t) \equiv t$, and the weights w_t are identically one. The solution is the least squares (LS) estimate of μ (i.e., the sample mean). The sample mean lacks robustness because the quadratic character of the least squares loss function causes outliers to have undue influence on the estimate. A robust estimate is obtained by using a ρ that grows more slowly than a quadratic function. The two main choices are: (a) a $\rho(t)$ that grows like $|t|$ for large t, and (b) a bounded $\rho(t)$. It is known that the former choice does not provide bias robustness (Martin, Yohai, and Zamar, 1989). Thus, we use a bounded $\rho(t)$ that was shown by Yohai and Zamar (1997) to be optimally bias-robust, subject to a constraint of specified efficiency, when the data are normally distributed. This $\rho(t)$ is implemented in the S-PLUS function `lmRob`. This type of $\rho(t)$ (RHO) is graphed in Figure 6.6 along with the corresponding $\psi(t)$ (PSI).

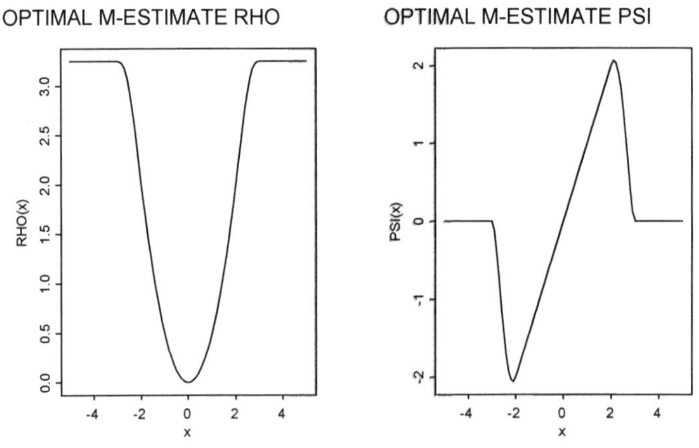

Figure 6.6 Optimal Bias Robust Rho and Psi Functions for 90% Gaussian Efficiency

The formula for the rho function $\rho(t)$ is

$$\rho(u) = \begin{cases} 0.5u^2, & |u| \le 2c \\ c^2\left(1.792 - .972\left(\frac{u}{c}\right)^2 + .432\left(\frac{u}{c}\right)^4 - .052\left(\frac{u}{c}\right)^6 + .002\left(\frac{u}{c}\right)^8\right), & 2c \le |u| \le 3c \\ 3.25c^2, & |u| \ge 3c. \end{cases}$$

The formula for the psi function $\psi(t)$ is easily obtained by differentiation. The role of the tuning constant c is to adjust the efficiency of the estimate to the desired level when the returns are normally distributed. **Efficiency when the returns are Gaussian** is defined to be the variance of the least squares (LS) estimator divided by the variance of the robust estimator. The graphs in Figure 6.6 show the ρ and ψ functions for an efficiency of 90%. This corresponds to an 11% increase in variance (only 3.3% in terms of increase in standard deviation) of the robust estimate over that of the least squares estimate (the sample mean in the present discussion). This increase in variance at nominal Gaussian returns is, in effect, a small "insurance premium" paid in exchange for protection against bias and inflated variance in the presence of outliers. The higher the premium paid, the more protection we get. If we require a Gaussian efficiency larger than 90%, the bound on the ρ grows, and the cutoff points where the ψ function goes to zero retreat further toward infinity. In the limit, when we require 100% Gaussian efficiency, the loss function becomes quadratic and we get the LS estimator.

The weight function associated with the optimal ψ function for 90% Gaussian efficiency has the shape shown in Figure 6.7. Note that the weight function is zero outside a finite interval; this means that the M-estimator will put zero weight w_t on any return r_t that has sufficiently large scaled residuals $\hat{\varepsilon} = (r_t - \hat{\mu}) / \hat{s}$ (i.e., the outliers will be totally rejected). Returns whose scaled residuals are sufficiently small (typically the bulk of them) will get weights w_t equal to one.

The S-PLUS function `lmRob` uses a sophisticated form of a nonlinear optimization method proposed by Yohai, Stahel, and Zamar (1991) to solve the M-estimate minimization problem (6.1). `lmRob` was designed for robustly fitting a linear regression (factor) model and computing associated robust statistical inference quantities such as robust standard deviations, t-statistics, and p-values. As a special case, `lmRob` can compute a robust estimate of mean returns (in which we have only an intercept term). We compute a robust M-estimate of the mean and its robust standard deviation, and a plot of the weights, using Code 6.2.

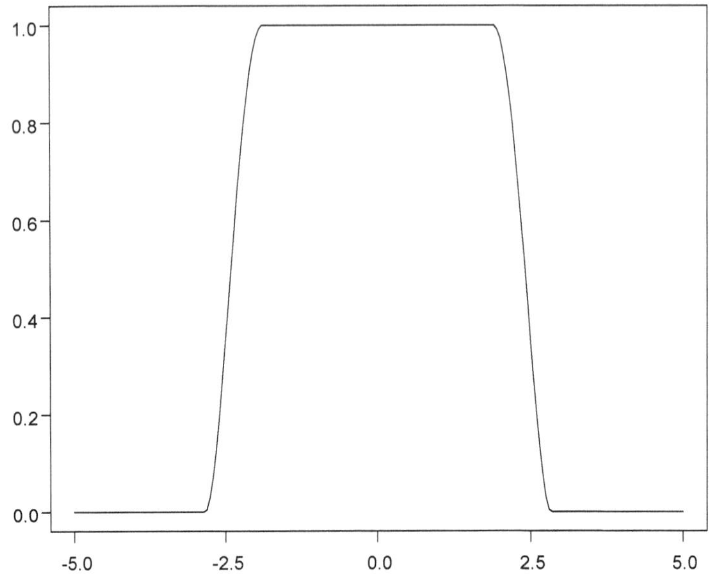

Figure 6.7 Optimal Weight Function for 90% Gaussian Efficiency

```
EVST.mean <- lmRob(EVST.returns~1, eff=.95)
coef(EVST.mean)
sqrt(EVST.mean$cov)
EVST.M.weights <- timeSeries(EVST.mean$M.weights,
  from="1/1/1997",by="months")
plot(EVST.M.weights,plot.args=list(type = "p"),
  reference.grid=F, xlab="TIME",ylab=WEIGHTS")
```

Code 6.2 Robust Location Estimate Weights

Use of the `eff = .95` argument gives us a robust estimate that would have 95% efficiency if the returns were normally distributed. In the weights plot of Figure 6.8, we see that the huge outlier in Q4 1988 gets zero weight, as do a few other large returns; this is not surprising in view of Figure 6.4.

We can repeat the commands above with a higher efficiency, say `eff =.98`, to perform a less severe down-weighting of data. This changes the robust mean returns estimate to –.066 and the standard deviation to .028 (which is not much of a change). The new weights, shown in Figure 6.9, are not very different from those for 95% Gaussian efficiency (shown in Figure 6.8). Note that as we change the efficiency, we change the cut-off values of the weight function in Figure 6.7. Higher efficiencies result in larger cut-off values and less severe rejection of outliers, while lower efficiencies result in smaller cutoff values and

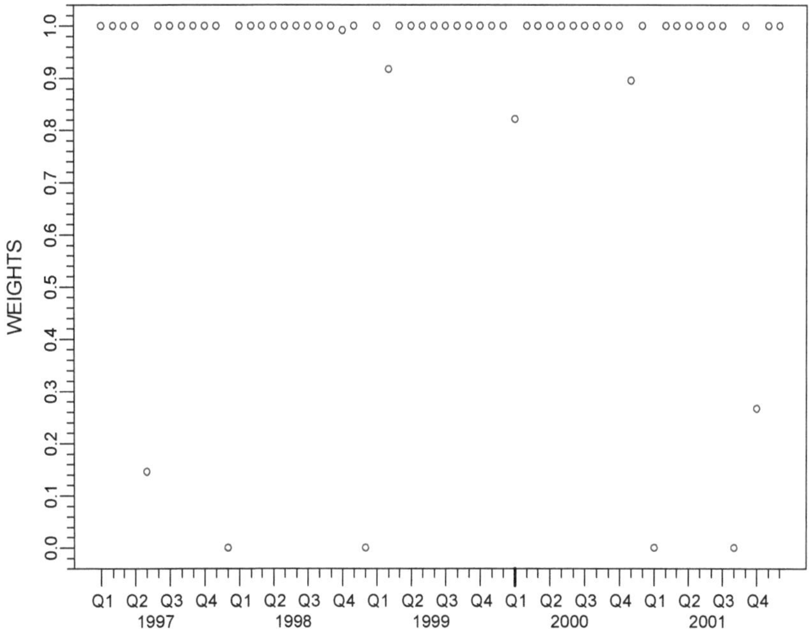

Figure 6.8 Robust M-Estimate Weights (95% Gaussian Efficiency)

more severe rejection of outliers. You are encouraged to try some lower values of efficiency, say .90 and .85. If you do so, you will see that while you get a few more zero weights and some other weights smaller than one, the robust estimate of mean returns and its standard deviation do not change much. This relative insensitivity of the robust estimate of mean returns with respect to the Gaussian efficiency/cutoff values is an attractive feature of the method.

There is also an S-PLUS function `location.m` that computes only a robust M-estimate of location (the mean); i.e., it does not compute a standard error like `lmRob`. It uses a somewhat different weight function (the Tukey biweight) that still gives weight zero to all sufficiently large residuals. With the previous data it gives a location estimate of –.06 instead of –.04, a difference less than one standard error (.28) of the sample mean without the large outlier:

```
> location.m(EVST.returns)
[1] -0.0597
```

Here are our recommendations on which estimator to use:

1. If experience with your data tells you that you have only a certain fraction of outliers, and you do not need a standard error estimate,

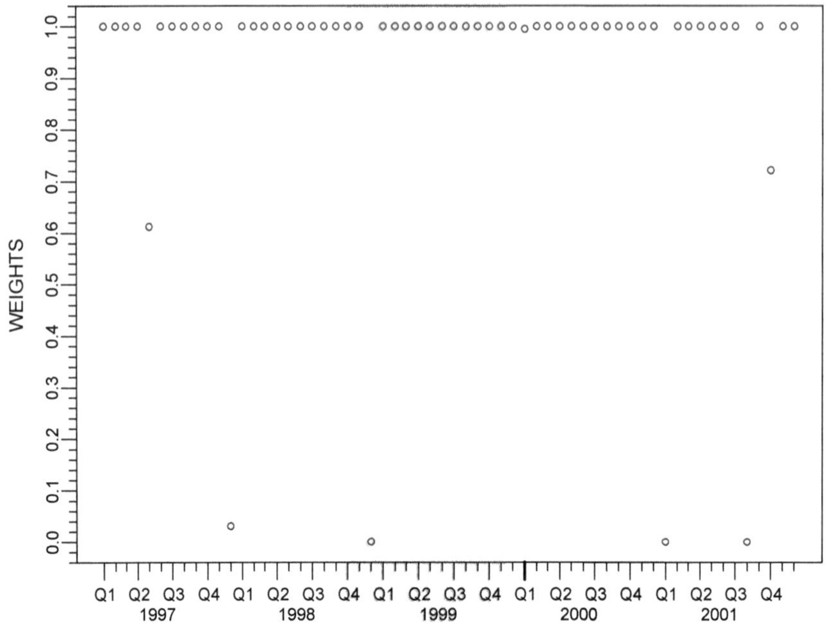

Figure 6.9 Robust M-Estimate Weights (98% Gaussian Efficiency)

use the trimmed mean `mean(returns, trim = alpha)` with trimming fraction `alpha` slightly greater than your worst case estimate of the fraction of outliers.

2. If you don't know much about the fraction of outliers and don't need a standard error, use `location.m`.
3. If you need a standard error estimate, use `lmRob` with the default Gaussian efficiency of 90%.

In S-PLUS, we can easily compute robust estimates of the mean returns for a collection of stocks with any of the robust mean (location) estimates above by using the `apply` function on the data frame of returns. To illustrate, we compute robust location estimates for five large cap stocks:

```
> rob.means <- apply(largecap.ts[, 3:8]@data,2,
     location.m)
> round(rob.means, 3)
   CAT     DD     G   GENZ     GM   HON
 0.007 0.003 0.01 0.038 0.009 0.02
```

6.4 Robust Estimates of Volatility

It is well-known that standard deviations (volatilities) of stock returns are usually estimated more accurately than means. Hence, there has been a tendency to assume that accuracy of estimation of standard deviations is not a problem. As the example in Section 6.1 clearly shows, however, the presence of even a single outlier can cause a dramatic change in the value of the standard deviation (in that particular case, a little over a fourfold increase from .22 if the outlier were not present to .92 with the outlier present).

There are several robust scale estimator functions in S-PLUS that provide robust estimates of returns volatilities and are approximately unbiased estimates of the standard deviation when the returns are normally distributed. These are the functions `mad` (median absolute deviation about the median), `scale.a`, and `scale.tau`. For the EVST returns, these functions give the following results:

```
> mad(EVST.returns)
[1] 0.1708058
> scale.a(EVST.returns)
[1] 0.1717936
> scale.tau(EVST.returns)
[1] 0.1864585
```

These values are all similar, and any one of the estimates can be used safely. Since the `mad` is the simplest and most transparent, it can be used as a default (though in some applications the smoothness properties and conceptual transparency of `scale.tau` might be preferred). See the S-PLUS help files for more details on these robust scale estimates.

6.4.1 Robustness Is Not Enough for Risk Management

Figure 6.10 shows the time series of ZIF returns along with a histogram and two normal density estimates of the returns. Clearly, the ZIF returns exhibit several outliers.

Code 6.3 gives the S-PLUS script to make the plot.

```
par(mfrow = c(2, 1))
ZIF.returns.ts = microcap.ts[,"ZIF"]
plot(ZIF.returns.ts, plot.args = list(type = "b",
  pch = "."), reference.grid = F, ylab = "RETURNS",
  main = "ZIF RETURNS")
```

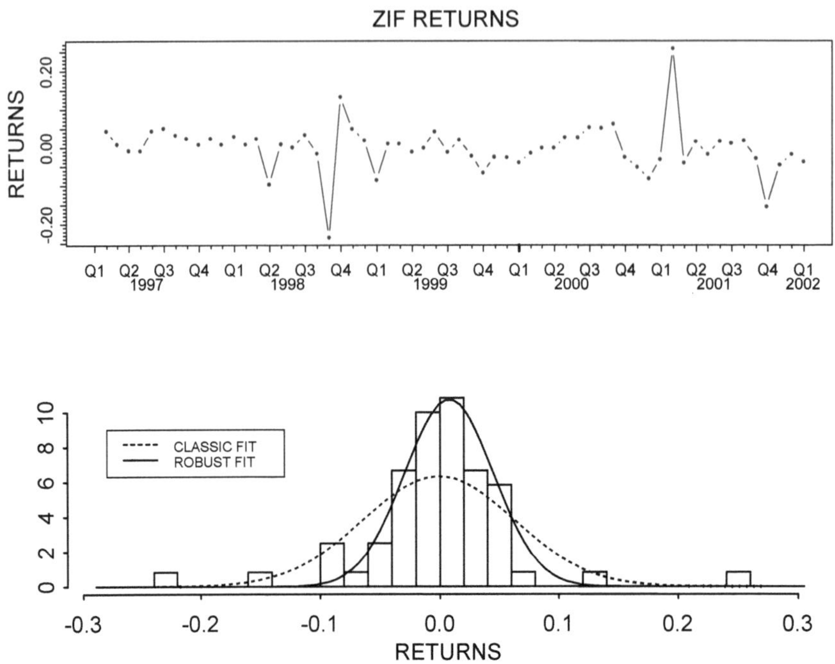

Figure 6.10 Time Series of ZIF Returns and Histogram and Two Density Estimates

```
returns = seriesData(ZIF.returns.ts)[,1]

mu = mean(returns)
sigma = stdev(returns)
mu.rob = median(returns)
sigma.rob = mad(returns)
k = 8

xlim = c(mu.rob - k * sigma.rob,
  mu.rob + k * sigma.rob)
hist(returns, nclass = "fd", col = 0, prob = T,
  xlim = xlim, xlab = "RETURNS")
x = seq(xlim[1], xlim[2], length = 100)
lines(x, dnorm(x, mu, sigma), lty = 8, lwd = 2)
lines(x, dnorm(x, mu.rob, sigma.rob), lwd = 2)
leg.names = c("CLASSIC FIT", "ROBUST FIT")
legend(-.28,9,legend=leg.names,lty=c(8,1),lwd=2)
par(mfrow = c(1, 1))
```

```
mu
mu.rob
sigma
sigma.rob
```

Code 6.3 Classical and Robust Normal PDF Fits

The classical and robust estimate values computed by the script above are:

```
> mu
[1] -0.0008619089
> mu.rob
[1] 0.007695105
> sigma
[1] 0.06311233
> sigma.rob
[1] 0.03714824
```

Suppose you want to calculate a 1% value-at-risk (VaR) (i.e., the lower 1% point of the fitted distribution). It is obvious from Figure 6.10 that neither the classical nor the robust fit of the normal distribution will suffice for this purpose. Let's examine the plots a bit more carefully.

Notice that although the mean and median (MED) differ, they are both rather close to zero because the positive and negative outliers tend to balance. On the other hand, the classical standard deviation is considerably larger than the median absolute deviation about the median (MAD). The dashed line is a normal density estimate based on the mean and standard deviation. The solid line is a normal density estimate based on MED and MAD. Because the mean and MED are quite close, both densities are well-centered on zero. But because of the difference between the standard deviation and MAD, the normal density based on the latter fits the bulk of the data well but fails to adequately describe the tails. This is to be expected since the data require a heavy-tailed density description. On the other hand, the normal density (fit with the classical estimates) is a very poor fit to the center of the data, but because of the inflated standard deviation estimate does a better job of estimating the tails (though it is still not good enough). This behavior of the normal fit explains why some value-at-risk (VaR) calculations are not so bad under normality at the 95% level but are quite inaccurate for 99% VaR. The robust MED and MAD indeed give you good estimates of the location and scale of the returns. But they are not a solution for getting better tail estimates! If you want good tail probability estimates, you need to fit a heavy-tailed distribution. Often for smallish sample sizes (such as five years of monthly data) a normal mixture with two or three components will do. For larger data sets (such as a year or more of daily data), one can do well by fitting stable distributions whose fat tails provide a good model for outliers (see Rachev and Mittnik, 2000).

6.4.2 Robust EWMA Estimates of Volatility

It is an overlooked fact that exponentially weighted moving average (EWMA) estimates often badly overestimate volatility following the occurrence of an isolated outlier return. Fortunately, it is quite easy to obtain a robust version of an EWMA volatility estimate by a simple modification of the classical EWMA algorithm, as we describe below. Figure 6.11 illustrates the dramatic difference between classical and robust EWMA estimates using two years of daily returns for a mid-cap stock with ticker ROH. The time series of returns, shown in the top panel of Figure 6.11, clearly reveals several outliers indicative of unusual movements in the price series. The middle panel shows a classical EWMA volatility time series estimate using a default smoothing parameter of .93. The classical EWMA estimate clearly grossly overestimates the volatility following the occurrences of the outliers, particularly after the two negative outliers in early Q2 and Q3 of 2000 and the outliers near the end of Q3 of 2001. The robust EWMA volatility estimate, shown in the bottom panel of Figure 6.11, does not suffer from this defect and produces much more reasonable-looking estimates of the volatility following the outliers.

The classical EWMA volatility estimate is the standard deviation series obtained as the square root of the variance estimates computed by the recursion

$$\hat{\sigma}_{t+1}^2 = \lambda \cdot \hat{\sigma}_t^2 + (1-\lambda) \cdot r_{t+1}^2, \qquad t \geq t_0. \tag{6.5}$$

Here t_0 is a starting time and $\hat{\sigma}_{t_0}^2$ is an initial variance estimate. We can easily construct a robust EWMA volatility estimate that in turn provides a robust unusual movement test (UMT) statistic with robustness of power as well as level of test. The robust EWMA algorithm is defined as follows:

$$\begin{aligned} \hat{\sigma}_{t+1}^2 &= \lambda \cdot \hat{\sigma}_t^2 + (1-\lambda) \cdot r_{t+1}^2, \quad \text{if } |r_{t+1}| \leq a \cdot \hat{\sigma}_t \\ &= \hat{\sigma}_t^2, \quad \text{if } |r_{t+1}| > a \cdot \hat{\sigma}_t. \end{aligned} \tag{6.6}$$

The parameter a is a rejection threshold. We recommend a default value of a = 2.5 for the rejection threshold; this results in using a pure prediction $\hat{\sigma}_{t+1}^2 = \hat{\sigma}_t^2$ about 1.2% of the time when the returns are normal and $\hat{\sigma}_t^2$ is equal to the true volatility of the return at time t.

In addition to having a good robust volatility estimate, one would sometimes like to have a good test statistic for providing an alert that a return is an outlier (and correspondingly that the asset price has made an unusually large movement). It is natural to use the following statistic, similar in form to a classical two-sided t-test:

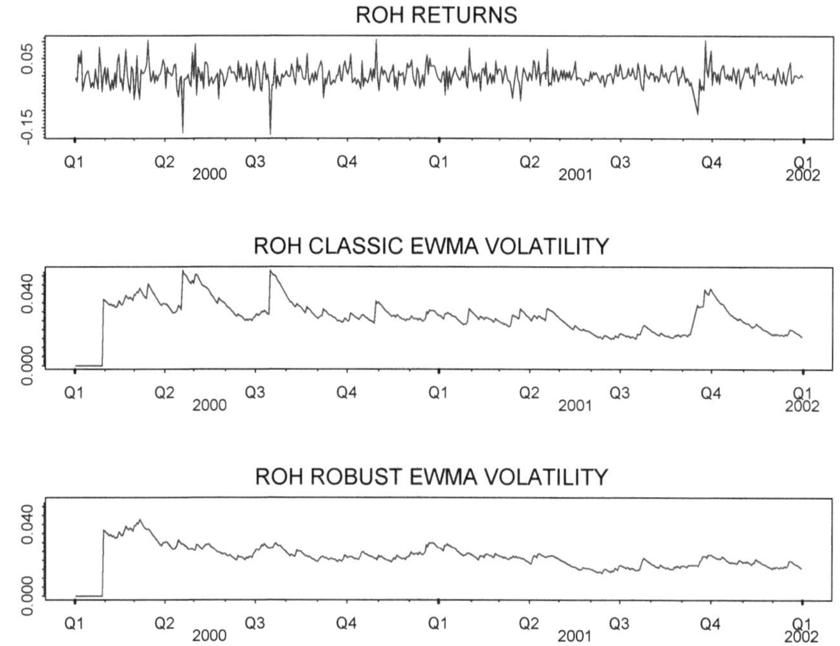

Figure 6.11 Time Series of Returns and Estimates of Volatility for ROH

$$UMT_t = \frac{|r_t|}{\hat{\sigma}_t}, \tag{6.7}$$

where r_t is the asset return at time t and $\hat{\sigma}_t$ is the classical EWMA estimate. This statistic is called the **classical UMT statistic**. However, since we know that the classical EWMA estimate overestimates volatility at times following an outlier, the denominator of the statistic will be larger than it would be without an outlier present. Thus we might anticipate it will result in a UMT with low power for detecting an unusual movement.[8] This suggests that one might obtain a robust UMT statistic by substituting the robust EWMA estimate $\hat{\sigma}_t$ in the denominator. Results for the classical and robust UMT's for the ROH returns are shown in Figure 6.12. The horizontal dashed line at $c = 3.5$ in Figure 6.12 is a test rejection threshold chosen to yield a false alarm rate of approximately .001. The classical UMT fails to detect any outlier returns or unusual movement in prices, while the robust UMT clearly detects the five largest outliers in Figure 6.11 and gives a weak indication of two others.

We remark that if the UMT test statistic had a standard normal distribution, a rejection threshold of c = 3.29 would yield a false alarm rate of .001. However, since this statistic is rather like a t-test, one has to naturally question the accuracy of a standard normal approximation. To answer this question, we make the Q-Q plot in Figure 6.13. This plot shows that the robust UMT statistic follows the normal distribution closely, with the exception of a few outliers and straggling values.

This leads us to use a normal distribution as a crude approximation for purposes of computing a threshold value. We fit this normal distribution to the data with robust location and scale estimates. The median estimate of location turns out to be exactly zero (and the mean is essentially equal to the median, with value .0002), while the robust scale estimate is 1.068 (as compared with the standard deviation value of 1.21, which is inflated by the outliers). The upper .0005 quantile of an $N(0,(1.068)^2)$ distribution is 3.51, giving a false alarm rate of .001 for the two-sided test. These values are not quite fair because they were computed post hoc. But in practice one could apply the same approach using data prior to the times at which the test statistics were computed.

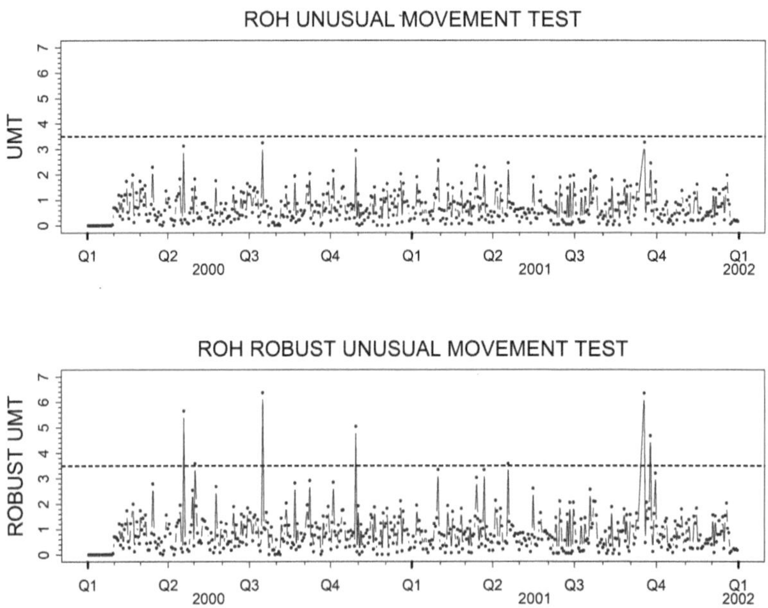

Figure 6.12 Classical and Robust UMT Test Statistics Time Series

Code 6.4 and Code 6.5 give an EWMA function and the script, respectively, for making the above computations and plots above.

```
ewma <- function(x, robust = T, lambda = 0.93,
  nstart = 20, a = 2.5)
{
  n <- length(x)
```

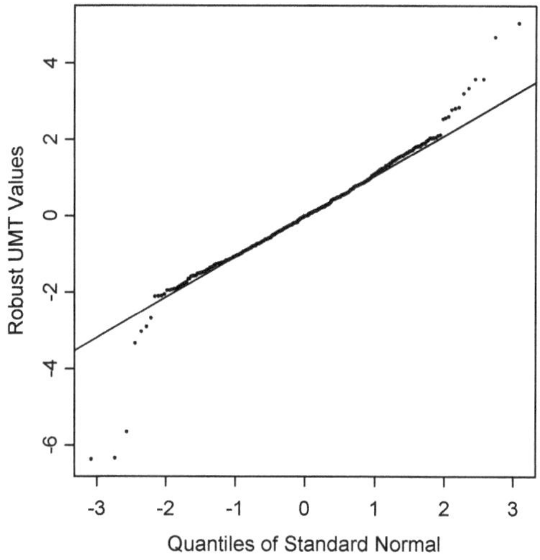

Figure 6.13 Robust EWMA and UMT for ROH

```
  # Compute initial variance estimate var.start
  var.start <- scale.tau(x[1:nstart])^2

  # Create output vector with padded zero's and
  # var.start
  varvec <- c(rep(0, nstart - 1), var.start,
     rep(0, n - nstart))

  # EWMA recursion
  var.old <- var.start
  ns1 <- nstart + 1
  for(i in ns1:n) {
     r2 <- x[i]^2
     if(robust && r2 > a^2 * var.old)
        r2 <- var.old
     var.new <- lambda * var.old +
        (1 - lambda) * r2
     var.old <- var.new
     varvec[i] <- var.new
  }
  varvec^0.5
}
```

Code 6.4 Function to Compute Classical and Robust EWMA

```
ticker <- "ROH"
tsdata <- midcapD.ts[,ticker]
returns <- tsdata@data[,1]
n <- length(returns)
lambda <-.93
nstart <- 20
a <- 2.5
thresh <- 3.5
vol.classic <- ewma(returns, robust=F,
     lambda=lambda, nstart=nstart,a=a)
vol.classic
vol.rob <- ewma(returns, lambda=lambda,
     nstart=nstart, a=a)
vol.rob
ylim <- range(vol.classic, vol.rob)
vol.classic.ts <- timeSeries(vol.classic,
  positions = tsdata@positions)
vol.rob.ts <- timeSeries(vol.rob,
  positions = tsdata@positions)
par(mfrow = c(3,1))
plot(tsdata[,ticker],reference.grid = F,
  main = paste(ticker,"RETURNS"))
plot(vol.classic.ts,ylim = ylim, reference.grid =
     F, main = paste(ticker,"CLASSIC EWMA
     VOLATILITY"))
plot(vol.rob.ts,reference.grid = F, ylim = ylim,
  main = paste(ticker,"ROBUST EWMA VOLATILITY"))
# Compute UMT
par(mfrow = c(2, 1))
umt.classic <-
  abs(returns[nstart:n])/vol.classic[nstart:n]
umt.classic <- c(rep(0,nstart-1),umt.classic)
umt.classic.ts = timeSeries(pos = tsdata@positions,
  data = umt.classic)
umt.robust <-
     abs(returns[nstart:n])/vol.rob[nstart:n]
umt.robust <- c(rep(0,nstart-1),umt.robust)
umt.robust.ts <- timeSeries(pos=tsdata@positions,
  data = umt.robust)
ylim <- 1.1*c(0,max(thresh,
  max(umt.classic,umt.robust)))
plot(umt.classic.ts, plot.args=list(type="b",
  pch="."), reference.grid=F, ylim=ylim,
  ylab="UMT",
  main=paste(ticker,"UNUSUAL MOVEMENT TEST"))
```

```
abline(thresh, 0,lty=8)
plot(umt.robust.ts, plot.args=list(type="b",
  pch="."), reference.grid=F, ylim=ylim,
  ylab="ROBUST UMT", main=paste(ticker,
  "ROBUST UNUSUAL MOVEMENT TEST"))
abline(thresh, 0,lty=8)
par(mfrow=c(1, 1))
# Q-Q plot and threshold estimate
par(pty="s")
umt.rob.signed <-
     returns[nstart:n]/vol.rob[nstart:n]
qqnorm(umt.rob.signed,
  ylab="Robust UMT Values",pch=".")
qqline(umt.rob.signed)
mean(umt.rob.signed)
sigma <- stdev(umt.rob.signed)
sigma
abs(qnorm(.0005,0,sigma))
sigma.rob <- scale.tau(umt.rob.signed)
mean(umt.rob.signed)
sigma.rob
abs(qnorm(.0005,0,sigma.rob))
par(pty = "")
```

Code 6.5 Compute and Plot Classical and Robust EWMA and UMT

We remark that the detection power of the classic UMT might be improved by using $\hat{\sigma}_{t-1}$ in the denominator instead of $\hat{\sigma}_t$. This should clearly improve the ability to detect isolated outliers, but it may not suffice to detect any additional outliers that follow in close time proximity to a first outlier. The reader could check this out by modifying Code 6.5 to use $\hat{\sigma}_{t-1}$ in place of $\hat{\sigma}_t$ in the test statistic (Exercise 5).

It is apparent that the robust EWMA volatility estimate has many potential uses in portfolio construction and risk management calculations, as would potential extensions to robust EWMA covariance matrix and mean return estimates. Clearly, in a time period subsequent to an isolated outlying return (positive or negative), one does not want to rebalance a portfolio based on a volatility estimate that grossly overestimates the true volatility and a covariance matrix estimate that is quite distorted by the outlying return. Generalizations of the robust UMT could be used to detect regime shifts from good times to bad times and vice versa.

6.5 Robust Betas

Beta estimates for assets are often used by portfolio managers to decide whether an asset should be added to a portfolio to increase or decrease its beta. It is therefore important to have reliable beta estimates that accurately reflect the risk and return characteristics of the assets under scrutiny. This section examines the impact of outliers on beta estimates and shows that even a single outlier in an asset's returns can adversely influence the conventional estimates of beta, which gives a completely misleading picture of the asset's risk and return characteristics.

CAPM[9] betas are typically estimated by fitting the single-factor market model

$$r_t = \alpha + r_{M,t}\beta + \varepsilon_t, \qquad t = 1,2,\cdots,n, \tag{6.8}$$

where r_t is the return on an asset or portfolio at time t, $r_{M,t}$ is the market return at time t, and ε_t is the error term in the model. In the United States, the market return is often taken to be the return on a value-weighted index of stocks from the NASDAQ, New York, and American stock exchanges. The parameter estimates $\hat{\alpha}$ and $\hat{\beta}$ are obtained by fitting a straight line to the scatterplot of r_t versus $r_{M,t}$ by some "good" method. The sanctified "good" method is that of (ordinary) least squares (LS), i.e., $\hat{\alpha}$ and $\hat{\beta}$ are obtained by minimizing the sum of squared residuals

$$\sum_{t=1}^{n}\left(r_t - \alpha - r_{M,t}\beta\right)^2. \tag{6.9}$$

This is often (but not always) good enough for large cap stocks, as the Microsoft example in Figure 6.14 and Figure 6.15 shows. The first of these figures shows the monthly time series of Microsoft returns and market returns, while the second displays both the LS and the robust straight-line fits and corresponding beta estimates. The LS and robust betas are quite close to one another, both on a relative basis and with respect to the ordinary LS standard error value of .28.

The robust beta is computed using the optimal bias robust regression M-estimate method described in Section 6.3, with $r_t - \mu$ replaced by $r_t - \alpha - r_{M,t}\beta$. This is the method implemented by the function `lmRob`. The time series plots of Figure 6.14 and the classic and robust beta computations for Figure 6.15 can be replicated using Code 6.6.

```
mkt.ret.ts <- largecap.ts[,"market"]
ticker <- "MSFT"
stock.ret.ts <- largecap.ts[,ticker]
par(mfrow = c(2,1))
plot(stock.ret.ts,plot.args = list(type = "b",
  pch = "."), reference.grid = F, ylab = "RETURNS",
  main = ticker)
plot(mkt.ret.ts,plot.args = list(type = "b",
  pch = "."), reference.grid = F, ylab = "RETURNS",
  main = "MARKET")
par(mfrow = c(1,1))
par(pty = "s")
mkt.ret <- mkt.ret.ts@data[,1]
stock.ret <- stock.ret.ts@data[,1]
plot(mkt.ret,stock.ret, xlab = "MARKET RETURNS",
  ylab = paste(ticker,"RETURNS"))
beta.ls <- lm(stock.ret ~ mkt.ret)
abline(beta.ls,lty = 8)
beta.rob <- lmRob(stock.ret ~ mkt.ret)
abline(beta.rob)
text.ls <- as.character(round(coef(beta.ls),2)[2])
text.ls <- paste("LS BETA =",text.ls)
text.rob <-
     as.character(round(coef(beta.rob),2)[2])
text.rob <- paste("ROBUST BETA =",text.rob)
legend(-.15,.37,c(text.ls,text.rob),lty = c(8,1))
par(pty = "")
```

Code 6.6 Classic and Robust Betas

It can happen, of course, that the returns of a stock contain one or more highly influential outliers that adversely influence the LS beta. This behavior is particularly prevalent in small cap and microcap stocks and is vividly illustrated in Figure 6.16 for the microcap stock EVST, whose time series of returns contains one very large outlier value representing a return of close to 700% (recall Figure 6.1). You can produce Figure 6.16 by replacing `largecap.ts` by `microcap.ts` and `ticker = "MSFT"` with `ticker = "EVST"` in Code 6.6.

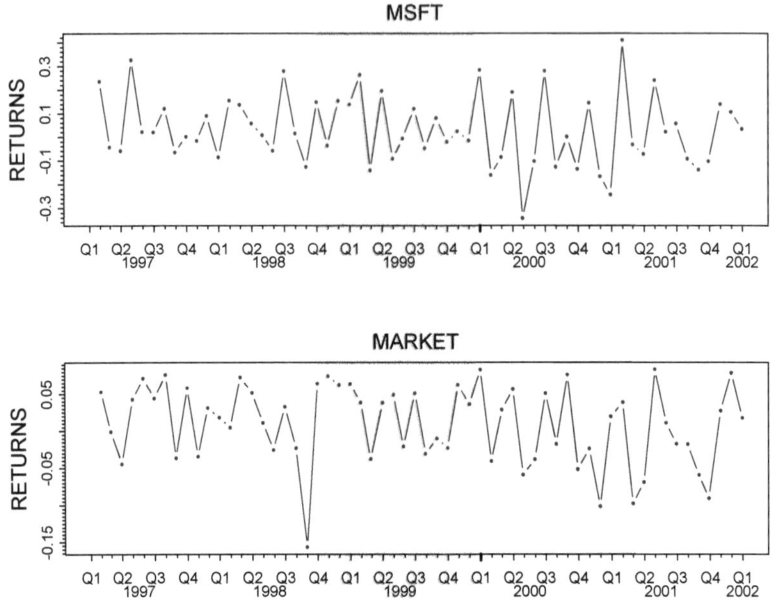

Figure 6.14 Microsoft and Market Monthly Returns for Five Years

Here the single outlier causes the LS line to fit the data quite poorly when the market returns are positive, whereas the robust line fit is not greatly affected by the outlier and fits the bulk of the data quite well. Note that in this example the outlier resulted in an LS beta of 3.17; a consumer of the LS beta would think that EVST has a high level of risk and high expected excess return relative to the market, even though this conclusion rests on a single data point (and one that was an amazingly high return at that). On the other hand, the robust beta value of 1.13 indicates that EVST behaves for the most part like the market, which is essentially the conclusion one would draw if the outlier were deleted.

Suppose that instead of the raw $\hat{\beta}$ estimates one computed an adjusted beta according to the fossilized shrinkage formula sometimes used by commercial financial data service providers,

$$\tilde{\beta} = .33 + .67 \cdot \hat{\beta}, \tag{6.10}$$

where $\hat{\beta}$ is either the LS or robust beta estimate.[10] This gives $\tilde{\beta}_{LS} = 2.45$ and $\tilde{\beta}_{ROBUST} = 1.09$, respectively. The influence of the outlier would be reduced, but this adjustment would not solve the problem with the LS estimate.

It is a rather surprising fact that most commercial providers of beta estimates appear to be totally unaware of the impact of outliers on the betas that they deliver. This is documented in Martin and Simin (2003), who found that out of

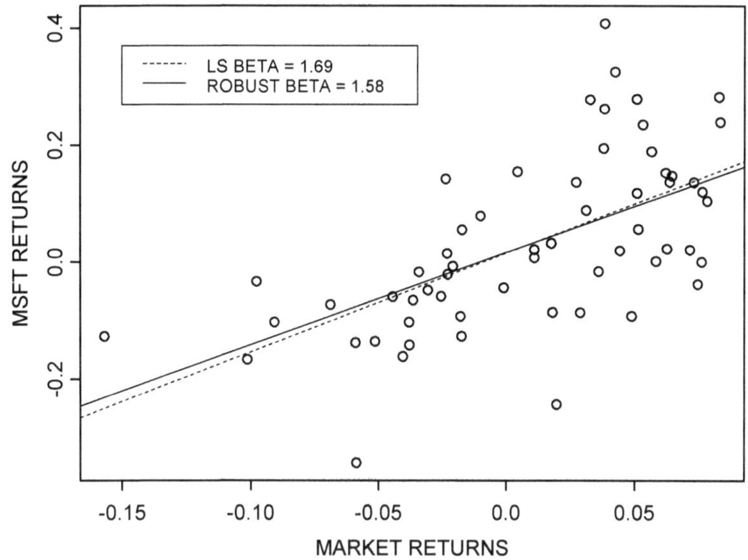

Figure 6.15 Least Squares and Robust Betas for Microsoft

the nine commercial providers of beta estimates that they surveyed, only two appeared to be aware of the issue.[11]

In order to more fully evaluate the value of a robust beta over an LS beta, one needs to know whether robust betas predict robust betas better than LS betas predict LS betas. Martin and Simin (2003) answer this question in the affirmative, giving further support to the use of robust betas in practice. Our recommendation is to compute both LS and robust betas and signal an alert that one or more outliers are probably influencing the LS estimate whenever the difference between the two estimates is larger than a user-supplied threshold. In this case, the provider should supply additional information such as a time series plot of returns, the time(s) of occurrence of the outlier(s), and potentially important related information such as corporate announcements, etc.

6.6 Robust Correlations and Covariances

In this section, we show that multivariate outliers in asset returns can have a substantial influence on correlation and covariance matrix estimates, and that one can use robust covariance matrix estimates to accurately measure the covariance and correlation structure of the bulk of the data. Figure 6.17 shows time series of 81 monthly returns for the following assets: U.S. alternative investments in DM (AI), high-quality German mortgage bonds (Pfand), U.S.

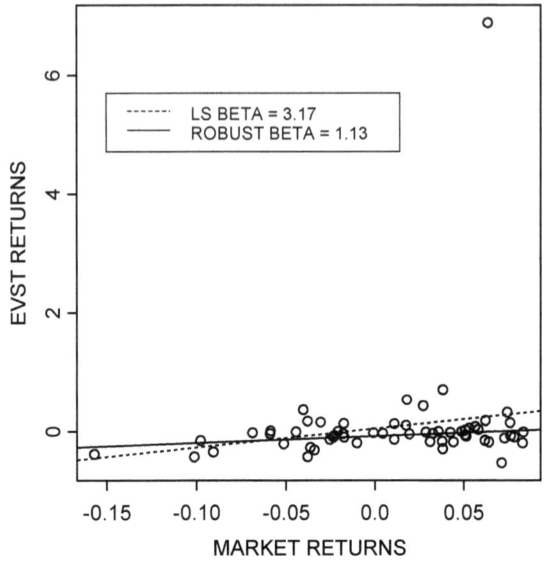

Figure 6.16 Least Squares and Robust Betas for EVST

private equity hedged in DM (PEHinDM), and U.S. high-yield bonds hedged in DM (USHYHinDM).[12]

The series of returns appear to have some distinct volatility regimes over time, and possibly a few outliers. For example, the PEHinDM and USHYHinDM series have relatively low volatility during the early time periods, while the Pfand series seems to have a lower volatility near the end of the series. The PEHinDM returns have an outlier during Q3 1998, and USHYHinDM appears to have an outlier in each of Q3 and Q4 1998 as well as an outlier in early 2001. The pairwise scatterplots in Figure 6.18 reveal some clear outliers and deviations from the elliptical shape of a multivariate normal distribution.

Figure 6.17 is a Trellis time series plot made with a modified version of the Trellis time series plotting function `seriesPlot` that comes with the S-PLUS add-on module S+FinMetrics.[13] Use the commands in Code 6.7 to make the Trellis time series plot and the pairwise scatterplots.

```
data.ts <- normal.vs.hectic.ts[-(1:60),2:5]
data <- seriesData(data.ts)
y.name <- colIds(data.ts)
seriesPlot(data.ts,one.plot=F,
  strip.text=y.name,col=1)
pairs(data)
```

Code 6.7 Trellis Time Series Plots and Pairwise Scatterplots

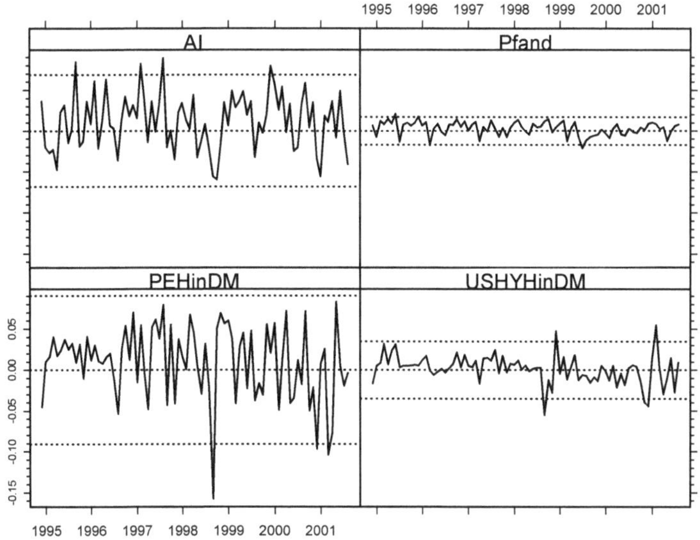

Figure 6.17 Monthly Time Series of Asset Returns

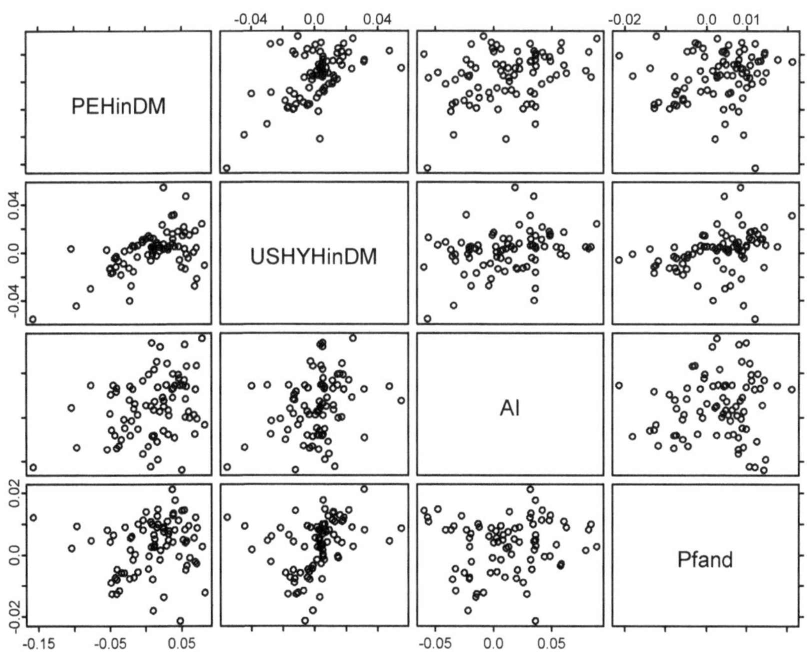

Figure 6.18 Pairwise Scatterplots of the Asset Returns

Figure 6.19 shows visual and tabular pairwise displays of classical and robust correlations for these returns. The robust correlations are obtained from a robust covariance matrix in a manner analogous to the way a classical correlation matrix is obtained from a classical covariance matrix: the elements of the robust covariance matrix are divided by the appropriate products of robust standard deviation (robust scale) estimates. For two of the pairs of returns, there are substantial differences between the classical and robust correlations: for Pfand and PEHinDM the classical correlation is .14 and the robust correlation is .49, and for Pfand and USHYHinDM, the classical correlation is .30 and the robust correlation is .66. These differences are consistent with the fact that the bulk of the data in the corresponding scatterplots clearly have a substantial positive correlation, while the outliers in these plots tend to make the data look more circular and hence less correlated.

Assuming you have already computed `data.ts` and `data` as in Code 6.7, Code 6.8 will produce Figure 6.19.

```
cov.fm <- fit.models(list(ROBUST = covRob(data),
  CLASSICAL = cov(data)))
plot(cov.fm,which.plots = 3)
```

Code 6.8 Robust Covariance Matrix and Correlation Display

The function `covRob`, appearing in Code 6.8, allows you to use any of several types of robust covariance matrices, with the default being the "Fast" Minimum Covariance Determinant (MCD) estimate of Rousseeuw and Van Driessen (1999). The MCD estimate computes the covariance matrix of the fraction `quan` of the data that yield the minimum covariance determinant, with the default `quan = .75`. The MCD estimate also returns a robust estimate of the multivariate mean (the mean returns in this application) consisting of the sample mean of the fraction `quan` of observations that yield the minimum covariance determinant. The reader is encouraged to experiment with different values of `quan` for the MCD estimate, and with the other robust covariance matrix estimates provided through `covRob` (see the online Robust Library User Guide (Insightful Corp., 2002) and help files for further details).

6.6.1 Uses of Robust Covariances and Correlations

There are at least three ways robust covariances and correlations can be used in portfolio construction:

1. As an exploratory data analysis (EDA) tool in order to discover whether the classical correlation and covariance estimates are influenced by outliers. In the case where the classical and robust methods agree, there is little need for concern, but when there are

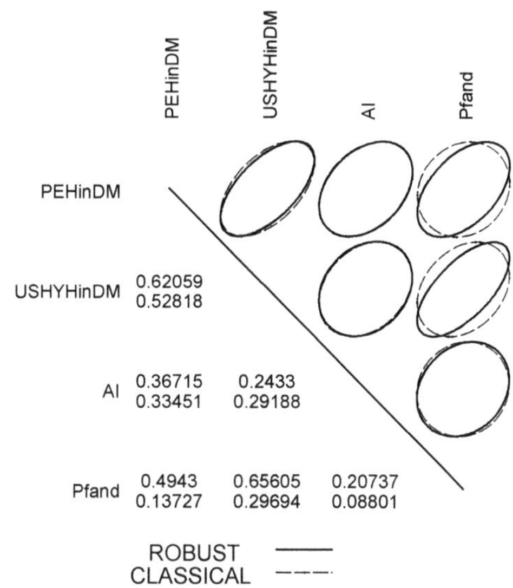

Figure 6.19 Classical and Robust Correlations for Asset Returns

substantial differences, one is well-advised to look more carefully at the data for possible explanations that will lead to better investment decisions. In some cases, influential outliers may be due to data errors, and in such cases the robust estimate will be more reliable than the classical estimate. In other cases, the influential outliers are valid data points, and the portfolio manager needs to decide whether they are representative of the future behavior of the returns, or are unique events that are unlikely to occur during the investment horizon under consideration and as such should be disregarded.

2. To construct robust multivariate distances for detecting unusual movements in multivariate returns (e.g., detecting normal times versus hectic times).
3. To obtain robust versions of Markowitz mean-variance optimal portfolios. By comparing the robust result with a classical mean-variance optimal portfolio, we will be alerted to the possibility that one or more outliers influence a particular optimal portfolio or Sharpe ratio.

The last two applications are described in the next two sections.

6.7 Robust Distances for Determining Normal Times versus Hectic Times

One way of defining "hectic" or "unusual" times, proposed by Scherer (2004), is based on the following statistical measure of the (squared) distance of a return vector $\mathbf{r}_t = (r_{t1}, r_{t2}, \cdots, r_{tk})$ from the vector of sample means $\hat{\boldsymbol{\mu}}$ over the history of interest:

$$d_t^2 = (\mathbf{r}_t - \hat{\boldsymbol{\mu}})' \hat{\boldsymbol{\Omega}}^{-1} (\mathbf{r}_t - \hat{\boldsymbol{\mu}}). \qquad (6.11)$$

See also Chow, Jacquier, Kritzman, and Lowry (1999). Here $\hat{\boldsymbol{\Omega}}$ is the classical sample covariance matrix and $\hat{\boldsymbol{\Omega}}^{-1}$ is its inverse, and we assume a history of length n. In the statistical literature, this distance is called the **Mahalanobis** distance. When the returns have a multivariate normal distribution and n is not too small, the distances above have a distribution that is well-approximated by a chi-squared distribution with k degrees of freedom (dof). By definition, "unusual" times are those that do not happen very often and so represent a smallish fraction of the returns history during which the data have considerably different behavior than during the remaining majority of "normal" times. Thus it is reasonable to define **unusual times** as those for which the values of d_t are larger than the square root of an upper-tail percentage point of this chi-squared distribution (e.g., the square root of the upper 1%, 2.5%, or 5% point).

Scherer (2004) provides a convincing example of this approach to detecting unusual times when the classical sample mean and sample covariance estimates are used in the squared distance above. In particular, the example shows that it is possible to separate unusual times from normal times. In general, however, the use of the classical sample mean and covariance matrix may not yield a highly reliable method of detecting unusual times: since outliers can distort the sample mean and covariance estimates, the resulting squared distance may not be very reliable. Robust mean and covariance matrix estimates do not suffer from this drawback and therefore are ideal alternatives to the classical sample means and covariances for detecting unusual times. Thus we compute robust Mahalanobis distances by replacing the classical mean and covariance matrix estimates in the Mahalanobis distance with robust estimates. We illustrate this approach using the data shown in Figure 6.20.

Figure 6.20 shows the classical and robust (Mahalanobis) distances for the time series of multivariate returns (i.e., the values of d_t). The horizontal dashed line in the figure is the upper 2.5% point of a chi-squared distribution with four degrees of freedom. This figure reveals that the classical distance only detects three unusual times (two of these are just barely detected), while the robust distance clearly detects thirteen unusual times in two distinct temporal clusters, a cluster of seven at the end of the series and a cluster of four near the middle of the series. There are also two other unusual times, month 56 and month 65.

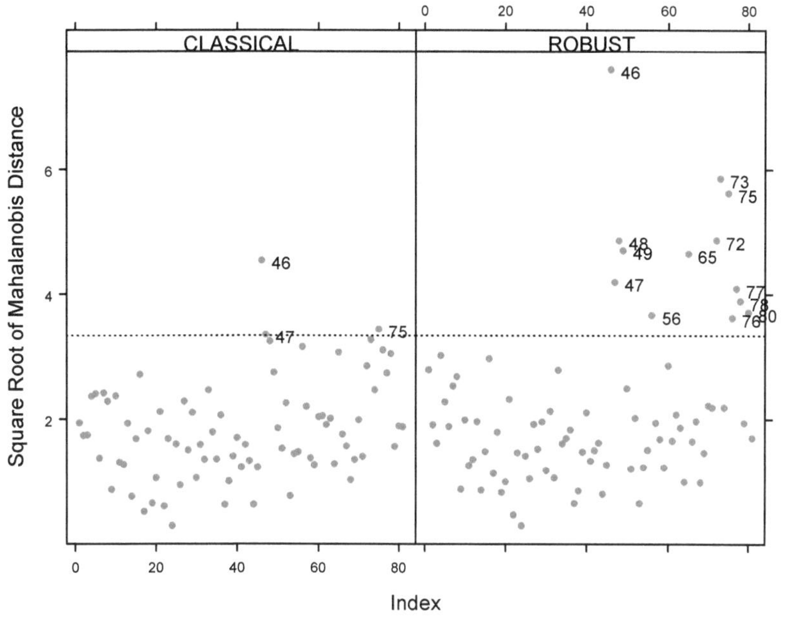

Figure 6.20 Robust and Classical Distances

Given the previous computation of `cov.fm` in Section 6.6, you can make the plot in Figure 6.20 with the command

```
plot(cov.fm,which.plots = 2,id.n = 14)
```

Now that we know the unusual times, we can repeat the application of classical and robust correlations and distances to the unusual portion of the data. The results are shown in Figure 6.21 and Figure 6.22 for the two clusters of unusual times.

The most striking bit of information revealed in Figure 6.21 is that the correlation between the German mortgage bond returns (Pfand) and returns on the other assets has switched from being positive during usual times to being substantially negative. This is natural during market downturns when there is a flight from equity investments. This behavior is also reflected in the pairwise scatterplots of Figure 6.23, which clearly reveal one or two outliers in the scatterplot of Al versus Pfand. The classical distances in Figure 6.22 completely miss the existence of these outliers, while the robust distances reveal two clear outliers and one marginal outlier.

Code 6.9 gives the S-PLUS code for producing the analysis of Figure 6.21, Figure 6.22, and Figure 6.23.

```
data <- data[c(46:49,72:80),]
cov.fm <- fit.models(list(ROBUST = covRob(data),
  CLASSICAL = cov(data)))
plot(cov.fm,which.plots = c(2,3),id.n = 14)
pairs(data)
```

Code 6.9 Analysis of Unusual Times Data

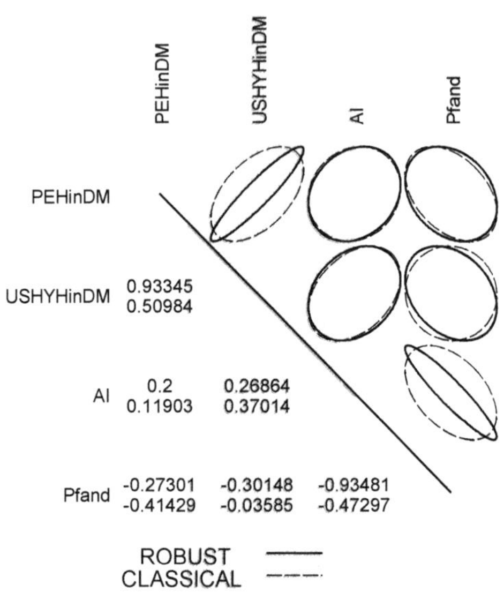

Figure 6.21 Classical and Robust Correlations for Two Clusters of Unusual Times

It is worth explaining why the distance measure (classical and robust) introduced above is the appropriate "statistical" distance. If you imagine an elliptical multivariate distribution for your returns (e.g., a multivariate normal or multivariate *t* distribution), then the right statistical distance is one that is the same for any data point lying along the same elliptical contour. This is what the Mahalanobis distance provides. A useful geometrical way to see what is going on with this distance is to consider the following re-expression of the (squared) distance. We assume that the true covariance matrix and mean return vector are $\mathbf{\Omega}$ and $\boldsymbol{\mu}$, and without loss of generality (by a shift of origin) assume that $\boldsymbol{\mu} = \mathbf{0}$. Then

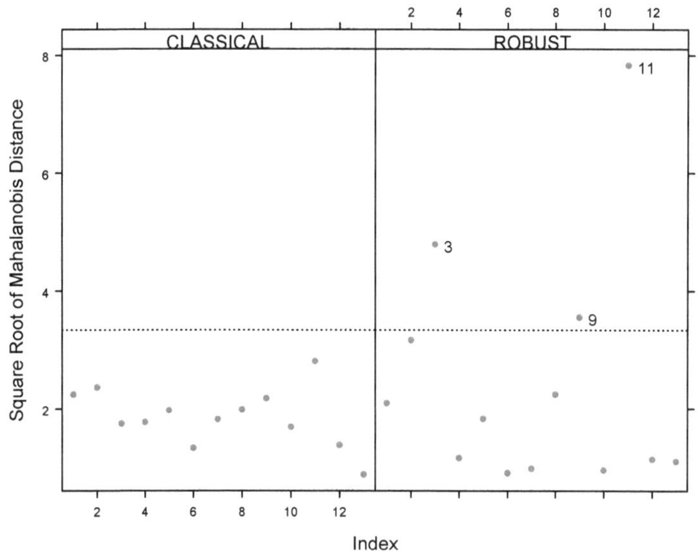

Figure 6.22 Robust and Classical Distances for Two Clusters of Unusual Times

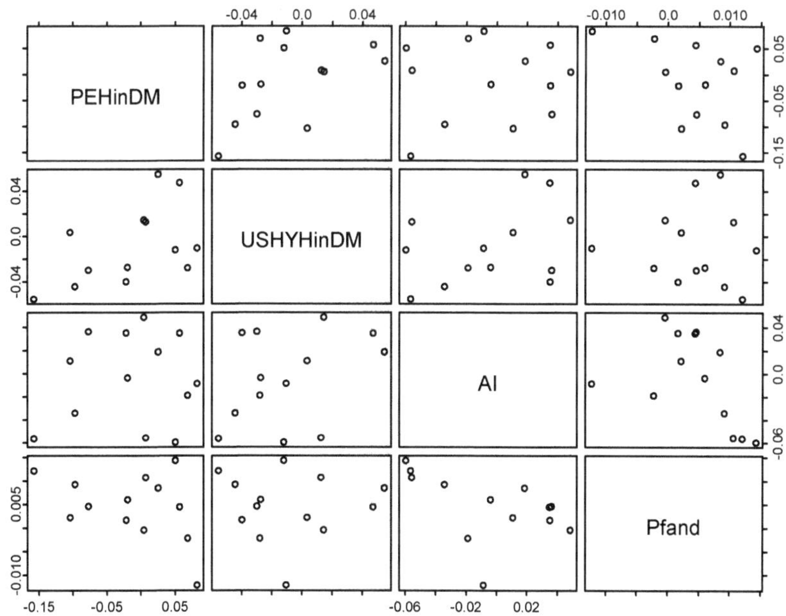

Figure 6.23 Pairwise Scatterplots for the Two Clusters of Unusual Times

$$\begin{aligned} d_t^2 &= \mathbf{r}_t{}'\boldsymbol{\Omega}^{-1}\mathbf{r}_t \\ &= \mathbf{r}_t{}'\boldsymbol{\Omega}^{-1/2}\cdot\boldsymbol{\Omega}^{-1/2}\mathbf{r}_t \\ &= \mathbf{z}_t'\cdot\mathbf{z}_t. \end{aligned} \tag{6.12}$$

You easily check that $\mathbf{z}_t'$ has the identity as its covariance matrix. Thus, using d_t is equivalent to making a transformation of the data so that the distribution is spherical rather than elliptical and then using the ordinary Euclidean distance in this new coordinate system.

The reason that a classical covariance matrix often fails to provide robust distances is that outliers often distort the estimated covariance matrix to such an extent that the transformation above does not result in a spherical scatter for the bulk of the data. Consequently, outliers are not reliably detected using the classical covariance matrix.

We emphasize the point that unusual times, consisting of locally extreme movements of one or more of the returns in a collection of returns, are frequently occurring behaviors by providing a second example, using monthly returns of four hedge fund indices: EMERGING MARKETS, EUROPE, EVENT DRIVEN, and EQUITY. The classical and robust correlations shown in Figure 6.24 clearly indicate that some outlying returns are influencing the classical covariance and correlation estimates.

Figure 6.25 shows that the classical distances give only a weak indication that there is something unusual going on at two or three time points, while the robust distances give a very strong indication of unusual movement at three to five time points in two clusters (time points 4 and 6 and time points 11, 12, 14, and 16).

A quick look at all pairwise scatterplots in Figure 6.26 reveals several multivariate outliers. The time series plots in Figure 6.27 reveal that the first cluster with unusual movement is in the emerging market returns in Q2 and Q3 of 1999, and the second cluster is joint unusual movements in the emerging market returns at the beginning of Q1 2000 and in the returns for Europe at the end of Q4 1999 and in the first and third months of Q1 2000.

Code 6.10 gives S-PLUS code for producing Figure 6.24 through Figure 6.27.

```
returns <- seriesData(hfunds.ts)
cov.fm <- fit.models(list(ROBUST = covRob(returns),
  CLASSICAL = cov(returns)))
plot(cov.fm,which.plots = c(2,3),id.n = 14)
pairs(returns)
par(mfrow = c(4,1))
y.name = colIds(hfunds.ts)
seriesPlot(hfunds.ts,one.plot=F,
  strip.text=y.name,col = 1)
```

Code 6.10 Robust Analysis for Hedge Fund Indices

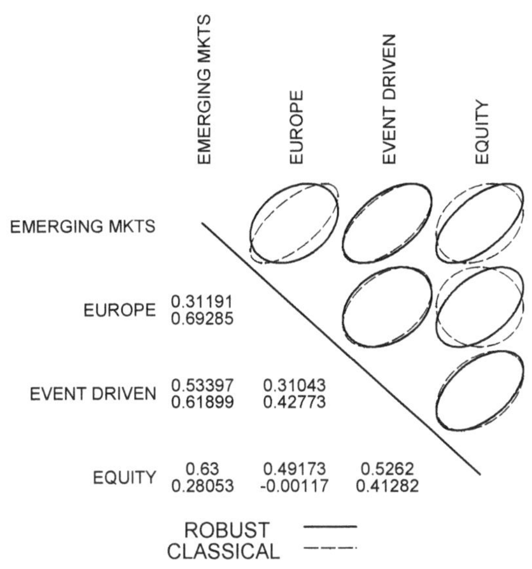

Figure 6.24 Classical and Robust Correlations for Four Hedge Fund Index Returns

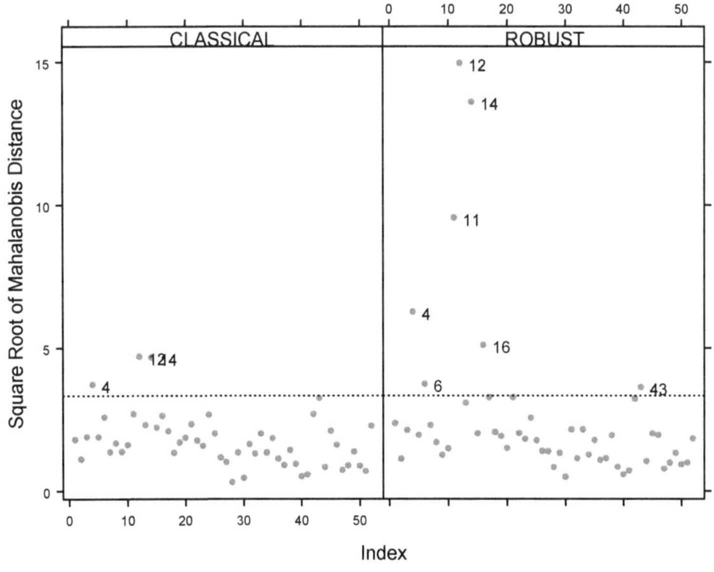

Figure 6.25 Robust and Classical Distances for Hedge Fund Index Returns

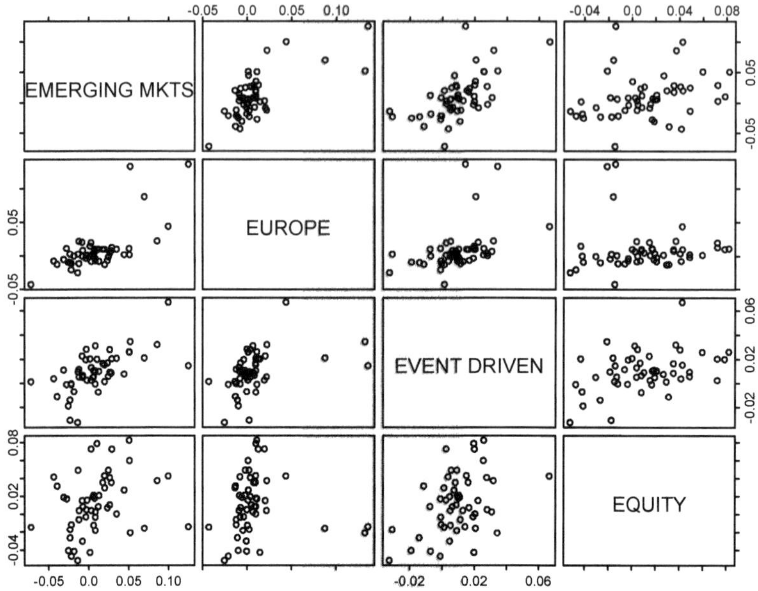

Figure 6.26 Pairwise Scatterplots for the Hedge Fund Index Returns

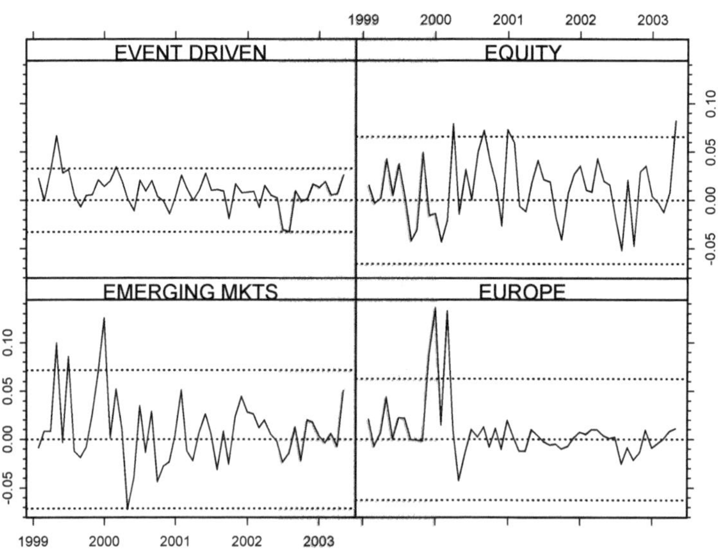

Figure 6.27 Time Series Plots of Hedge Fund Index Returns

6.8 Robust Covariances and Distances with Different Return Histories

It often happens that histories of returns for a collection of portfolio assets cover different periods of time. This situation is quite prevalent in "funds of funds" contexts where a manager is trying to select and optimize a portfolio of funds from a pool in which some funds have existed for only a few years while others have existed for ten years or more. In such a case one does not, at first blush, have an obvious way to compute classical or robust covariance matrices using all the data available. When confronted with this situation, most managers will opt to use the longest common history of the data by truncating all the data to the length of the shortest history available, a practice that often wastes useful information in the asset returns with longer histories. For example, Figure 6.28 shows five sets of hedge fund index returns, with the Emerging Markets (EM) index having the longest history (January 29, 1993 to March 31, 2003), and the High Yield (HY) and Health indices having common shortest histories (January 31, 1997 to March 31, 2003). All of the returns above exhibit a clear negative outlier in 1998, when markets took a dive following the Russian credit default. In addition, Health (a Health and Biotech index) exhibited a wild positive swing in 1999 prior to the dot-com collapse, as well as a wild negative swing following the dot-com collapse in spring of 2000, and Events exhibits a large positive outlier in 1995. A good detection method should reveal these unusual movements, along with others that may not be so apparent, at any time in the entire history of the series (from the earliest date of the longest series to the end of the series). It would be wasteful to throw away the Equity, EM, and Events returns prior to January 31, 1997, in order to compute robust covariance matrices and robust distances. To detect unusual times (or simply unusual data) at every instance over the entire time span of the data, we need a method to compute a robust covariance matrix of appropriate dimension and the associated robust distances.

Effective use of all the data is a classical missing data problem for which there exists a solution in the context of maximum likelihood estimation under multivariate normal returns (Stambaugh, 1997). Here we briefly explain the method in detail for the special case of two groups of assets, where each asset within a group has the same history, and indicate how the method is generalized to more than two groups.

Let the first group have k_1 assets and let the second group have k_2 assets, where the first group has the longer history, $t = 1, 2, \cdots, T$, and the second group has the shorter history, $t = s, s+1, \cdots, T$, with $s > 1$. Let $\hat{\boldsymbol{\mu}}_{long}^{ML}$ and $\hat{\boldsymbol{\Omega}}_{long,long}^{ML}$ be the Gaussian maximum likelihood estimators of the mean vector and covariance matrix of the first group with the long history (i.e., the usual sample mean vector

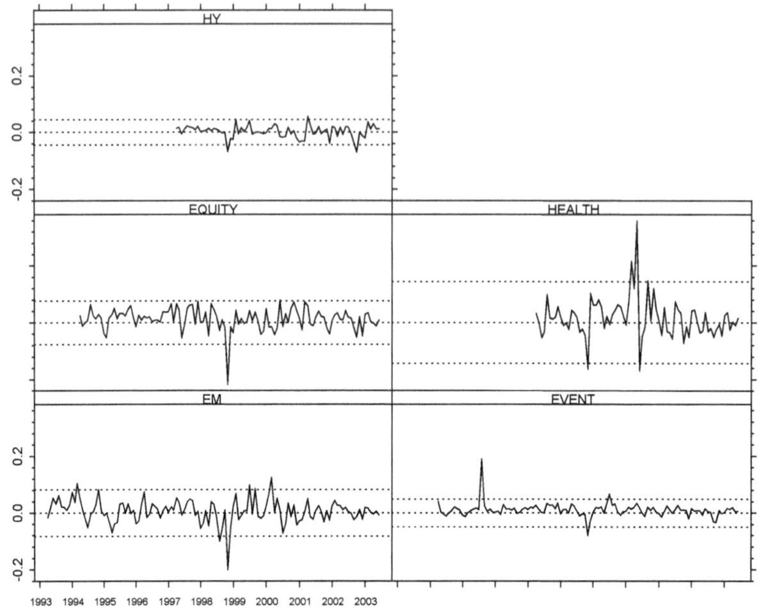

Figure 6.28 Hedge Fund Index Returns with Different Starting Dates

and sample covariance matrix with divisor T). Let $\hat{\boldsymbol{\mu}}_{long}^{truncated}$ be the sample mean vector of the longer group after truncating the returns to make their history have the same period as the shorter group, and let $\hat{\boldsymbol{\mu}}_{short}$ be the sample mean vector of the shorter group. It is important to note that this is not the maximum likelihood estimator of the shorter group's mean vector: the longer series is generally correlated with the shorter series and therefore contains information about the mean vector of the shorter series. Let $\mathbf{r}_{long,t}$, $t = 1, 2, \cdots, T$, be the k_1-dimensional column vectors of returns of the first group, and let $\mathbf{r}_{short,t}$, $t = s, s+1, \cdots, T$, be the k_2-dimensional column vectors of the second group. Consider the multivariate linear regression model

$$\mathbf{r}_{short,t} = \boldsymbol{\alpha} + \mathbf{B} \cdot \mathbf{r}_{long,t} + \boldsymbol{\varepsilon}_t, \quad t = s, s+1, \cdots, T \tag{6.13}$$

of the shorter set of returns on the longer set of returns over the shorter history. Let $\hat{\boldsymbol{\alpha}}$ and $\hat{\mathbf{B}}$ be the Gaussian maximum likelihood (least squares) estimates of the regression coefficients, and let $\hat{\boldsymbol{\Omega}}_{\varepsilon}$ be the sample covariance matrix of the residuals $\hat{\boldsymbol{\varepsilon}}_t$, $t = s, s+1, \cdots, T$, from the maximum likelihood fit.

We can now summarize the overall maximum likelihood estimation results. The maximum likelihood estimate (MLE) of the mean vector of the shorter group series is

$$\hat{\boldsymbol{\mu}}_{short}^{ML} = \hat{\boldsymbol{\mu}}_{short} + \hat{\mathbf{B}} \cdot \left(\hat{\boldsymbol{\mu}}_{long}^{ML} - \hat{\boldsymbol{\mu}}_{long}^{truncated} \right). \tag{6.14}$$

The overall mean vector MLE is

$$\hat{\boldsymbol{\mu}}^{ML} = \left(\hat{\boldsymbol{\mu}}_{long}^{ML}, \hat{\boldsymbol{\mu}}_{short}^{ML} \right). \tag{6.15}$$

The maximum likelihood estimate of the overall covariance matrix is

$$\hat{\boldsymbol{\Omega}}^{ML} = \begin{pmatrix} \hat{\boldsymbol{\Omega}}_{long,long}^{ML} & \hat{\boldsymbol{\Omega}}_{long,short}^{ML} \\ \hat{\boldsymbol{\Omega}}_{short,long}^{ML} & \hat{\boldsymbol{\Omega}}_{short,short}^{ML} \end{pmatrix}, \tag{6.16}$$

where

$$\hat{\boldsymbol{\Omega}}_{short,short}^{ML} = \hat{\boldsymbol{\Omega}}_{\varepsilon} + \hat{\mathbf{B}} \cdot \hat{\boldsymbol{\Omega}}_{short,short}^{ML} \cdot \hat{\mathbf{B}}' \tag{6.17}$$

and

$$\hat{\boldsymbol{\Omega}}_{short,long}^{ML} = \hat{\mathbf{B}} \cdot \hat{\boldsymbol{\Omega}}_{long,long}^{ML}. \tag{6.18}$$

In applications where there are more than two groups and more than two sets of common histories with different starting dates, the method above can be applied recursively to compute an overall Gaussian maximum likelihood estimate of the mean vector and covariance matrix. Details may be found in Section 4 of Stambaugh (1997).

6.8.1 Robustifying the Stambaugh Method

The Stambaugh method relies heavily on a multivariate Gaussian assumption for the returns. As we have seen, this is not a very safe assumption when dealing with asset returns. Furthermore, we need a robust version of the method that is not much influenced by a few outliers. Fortunately, it is rather straightforward to create a robust version by making the following three modifications: (a) replace sample mean estimates by robust location estimates, (b) replace the least squares multivariate regression estimates $\hat{\boldsymbol{\alpha}}$ and $\hat{\mathbf{B}}$ with robust regression estimates, and (c) replace each of the Gaussian MLE sample covariance matrix estimates above (including $\hat{\boldsymbol{\Omega}}_{\varepsilon}$) with a robust covariance matrix estimate. In the example below, we use (a) `location.m` for the robust location estimates, (b) `lmRob` to obtain the robust multivariate regression by computing a set of robust univariate regressions, and (c) `covRob`, with the default fast MCD method and setting `quan = .9`. We note that the method is such that when all the component covariance matrix estimates below are positive definite, the overall robust covariance matrix

$$\hat{\Omega}^{ROB} = \begin{pmatrix} \hat{\Omega}^{ROB}_{long,long} & \hat{\Omega}^{ROB}_{long,short} \\ \hat{\Omega}^{ROB}_{short,long} & \hat{\Omega}^{ROB}_{short,short} \end{pmatrix} \tag{6.19}$$

will be positive definite.

6.8.2 Robust Distances and Degrees of Freedom at Different Time Points

In order to compute a robust distance at each point in time, one needs to use the appropriate robust covariance matrix. Suppose that as you move through the history of a set of returns with different starting dates you encounter returns of dimensions $k_1, k_2, \cdots, k_M$. Then at a time point where there exist k_i returns, you use the corresponding $k_i \times k_i$ robust covariance matrix. Then the corresponding degrees of freedom for the chi-squared upper 2.5% cutoff point is k_i.

6.8.3 The Hedge Fund Indices Example

We used the robustified Stambaugh method[14] to obtain robust covariance matrices $\hat{\Omega}_i^{ROB}$, $i = 1,2,3$, and associated robust distances for the hedge fund indices whose time series were displayed at the beginning of this section (Figure 6.28). The results are shown in Figure 6.29. Note the increasing staircase behavior of the chi-squared upper 2.5% thresholds for each of the three groups with common starting dates owing to the increase in chi-squared degrees of freedom as more assets came online in 1994 and 1997. The robust distances detect outliers with greater power than the classical distances, thereby clearly revealing multivariate outliers that the classical method detects only weakly or not at all.

The robust Stambaugh method code is long, and is not included with this book.

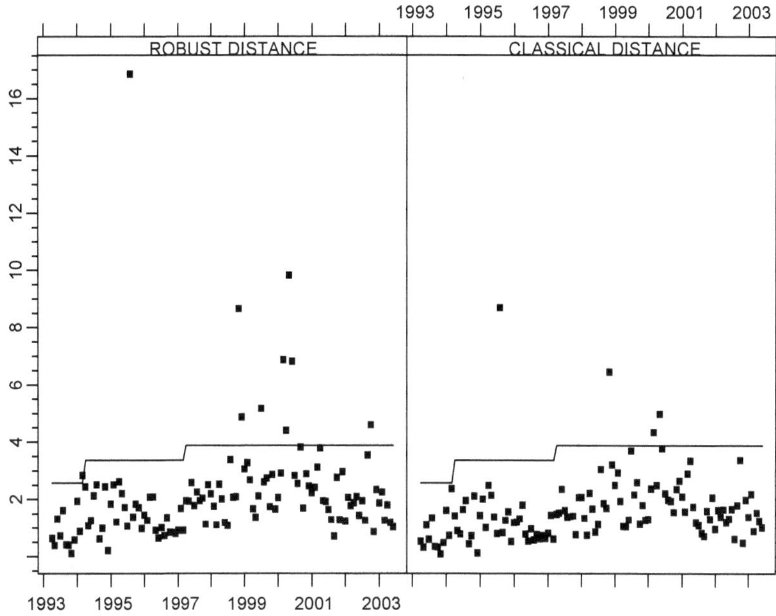

Figure 6.29 Classical and Robust Distances for Hedge Fund Indices

6.8.4 Comment on Using Chi-Square Percentage Points

One should be aware of several points concerning use of the (square root of the) 2.5% upper percentage point of the chi-squared distribution as a detection threshold. First, the threshold is somewhat arbitrary, and one could equally well use an upper 5% point or upper 1% point, the former yielding a larger false alarm rate and the latter yielding a smaller false alarm rate than the 2.5% point. We advise against using anything smaller than the upper 5% point since false alarm rates that are too high can lead to detection of outliers and unusual times even when the data are perfectly stationary and normally distributed (e.g., the upper 25% point recommended by Chow, Jacquier, Kritzman, and Lowry (1999) would exhibit such behavior). As the chi-squared approximation is not very reliable under non-normality (see, for example, Rocke and Woodruff, 1996), one may prefer to make a kernel density estimate of the classical and robust distances and look for clustering of unusual times in terms of multimodality of the density estimates.

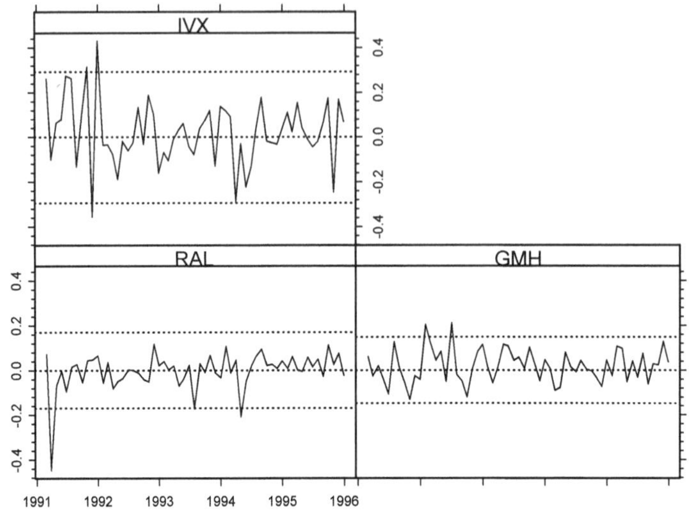

Figure 6.30 Time Series of RAL, GMH, and IVX Returns

6.9 Robust Portfolio Optimization

Since classical estimates of mean returns and covariances can be adversely influenced by the presence of one or more outliers, it should not be surprising to find that Markowitz mean-variance optimal portfolios based on these classical estimates can also be adversely influenced by such outliers. As an example of the extent to which outliers can influence the estimated Markowitz efficient frontier, consider the time series of highly volatile monthly stock returns for RAL, GMH, and IVX from February 28, 1991 to December 29, 1995, shown in Figure 6.30.

RAL is distinguished by having a single negative outlier at the beginning of the series, while GHM and IVX have a relatively high volatility in the early time periods when compared with the rest of the series. The values of robust and classical sample means and standard deviations for these returns are shown in Table 6.1.

It is evident that the outliers have the largest impact on RAL, where the sample mean and robust mean values are .003 and .009, respectively, and the classical and robust standard deviations are .085 and .055. The differences in correlations between the two estimates are shown in Figure 6.31.

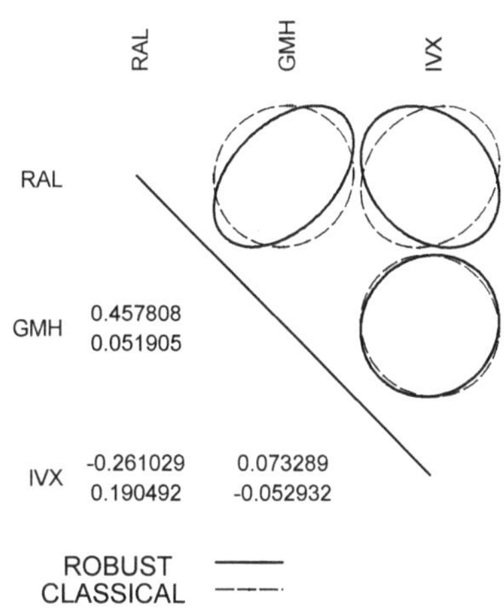

Figure 6.31 Classical and Robust Correlations for RAL, GHM, and IVX

Table 6.1 Means and Standard Deviations for RAL, GHM, and IVX

	RAL	GMH	IVX
Classic Mean	–.003	.019	.021
Robust Mean	.009	.018	.020
Classic Std. Dev.	.085	.074	.147
Robust Std. Dev.	.055	.082	.131

We see a substantial shift of +.41 for the RAL/GHM correlation and –.45 for the RAL/IVX correlation when substituting a robust correlation for a classical correlation. The S-PLUS code for the plots above is similar to Code 6.7 and Code 6.8 provided in Section 6.6.

We made the plots in Figure 6.30 and Figure 6.31 with Code 6.7 and Code 6.8 using `returns.three.ts` in place of `normal.vs.hect.ts` and deleting the `pairs` command in Code 6.7.

Now we use NUOPT to compute a robust efficient frontier with a constraint of no short-selling by simply replacing the classical sample mean returns and sample covariance estimates with robust estimates. The resulting efficient frontier is displayed in Figure 6.32 along with the classical efficient frontier and the maximum Sharpe ratios based on a monthly risk-free rate of .003. The display also shows the classical and robust means and standard deviations of each of the three stocks along with their ticker symbols.

The (maximum) Sharpe ratios are approximately the same for both frontiers. However, the classical efficient frontier indicates that the investor can achieve about 10 to 20 basis points (monthly) more than with the robust frontiers for sufficiently high levels of risk. On the other hand, the robust efficient frontier

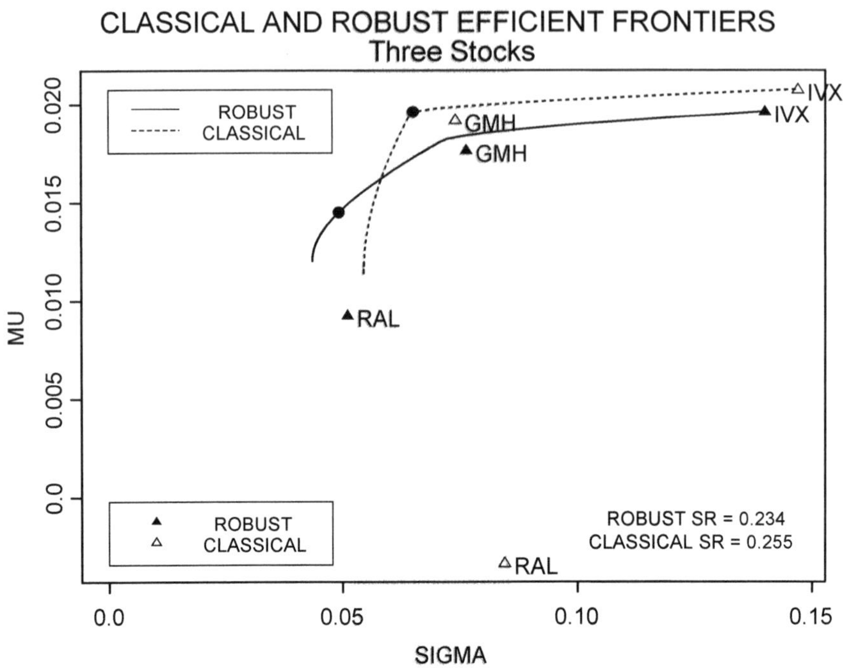

Figure 6.32 Classical and Robust Efficient Frontiers for Three-Stock Portfolio

offers higher levels of return than the classical frontier for lower levels of risk and a considerably lower risk for the (global) minimum variance portfolio. The means and standard deviations of the individual stocks in the plot above are the classical sample mean and sample standard deviation. Note that the main difference between the values of the classical and robust means and standard deviations are for RAL. The optimal weights for the classical and robust efficient frontiers are displayed in Figure 6.33.

It should not be surprising to see that, for small levels of risk, the classical portfolio gives considerably less weight to RAL than does the robust portfolio (recall that RAL has one large negative outlier at the beginning of the series) and that for the largest levels of risk both portfolios give about the same relative weights to GMH and IVX. We also see that the robust portfolio gives no weight to GMH in the minimum variance portfolio and also gives considerably less weight to GMH than does the classic portfolio for lower levels of risk.

Code 6.11 creates the function `rob.mv.efronts` for computing and displaying mean-variance and robust efficient frontiers and optionally plotting the portfolio weights for each. The last line of Code 6.11 executes the function on the three-asset time series object `returns.three.ts`:

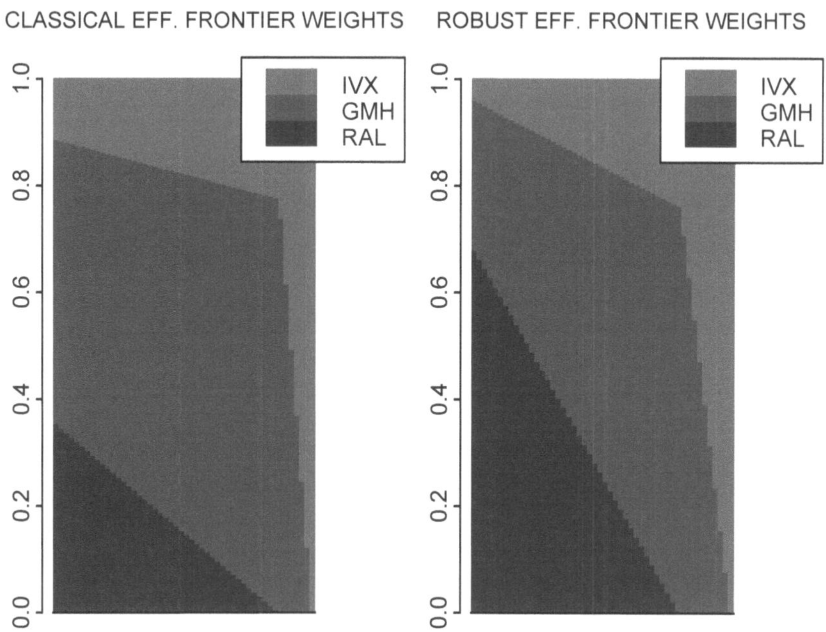

Figure 6.33 Weights for Classical and Robust Efficient Frontier Portfolios

```
rob.mv.efronts <- function(returns.ts, rf=.003,
  n.ret=50, plot.weights=F, a=1, sharpe=T,
  display.points=T, display.names=T,
  display.letters=F)
{
  returns <- seriesData(returns.ts)
  p <- ncol(returns)
  # Parameter comments
  # Use a= 1 for no short selling, and
  # adjust a > 1 for short selling
  # If using display.letters=T, set
  # display.points=F and display.names=F)
  # Compute Classical Efficient Frontier
  meanVec1 <- apply(returns,2,mean)
  covMat1 <- var(returns)
  sigma1 <- diag(covMat1)^.5
  max.ret1 <- max(meanVec1)*a
  ef.classic <- portfolioFrontier(covMat1,
     meanVec1,
     wmin=0, max.ret=max.ret1, n.ret=n.ret)
  # Compute Robust Efficient Frontier
```

```
meanVec2 <- apply(returns,2,location.m)
covMat2 <- covRob(returns,estim = "mcd",
   quan = .9)$cov
sigma2 <- diag(covMat2)^.5
max.ret2 <- max(meanVec2)*a
ef.robust <- portfolioFrontier(covMat2, meanVec2,
   wmin=0, max.ret=max.ret2, n.ret=n.ret)
# Plot Efficient Frontiers
xlim <- range(ef.classic$sd,ef.robust$sd,
   sigma1,sigma2,0)
ylim <- range(ef.classic$ret,ef.robust$ret,
   meanVec1,meanVec2,0)
plot(ef.classic$sd,ef.classic$ret,xlim=xlim,
   ylim=ylim, type = "n" ,xlab="SIGMA",ylab="MU")
lines(ef.classic$sd,ef.classic$ret,lty = 8)
lines(ef.robust$sd,ef.robust$ret,lwd=2)
# Plot Stock Mu's and Sigma's and Add Legend
title(main="CLASSICAL AND ROBUST EFFICIENT
   FRONTIERS\n Three Stocks")
if(display.letters) {
   for(i in 1:p) {
      points(sigma1[i],meanVec1[i],
         pch = letters[i])
      points(sigma2[i],meanVec2[i],
         pch = LETTERS[i])
   }
   if(display.names){
      text(sigma1 + 0.002, meanVec1,
         names(meanVec1), adj=0)
      text(sigma2 + 0.002,meanVec2,
         names(meanVec2), adj=0)
   }
   x = xlim[1]+.15*(xlim[2]-xlim[1])
   y = ylim[1]+.10*(ylim[2]-ylim[1])
   leg.names = c("A,B,.. Robust Mu's,Sigma's",
      "a,b,. Classical Mu's,Sigma's")
   text(x,y,leg.names[1])
   text(x,y-.002,leg.names[2])
} # endif display.letters

if(display.points){
   points(sigma1,meanVec1,pch = 2)
   points(sigma2,meanVec2,pch = 17)
   if(display.names){
      text(sigma1 + 0.002, meanVec1,
```

```
            names(meanVec1), adj= 0)
        text(sigma2 + 0.002, meanVec2,
            names(meanVec2), adj= 0)}
        x = xlim[1]+.00*(xlim[2]-xlim[1])
        y = ylim[1]+.13*(ylim[2]-ylim[1])
        leg.names = c("   ROBUST","CLASSICAL")
        legend(x,y,leg.names, marks = c(17,2))
} #endif display.points

# Add legend
x = xlim[1]+.0*(xlim[2]-xlim[1])
y = ylim[2]-.0*(ylim[2]-ylim[1])
leg.names = c("   ROBUST","CLASSICAL")
legend(x,y,leg.names,lty=c(1,8))

# Compute and Display Maximum Sharpe Ratio's,
# and Add Bullets
if(sharpe) {
    i.maxsr.classic = order((ef.classic$ret-
        rf)/ef.classic$sd)[n.ret]
    i.maxsr.robust = order((ef.robust$ret-
        rf)/ef.robust$sd)[n.ret]
    sr.classic = ((ef.classic$ret-
        rf)/ef.classic$sd)[i.maxsr.classic]
    sr.robust = ((ef.robust$ret-
        rf)/ef.robust$sd)[i.maxsr.robust]
    points(ef.classic$sd[i.maxsr.classic],
        ef.classic$ret[i.maxsr.classic],pch = 16)
    points(ef.robust$sd[i.maxsr.robust],
        ef.robust$ret[i.maxsr.robust],pch = 16)
    x = xlim[1]+.85*(xlim[2]-xlim[1])
    y = ylim[1]+.1*(ylim[2]-ylim[1])
    text(x,y,paste("  ROBUST SR =",
        round(sr.robust,3)))
    y = ylim[1]+.05*(ylim[2]-ylim[1])
    text(x,y,paste("CLASSICAL SR =",
        round(sr.classic,3)))
} #endif sharpe

# Plot Portfolio Weights for Both Efficient
# Frontiers
if(plot.weights) {
    par(mfrow = c(1,2))
    barplot(ef.classic$weights,
        legend = names(returns))
```

```
      title(main = "CLASSICAL EFF. FRONTIER
         WEIGHTS")
      barplot(ef.robust$weights,
         legend = names(returns))
      title(main = "ROBUST EFF. FRONTIER WEIGHTS")
      par(mfrow = c(1,1))
   }#endif plot.weights
} # end function definition

rob.mv.efronts(returns.three.ts,plot.weights = T)
```

Code 6.11 Robust Efficient Frontiers

6.9.1 Effect of Outliers on the Sample Mean versus the Sample Covariance Matrix

By making small modifications to Code 6.11 above, you can easily do a sensitivity analysis to see whether the influence of the outliers on the classical efficient frontier is primarily through distortion of the mean estimate or primarily through distortion of the covariance matrix estimate. To use only a robust covariance estimate, replace `location.m` with `mean` in the expression

```
meanVec2 <- apply(returns,2,location.m)
```

in the Code 6.11 function. To use only a robust mean estimate (and the classical covariance estimate), change the code line

```
covMat2 <- covRob(returns)$cov
```

to

```
covMat2 <- var(returns).
```

The results of making these two changes separately are shown in Figure 6.34 and Figure 6.35, respectively.

These displays indicate that it is not enough to use only robust means or only robust covariances. Combining the information in Figure 6.33, Figure 6.34, and Figure 6.35, it appears that the difference between the classical and robust efficient frontiers is a result of outliers influencing both the sample mean and sample covariance estimates.

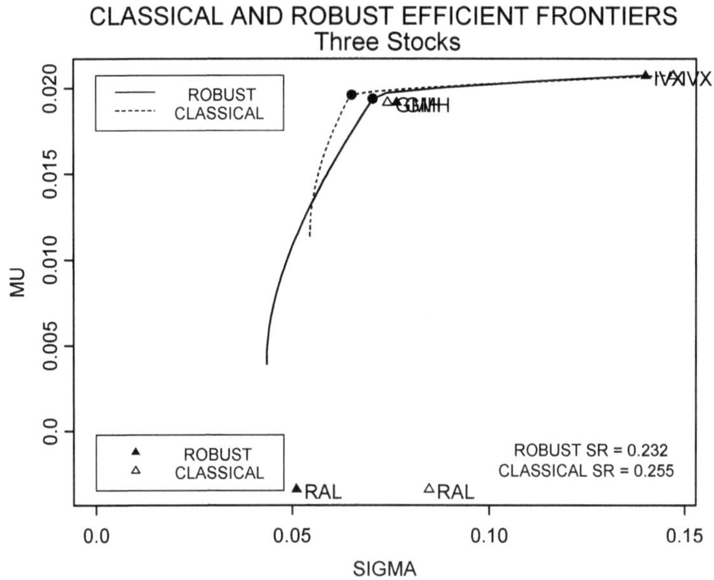

Figure 6.34 Robust Efficient Frontier with Robust Covariance Estimate Only

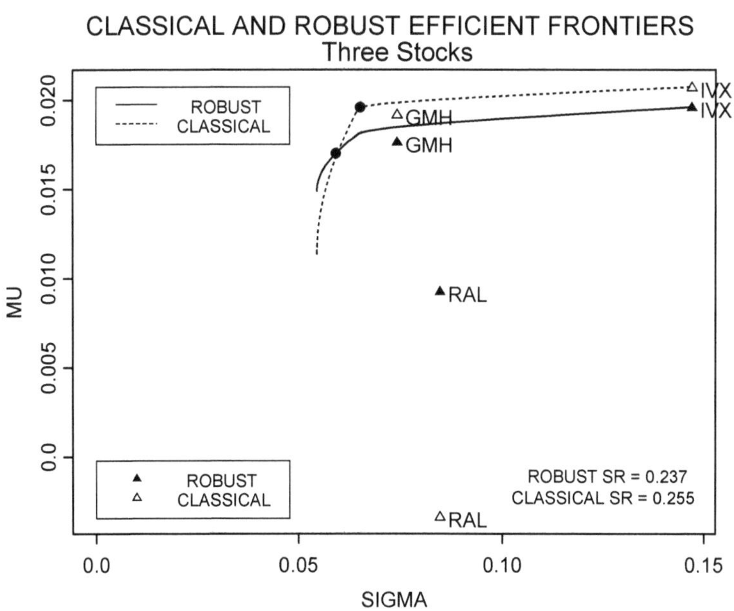

Figure 6.35 Robust Efficient Frontier with Robust Mean Only

6.9.2 Other Examples and Alternative Asset Plot Labels

You can easily find many examples where the classical and robust efficient frontiers differ by a considerable amount. This is particularly true for microcap and small cap stocks, but you can also find examples of this type for mid-cap and large cap stocks. We give three examples in Figure 6.36 through Figure 6.38 in support of this claim using options in Code 6.11 that provide for alternative displays of both classical and robust means and standard deviations.

In Figure 6.36, we display the two efficient frontiers for a group of five small cap stocks with a solid (open) triangle symbol for the robust (classical) means and standard deviations of the returns of the individual stocks. Although we can see that there are substantial differences in the robust and classical means and standard deviations, we cannot see their individual changes.

We make the plot of Figure 6.36 with the commands

```
tickers <- c("TOPP","KWD","HAR","RARE","IBC")
returns.ts <- smallcap.ts[,tickers]
rob.mv.efronts(returns.ts)
y.name <- colIds(returns.ts)
seriesPlot(returns.ts,one.plot=F,strip.text=y.name,
  col = 1)
```

Figure 6.37 and Figure 6.38 are made by modifying the code above in obvious ways.

We note that, in general, mid-cap and large cap stocks are less prone to having large outliers than small caps and microcaps, and when there are no influential outliers in returns, the values for classical and robust means, standard deviations, and covariances will be close to one another. In such situations, the classical and robust efficient frontiers will be very similar, as in Figure 6.38, and one need not worry about influential outliers.

In the case of many stocks, the ticker symbols may overlap a lot, and you may prefer to use uppercase and lowercase letters in order to visualize the changes in individual means and standard deviations, as in Figure 6.39. You can get these kinds of labels in your efficient frontier plot by using the optional arguments `display.points = F`, `display.names = F`, and `display.letters = T` in the function `rob.mv.efronts`.

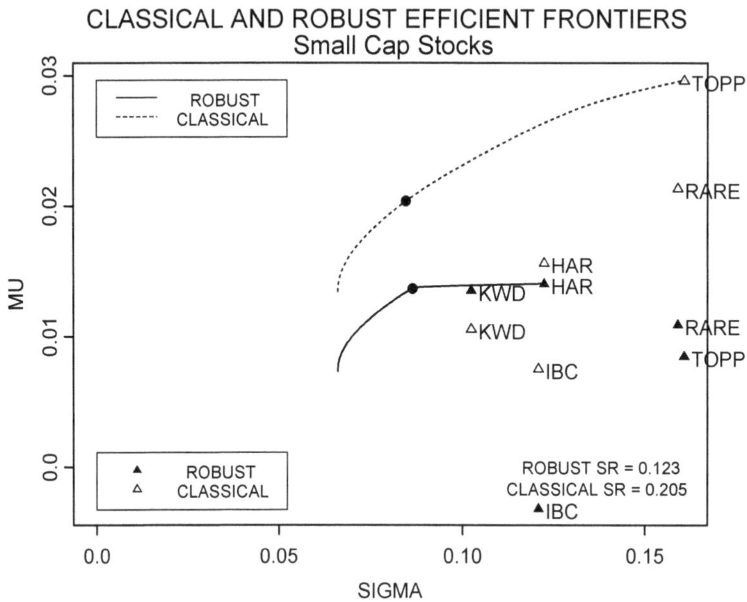

Figure 6.36 Efficient Frontiers for Small Cap Stocks with Classical and Robust Mu and Sigma

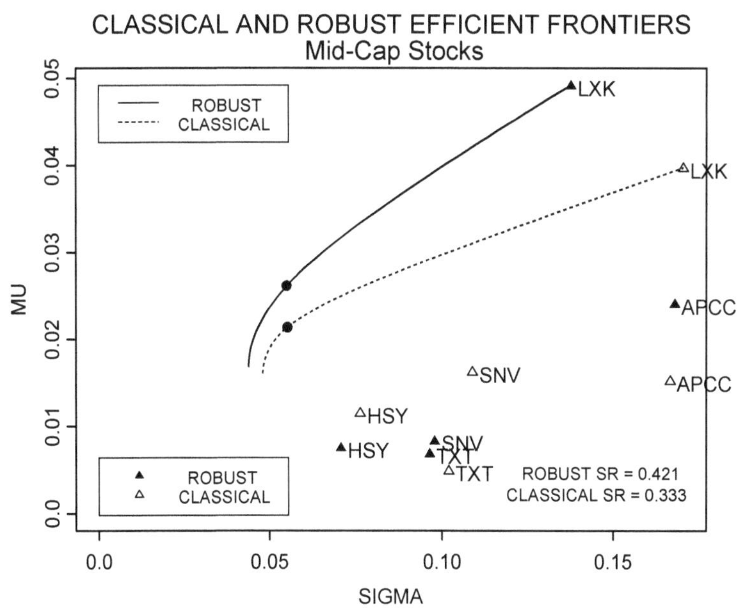

Figure 6.37 Efficient Frontiers for Mid-Cap Stocks with Ticker Symbols

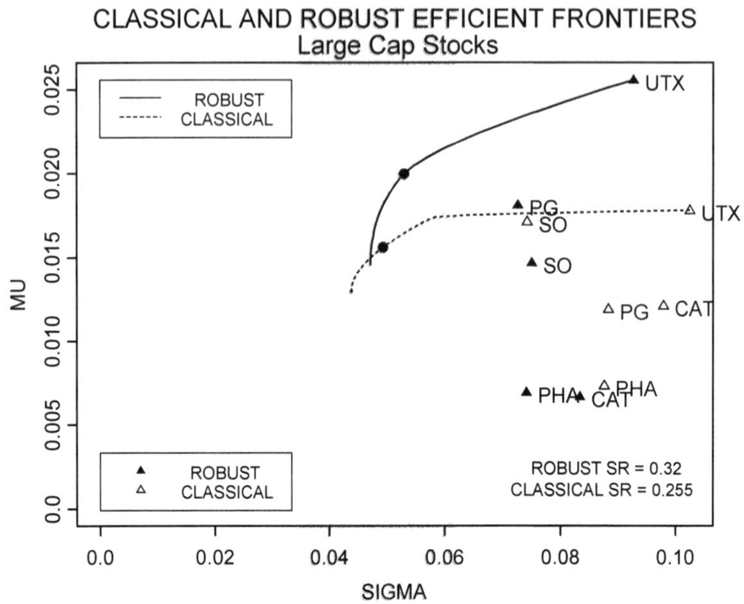

Figure 6.38 Efficient Frontiers for Large Cap Stocks with Letters for Mu and Sigma

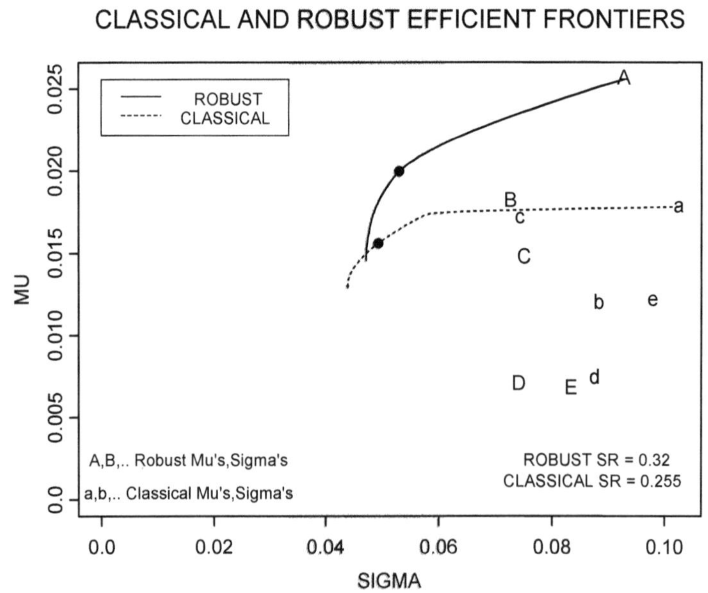

Figure 6.39 Typical Efficient Frontiers for Large Cap Stocks

6.9.3 Classical or Robust Efficient Frontier: Which to Use?

It is important to keep in mind that a robust efficient frontier is based on robust estimates of the means and covariance matrix, which themselves represent the mean and covariance of the bulk of the returns. As such, a robust efficient frontier represents the bulk of the data. Whether this is an adequate representation of the future behavior of your returns is open to serious question. Therefore, at this point it is not clear whether you should prefer making an investment decision based on a robust efficient frontier rather than a classical efficient frontier.

Of one thing we are sure: a robust efficient frontier is a valuable diagnostic tool. When the robust and classical frontiers are quite close to one another, as in Figure 6.39, the returns are quite likely to be free of influential outliers and well-approximated by a multivariate normal distribution. In this case, one can feel reasonably confident in using the mean-variance efficient frontier. When the two efficient frontiers differ by a significant amount, as in Figure 6.36 though Figure 6.38 above, there are likely to be influential outlying returns, and it is unlikely that the returns are well-approximated by a multivariate normal distribution. In such cases one is alerted to the need to carry out some exploratory data analysis (EDA) of the returns data and think carefully about what to do. One way to start such an EDA is by making time series plots of your returns to see if there are any obvious outliers, whether those outliers are positive or negative, and where they occur in the series of returns (e.g., early, middle or late in the period of interest). We illustrate what this initial step can reveal in the examples above.

The Trellis time series plots of the stocks in the small cap portfolio above, provided in Figure 6.40, reveal that three of the series, TOPP, RARE, and IBC, have one or more dominant positive returns outliers and that there are no dominant negative outliers in any of the series of returns. In the efficient frontiers display of Figure 6.36, you see that the robust means and standard deviations of the returns for TOPP, RARE, and IBC are substantially smaller than those of classical means and standard deviations, as might be anticipated. Correspondingly, the robust efficient frontier is lower and slightly to the left of the classical frontier. An investor who uses the robust efficient frontier is taking a conservative view with regard to the potential occurrence of future positive outliers in one or more of the series TOPP, RARE, and IBC (i.e., he is not betting on such occurrences in the future). Such an investor is, in a sense, implementing a Bayesian approach based on his own subjective prior distributions about the probability of future positive outliers.

The situation for the time series plots of the mid-cap stocks in Figure 6.41 is different. APCC has a number of substantially negative values but no clearly dominant outliers, while LXK has three or four dominant negative outliers. Furthermore, in each case there are no equivalent offsetting positive values. For

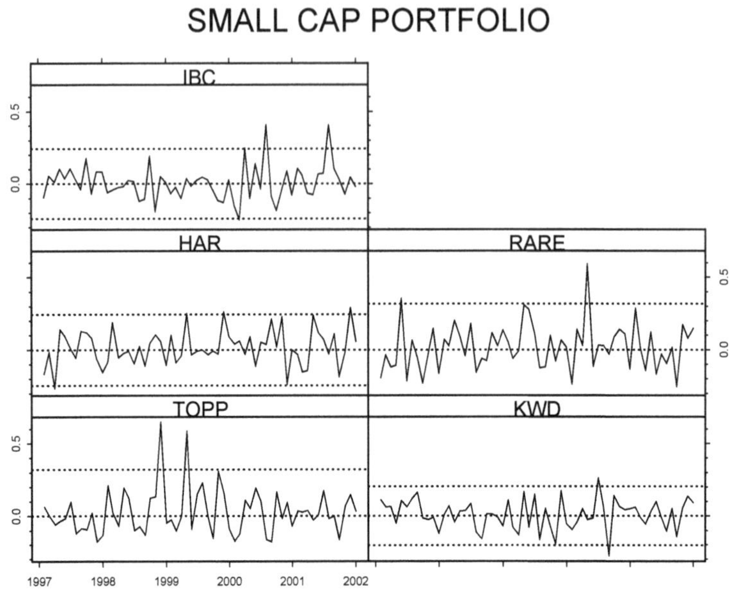

Figure 6.40 Small Cap Portfolio Time Series

TXT, the positive and negative extreme values appear to be roughly offsetting, and SNV has a single positive outlier near the beginning of the series. This information is depicted in the efficient frontiers of Figure 6.36: the robust means of the APCC and LXK returns are larger than their sample means; the robust standard deviation of LXK is smaller than its classical sample standard deviation, while the robust and classical standard deviations of APCC have almost the same values; the robust mean and standard deviation of SNV are smaller than the classical sample mean and standard deviation; and there is a small difference between the robust and classical means and standard deviations for TXT. Given the different behaviors of these means and standard deviations, one might not expect the robust efficient frontier to dominate the classical frontier at all levels of risk. That this is the case reflects the fact that the larger values of the robust means along with the large values of the robust standard deviations result in considerable "leverage" effects in determining the location of the robust efficient frontier relative to the classical efficient frontier.

A word of caution is in order: in this situation, would a wise investor trust the higher returns achievable with the robust efficient frontier? Since the negative returns for APCC and LXK are near the end of the series, an investor may well be wary of assuming that such returns will not occur again in the near future and therefore reject the optimism of the robust efficient frontier.

Figure 6.42 shows the Trellis time series plots of returns for the stocks of the large cap portfolio whose efficient frontiers are shown in Figure 6.38. The tickers UTX, PG, PHA, SO, and CAT correspond to the letters "A," "B," "C,"

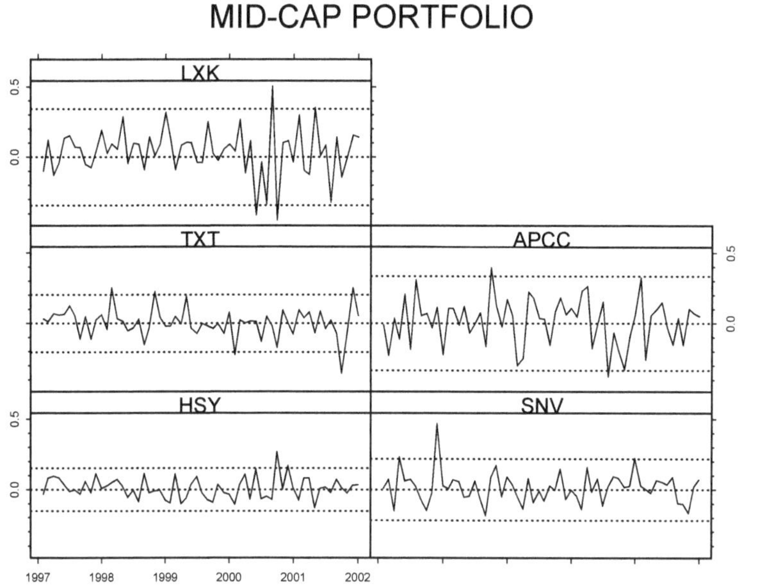

Figure 6.41 Mid-Cap Portfolio Time Series

"D," "E" (and "a," "b," "c," "d," "e"). The time series plots reveal that UTX has one or two possible negative outliers, PG has one large negative outlier, PHA has one positive outlier and one negative outlier, and CAT has one large positive outlier. The corresponding differences in locations of the robust means and standard deviations (the points labeled "A," "B," "C," and "E," respectively) and the classical means and standard deviations (the points labeled "a," "b," "c," and "e," respectively) are what one would expect. In this case the overall configuration of the outliers in the returns and the resulting robust versus classical means and standard deviations is rather complex, and one cannot easily guess the relative positioning of the robust and classical efficient frontiers.

While the analysis above may provide some guidance in choosing between a robust and classical efficient frontier when making an investment decision, better tools are needed to determine the relative performance of investments made with these two approaches. One such tool is a **bootstrapped efficient frontier** that can help determine whether the difference between a robust and classical efficient frontier is "real" or whether it is just a result of statistical variability.

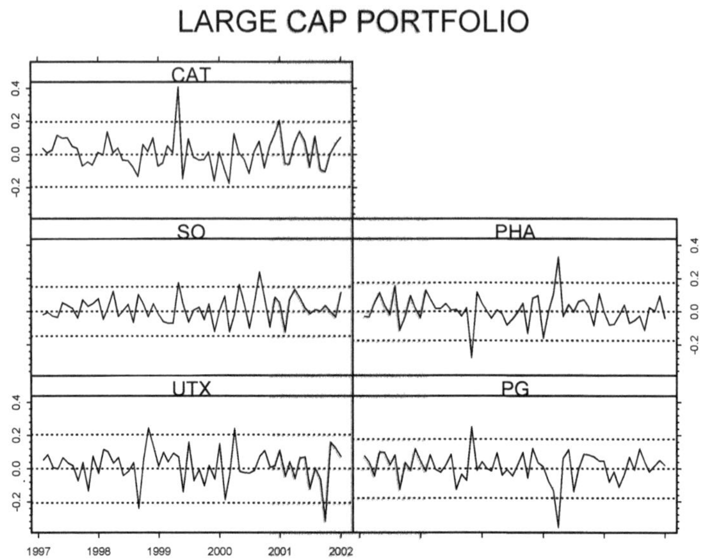

Figure 6.42 Large Cap Portfolio Time Series

6.9.4 Bootstrapped Efficient Frontiers and Sharpe Ratios

Classical and robust efficient frontiers are complicated functionals of the underlying distribution of the returns, and the exact distributions of efficient frontiers (even the mean-variance efficient frontier) are generally intractable. In Chapter 4, we discussed parametric portfolio resampling in which multivariate normal samples are generated based on the sample mean and covariance of the returns. In Section 4.5.4, we noted the lack of statistical foundation of the variant proposed by Jorion (1992) and Michaud (1998), and in Section 4.6 we applied nonparametric bootstrap methods to estimate confidence intervals for the Sharpe ratio.[15] In this section we continue to use the nonparametric bootstrap to calculate and visualize the variability of both robust and classical mean-variance efficient frontiers and their maximum Sharpe ratios.

A primary advantage of using the nonparametric bootstrap to assess the average behavior and variability of these two types of efficient frontiers is that the results tell us all that the data have to say about the unknown distribution of the multivariate returns. In this kind of bootstrap sampling, some samples will have fewer outliers (sometimes zero outliers) than in the original sample and some will have more outliers than in the original sample. In this way, the efficient frontier variability is reflecting the various possible future efficient frontier curves based on samples coming from a nonparametric estimate of the unknown returns distribution.

The dashed lines in Figure 6.43 show 30 bootstrapped classical mean-variance efficient frontiers for the small cap portfolios of Figure 6.36 and Figure 6.40, and the dashed lines in Figure 6.44 show 30 bootstrapped robust efficient frontiers for the same portfolio. The solid line in each figure is the efficient frontier based on the original set of returns. The same bootstrap replicates are used for each figure (i.e., for each mean-variance efficient frontier there is a corresponding robust efficient frontier computed with the same bootstrap sample). Each solid dot is the "bullet point" representing the (global) minimum variance portfolio for the corresponding bootstrap sample and associated efficient frontier. It may be noted immediately that both the original efficient frontiers are biased in that they are not centrally located in the scatter of bootstrap efficient frontiers. Since the horizontal and vertical ranges of both axes are the same in the two figures, you can deduce that the mean return of the robust minimum variance portfolio appears to be less than that of the mean-variance minimum variance portfolio, while the risk of the former is at least as small as the risk of the latter.

Figure 6.45 shows boxplots of the differences for each bootstrap sample between the mean returns and risks of the robust and mean-variance minimum variance portfolios along with the differences between the Sharpe ratios. The

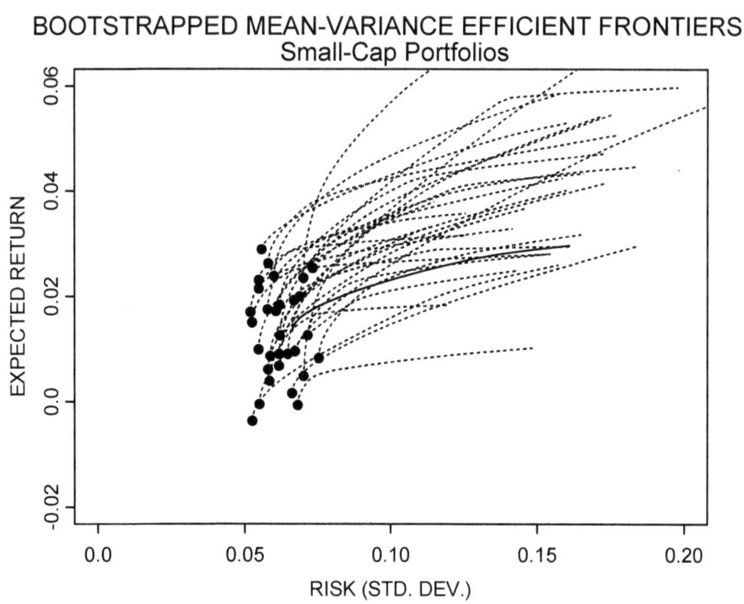

Figure 6.43 Bootstrapped Mean-Variance Efficient Frontiers for Small Cap Portfolio

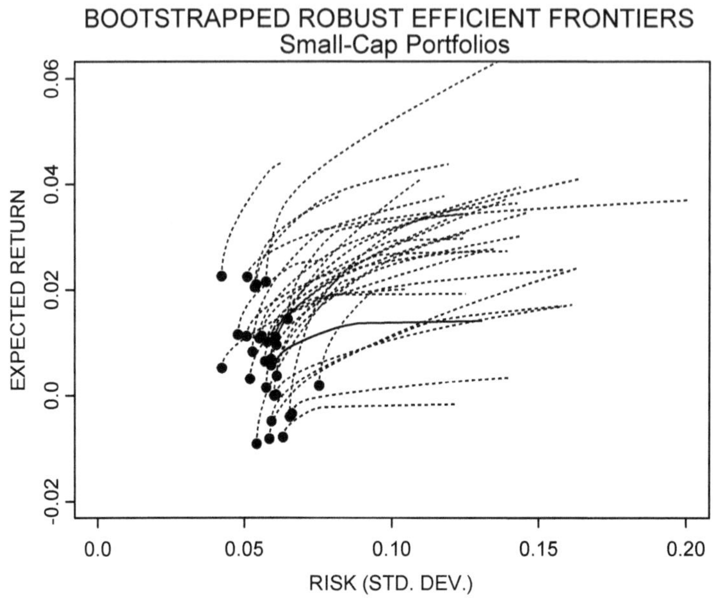

Figure 6.44 Bootstrapped Robust Efficient Frontiers for Small Cap Portfolio

notches in the boxplots correspond to approximate 95% confidence intervals for the median differences. None of these confidence intervals contains zero, indicating that the performance of the robust optimal portfolio is significantly lower than that of the mean-variance portfolio at a level of 5%. The median difference between the mean-variance and robust Sharpe ratios is only about –.06, which is not likely to be of much financial consequence.

The choice $B = 30$ for the number of bootstrap replicates may well be too small to draw firm conclusions. To get an idea of how things will change with an increasing number of replicates, we ran the bootstrap program with $B = 100$ replicates (without plotting efficient frontiers); the resulting boxplots are shown in Figure 6.46. For this particular example, the results are not substantially different from those of the bootstrap with $B = 30$.

The S-PLUS and NUOPT code for the above computations is provided below in the form of the two functions `boot.efronts` (Code 6.12) and `efront.nuopt.forboot` (Code 6.13) and a short script (Code 6.14) for calling `boot.efronts` using the small cap returns. When running Code 6.12 and Code 6.13, one must allow sufficient time for the bootstrap computations to finish.

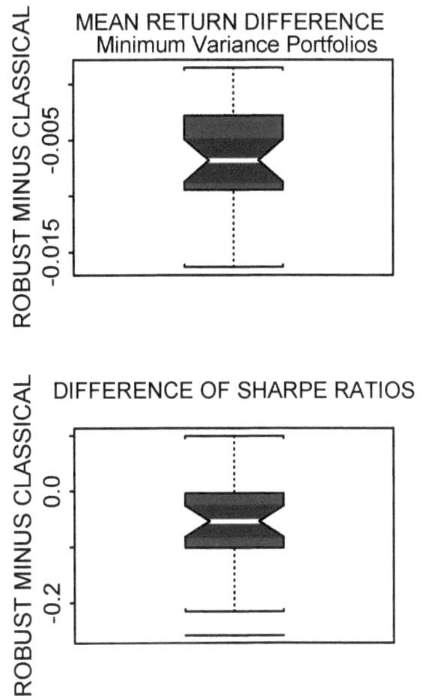

Figure 6.45 Bootstrap Portfolio Performance Differences for $B = 30$

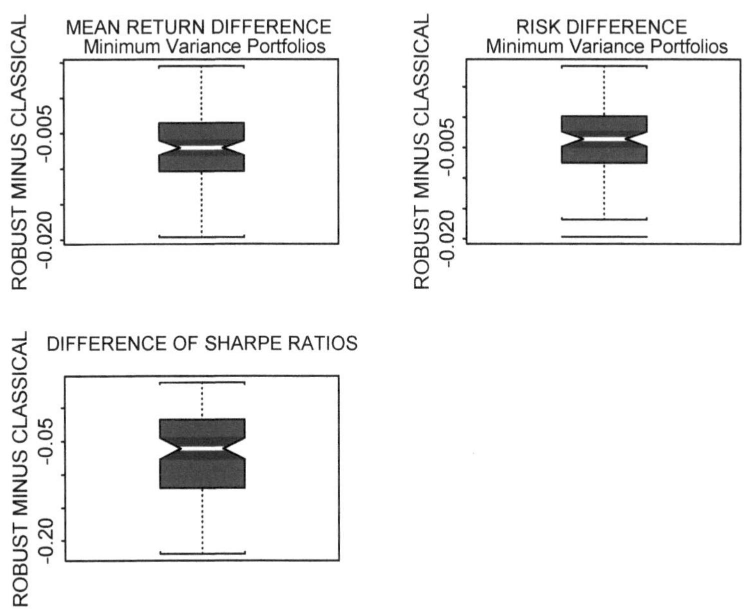

Figure 6.46 Bootstrap Portfolio Performance Differences for $B = 100$

```
boot.efronts <- function(returns.ts, B=30, rf=.03,
  npoints=20, k.mu=4, k.sigma=1.5, mv=T, tan=F,
  plotit=F, estim = "mcd", quan=.9)
{
  # B is the number of bootstrap samples
  # k.mu controls vertical axis plotting range
  # k.sigma controls horizontal axis plotting range
  # Adjust k.mu, k.sigma to minimize plot
  # "Line out of bounds" Warnings
  # mv = T to display bullet at minimum var.
     portfolio
  # tan = T to display bullet at tangency portfolio
  # plotit = T to plot efficient frontiers
  # estim = "mcd" to use MCD est. (avoid auto.
  # default choice)
  # quan is the fraction of data used by the MCD
  # Compute Bootstrap Samples Indices
  returns <- seriesData(returns.ts)
  n <- nrow(returns)
  m <- ncol(returns)
  B <- 30
  boot.index <- samp.boot.mc(n,B)
  # Compute Classic Efficient Frontier
  covmat <- var(returns)
  mu <- apply(returns,2,mean)
  max.ret <- max(mu)
  ef <- portfolioFrontier(covmat, mu, wmin=0,
     max.ret=max.ret,n.ret=npoints)
  sd = ef$sd
  ret = ef$returns
  # Compute Robust Efficient Frontier
  cov.rob <- covRob(returns,estim = estim,
     quan = quan)
  covmat.rob <- cov.rob$cov
  #mu.rob <- cov.rob$center
  mu.rob <- apply(returns,2,location.m)
  max.ret <- max(mu.rob)
  ef.rob <- portfolioFrontier(covmat.rob, mu.rob,
     wmin=0, max.ret=max.ret, n.ret=npoints)
  sd.rob <- ef.rob$sd
  ret.rob <- ef.rob$returns
  # Set Axis Limits
  xlim <- k.sigma*c(0,max(sd,sd.rob))
  ylim <- k.mu*c(0,max(ret,ret.rob))
  xlim <- c(0,.2)
```

```
ylim <- c(-.02, .06)
# Plot Original Classic Efficient Frontier
if(plotit) {
   plot(sd, ret, xlim = xlim, ylim = ylim,
      type = "l", xlab ="RISK (STD. DEV.)",
      ylab ="EXPECTED RETURN")
   lines(sd,ret,lwd = 2)
   title(main="BOOTSTRAP MEAN-VARIANCE EFFICIENT
      FRONTIERS \n Small cap Portfolios")
} #endif plotit

# Compute and Plot Classic Bootstrapped Frontiers
names <- c("SD.MV","MU.MV","SD.TAN","MU.TAN",
   "SHARPE")
out <- matrix(rep(0,5*B),ncol = 5)
dimnames(out) <- list(NULL,names)

for(i in 1:B) {
   ef.classic = efront.nuopt.forboot(
      returns[boot.index[,i],],plotit,
      robust = F, estim = estim, quan = quan,
      rf = rf, mv = mv, tan = tan)
   out[i,] = ef.classic
} # endfor i in 1:B
round(out,3)
# Plot Original Robust Efficient Frontier
if(plotit) {
   plot(sd.rob, ret.rob, xlim = xlim, ylim =
   ylim,
      xlab = "RISK (STD. DEV.)",
      ylab = "EXPECTED RETURN",type = "l")
   lines(sd.rob,ret.rob,lwd = 2)
   title(main="BOOTSTRAP ROBUST EFFICIENT
   FRONTIERS
      \n Small cap Portfolios")
} # endif plotit

# Compute and Plot Robust Bootstrapped Frontiers
names <- c("SD.MV","MU.MV","SD.TAN","MU.TAN",
   "SHARPE")
out.rob <- matrix(rep(0,5*B),ncol = 5)
dimnames(out.rob) <- list(NULL,names)
for(i in 1:B){
   ef.rob = efront.nuopt.forboot(
      returns[boot.index[,i],],plotit,
```

```
        robust = T, estim = estim, quan = quan,
        rf = rf, mv = mv, tan = tan)
     out.rob[i,] = ef.rob
}
  round(out.rob,3)
  par(mfrow = c(2,2))
  boxplot(out.rob[,2]-out[,2],notch = T,
     ylab = c("ROBUST MINUS CLASSICAL"))
  title(main = "MEAN RETURN DIFFERENCE \n Minimum
     Variance Portfolios")
  boxplot(out.rob[,1]-out[,1],notch = T,
     ylab = c("ROBUST MINUS CLASSICAL"))
  title(main = "RISK DIFFERENCE \n Minimum
     Variance Portfolios")
  boxplot(out.rob[,5]-out[,5],notch = T,
     ylab = c("ROBUST MINUS CLASSICAL"),
     ylim = range(out.rob[,5]-out[,5]))
  title(main = "DIFFERENCE OF SHARPE RATIOS")
  par(mfrow = c(1,1))
 }
```

Code 6.12 Bootstrapped Efficient Frontiers and Sharpe Ratios

```
efront.nuopt.forboot <- function(returns, plotit =
  T, robust, estim, quan, rf = 0.005, mv = T, tan =
  F, npoints = 50)
{
  if(robust) {
     covmat <- covRob(returns,estim = estim,
        quan = quan)$cov
     mu <- apply(returns,2,location.m)
  }
  else {
     covmat <- var(returns)
     mu <- apply(returns,2,mean)
  }
  #sd <- apply(returns, 2, stdev)
  ef <- portfolioFrontier(covmat, mu,wmin = 0,
     max.ret = max(mu), n.ret = npoints)
  sdopt <- ef$sd

  muopt <- ef$returns
  # Compute minimum variance portfolio
  port.mv <- c(sdopt[1], muopt[1])
  names(port.mv) <- c("SD.MV", "MU.MV")
  # Compute tangency portfolio
```

```
    sharpe <- (muopt - rf)/sdopt
    iopt <- order(sharpe)[npoints]
    sharpe.max <- sharpe[iopt]

    names(sharpe.max) <- "SHARPE"
    port.tan <- c(sdopt[iopt], muopt[iopt])
    names(port.tan) <- c("SD.TAN", "MU.TAN")
    # Plot results
    if(plotit) {
       lines(sdopt, muopt, lty = 8)
       #points(max(sdopt), max(muopt), pch = ".")
       if(mv == T)
          points(port.mv[1], port.mv[2], pch = 16)
       if(tan == T)
          points(port.tan[1], port.tan[2], pch = 18)
    }
    c(port.mv, port.tan,sharpe.max)
}
```

Code 6.13 NUOPT Efficient Frontiers for Bootstrap Function

```
tickers <- c("TOPP","KWD","HAR","RARE","IBC")
returns.ts <- smallcap.ts[,tickers]
boot.efronts(returns.ts,plotit = T)
```

Code 6.14 Bootstrap Efficient Frontiers Example

The reader is encouraged to experiment with Codes 6.12–6.14 on a variety of portfolios using the returns data set included with this book.

6.9.5 Efficient Frontiers Based on the Classical and Robust Stambaugh Methods

In Section 6.7, we discussed the Stambaugh normal distribution maximum likelihood method of estimating a mean vector and covariance matrix for asset returns having unequal histories and showed how to make the method robust. With these two types of mean vector and covariance matrix estimates in hand, we can proceed as usual to compute both a classical and a robust efficient frontier. Figure 6.47 and Figure 6.48 show the results of doing this using the returns pictured in Figure 6.28. Note the dominant role of Health along with Events in determining the limits of the classical efficient frontier as compared with the dominant role of Equity along with Events in determining the robust efficient frontier. Note also that the reduction in risk when moving from classical sample standard deviations to robust standard deviations is roughly the same for all indices except Health, which exhibits a much more substantial

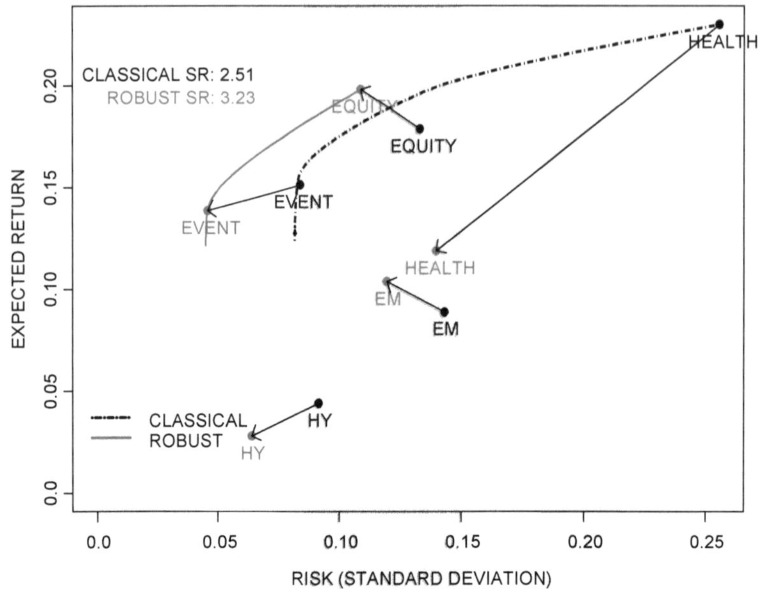

Figure 6.47 Efficient Frontiers for Index Returns with Unequal Histories

reduction in risk. Health also exhibits a substantial reduction in mean return when moving from the sample mean estimate to a robust location estimate. The distribution of portfolio weights along the two efficient frontiers is quite different, with the classical portfolio weights relying more heavily on Health at higher levels of risk and return. By way of contrast, the robust portfolio relies more heavily on Equity at higher levels of return (and risk) and gives Health a zero weight for all possible portfolios.

A glance at the time series of returns for the indices in Figure 6.28 reveals that Health was giving exceptional gains during 1999 (probably by riding the dot-com bubble) and exhibited exceptional losses during the dot-com crash in 2000, followed by relatively lower volatility and unexceptional returns during 2001, 2002, and early 2003. Therefore, it would not be surprising to find many investors preferring the robust efficient frontier for making their investment decision. One can say the robust approach is an automatic method for down-weighting the anomalous returns in the data, thereby calculating an efficient frontier that represents the "normal" behavior of the data. Lacking special information, the "normally" behaving data are the only part of the data that is predictable.

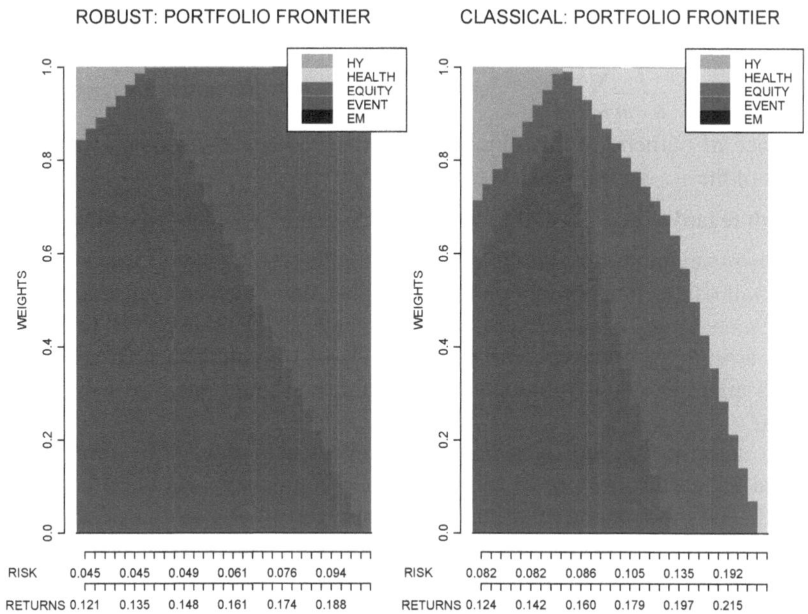

Figure 6.48 Classic and Robust Portfolio Weights for Hedge Fund Indices

6.10 Conditional Value-at-Risk Frontiers: Classical and Robust

A definition of coherent risk measure introduced by Artzner et al. (1997, 1999) was discussed in Section 5.6.2, where it was pointed out that value-at-risk (VaR) is not a coherent risk measure.[16] It is also the case that standard deviation is not a coherent risk measure, which makes the classical Markowitz mean-variance method suspect. On the other hand, it was pointed out in Sections 5.6.2 and 5.6.3 that conditional value-at-risk (CVaR) is a coherent risk measure that leads to a computationally attractive portfolio optimization approach. The question, therefore, is how one might make the CVaR method of portfolio optimization robust. We note that a CVaR optimal portfolio (CVaR portfolio for short) does not involve an estimate of the covariance matrix; it only involves estimation of the mean returns. Thus, at first glance one can make a CVaR robust by simply replacing the sample mean estimates of the μ_i by one of the robust location estimates in Section 6.2. However, outliers can also influence the individual scenario returns $\sum_{i=1}^{n} w_i r_{i,s}$ in $e_s = \max\left[0,\ VaR - \sum_{i=1}^{n} w_i r_{i,s}\right]$ and hence can influence the value of the objective function $CVaR = VaR - \frac{1}{\alpha \cdot m}\sum_{i=1}^{m} e_s$ (see Section 5.6.3). In order to control the influence of outliers on this objective function, one needs to down-weight the values of the $r_{i,s}$ appearing above.

With regard to down-weighting the $r_{i,s}$, there are two questions. First, should one down-weight these values at all? After all, the whole point of using CVaR is to use tail risk to find optimal portfolio weights, and in general one expects that one or more large negative returns in a given asset will tend to reduce the weight in that asset in the optimal CVaR portfolio. Down-weighting such large negative returns might be counterproductive. Second, how should one down-weight the values?

We defer the second question for a moment and address the first. As we pointed out earlier, the use of robust portfolio computations is not a be-all and end-all. The most important value of a robust portfolio is its diagnostic value: when the robust and classical efficient frontiers agree, there is no need to worry, and when they differ the portfolio manager needs to make a decision on which to use based on all the other information available. In the end, the manager may be inclined to take one of the following positions:

> The *robust view,* in which the manager does not trust that *any* past outlier returns will repeat themselves with any degree of predictability and therefore uses a robust portfolio solution since it reflects the behavior of the bulk of the data, excluding outliers;
>
> The *pessimistic view,* in which the manager does not trust that past *positive* outliers are to be expected over the investment horizon, but that past negative returns outliers indicate possible future negative returns outliers, and therefore *down-weights only positive returns outliers*;
>
> The *optimistic view,* in which the manager does not believe that past *negative* returns outliers will repeat themselves over the investment horizon, but that past positive returns outliers indicate possible future positive returns outliers, and therefore *down-weights only negative returns outliers.*

The optimistic or pessimistic views might be taken, for example, when the corresponding outlier or outliers occur only during the early part of the history used to optimize the portfolio or when the manager has other information about some or all of the portfolio assets under consideration.

With these three managerial views in mind, one might first think to use robust distances based on a robust covariance matrix estimate as in Section 6.6 to create weights for each scenario vector $r_s = (r_{1,s}, r_{2,s}, \cdots, r_{n,s}),\ s = 1, 2, \cdots, m$. There are at least two reasons why this is not a highly appealing approach. First, it is not clear how to modify the robust distance approach in a simple manner to accommodate the three distinct manager views. Second, since CVaR portfolio optimization does not require computation of a covariance matrix (which can be computationally burdensome when dealing with a large portfolio), it is attractive to avoid this approach. Consequently, we propose a much simpler approach based on down-weighting outliers in each set of asset returns, one at a time, according to which of the views above the manager takes. While this simpler approach has the deficiency that it is not able to detect and down-weight potentially influential multivariate returns outliers that are not univariate outliers, it has the virtue of simplicity and appears to help in many situations occurring in practice.

We use a special form of the outlier-down-weighting approach, often called **trimming**, which is done as follows: for each set of returns $r_{i,s},\ s = 1, 2, \cdots, m$, compute a robust location estimate $\hat{\mu}_i$, a robust scale estimate $\hat{\sigma}_i$, and the resulting residuals $res_{i,s} = r_{i,s} - \hat{\mu}_i$. For a manager with a robust view, compute the symmetrically trimmed returns

$$\tilde{r}_{i,s} = \begin{cases} r_{i,s}, & |res_{i,s}| \le a \cdot \hat{\sigma}_i \\ \hat{\mu}_i, & |res_{i,s}| > a \cdot \hat{\sigma}_i . \end{cases} \tag{6.20}$$

For a manager with a pessimistic view, compute the positive-trimmed returns

$$\tilde{r}_{i,s} = \begin{cases} r_{i,s}, & res_{i,s} \le a \cdot \hat{\sigma}_i \\ \hat{\mu}_i, & res_{i,s} > a \cdot \hat{\sigma}_i , \end{cases} \tag{6.21}$$

and for a manager with an optimistic view, compute the negative-trimmed returns

$$\tilde{r}_{i,s} = \begin{cases} \hat{\mu}_i, & res_{i,s} \le -a \cdot \hat{\sigma}_i \\ r_{i,s}, & res_{i,s} > -a \cdot \hat{\sigma}_i . \end{cases} \tag{6.22}$$

We use a default value $a = 3$, which results in about 2.6 symmetric trimmed residuals out of 1000 for normally distributed returns and about 1.3 out of 1000 for the two other cases.

Code 6.15 gives the S-PLUS code for the function `trimmed.returns` that computes the above trimmed returns:

```
trimmed.returns <- function(x, view = "robust",
  a = 3)
```

```
{
  p <- ncol(x)
  n <- nrow(x)
  ind <- matrix(0,nrow = n, ncol = p)
  for(j in 1:p) {
     mu <- location.m(x[,j])
     scale <- scale.tau(x[,j])
     resid <- x[,j]- mu
     if(view == "pessimistic") {
        x[resid > a*scale,j] <- mu
        ind[resid > a*scale,j] <- 1
     }
     else if(view == "optimistic") {
        x[resid < -a*scale,j] <- mu
        ind[resid < -a*scale,j] <- 1
     }
     else if(view == "robust") {
        x[abs(resid) > a*scale,j] <- mu
        ind[abs(resid) > a*scale,j] <- 1
     }
     else
        stop("view must be \"pessimistic\",
           \"optimistic\", or \"robust\"")
     ind <- data.frame(ind)
     names(ind) <- names(x)
  }

  list(returns.trimmed = x,ind = ind)
}
```

Code 6.15 Trimmed Returns

Once the matrix of returns has been trimmed using the function above, you compute a CVaR efficient frontier in a manner similar to that used in Section 5.6.3. We note that in Section 5.6.3 the function `CVaR.frontier` computes a frontier for long-only portfolios and for target returns ranging from the minimum return sample mean return to the maximum return sample mean. Consequently, the resulting frontier typically contains inefficient positions. In order to compute a CVaR efficient frontier, we first need to find the global minimum CVaR portfolio. This is easily done as follows. Remove the line of code that specifies a return constraint from the function `CVaR.model` in Section 5.6.3:

```
Sum(mu.bar[i]*w[i],i) == mu.target.
```

Name the resulting function `CVaR.globalmin.model`. Now you need to modify the function `CVaR.portfolio` by replacing the code line

```
call(CVaR.model)
```

in that function with

```
call(CVaR.globalmin.model)
```

Name the new function `CVaR.globalmin.portfolio`. You should also add the lines

```
S <- as.matrix(S)
dimnames(S) <- NULL
```

at the beginning of these two functions, the first to allow a data frame as an argument to the function and the second to remove the column names, which the current version of NUOPT does not accept. Since the function `CVaR.globalmin.portfolio` differs in a few other places from `CVaR.portfolio`, Code 6.16 gives the code for the revised version of `CVaR.globalmin.portfolio`:

```
CVaR.globalmin.portfolio <- function(S, alpha)
{
  S <- as.matrix(S)
  dimnames(S) <- NULL
  call(CVaR.globalmin.model)
  CVaR.system <-
     System(CVaR.globalmin.model,S,alpha)
  solution <- solve(CVaR.system, trace=T)
  weight <- solution$variable$w$current
  w <- as.matrix(weight)
  mu <- as.matrix(apply(S,2,mean))
  mu.min <- as.numeric(t(w)%*%mu)
  risk <- solution$objective

  return(mu.min,risk)
}
```

Code 6.16 Global Minimum CVaR Portfolio

The argument `alpha` above, and in what follows, specifies the tail probability for CVaR.

Code 6.17 provides `CVaR.eff.frontier`, a slightly modified version of the function `CVaR.frontier` in Section 5.6.3, that computes and optionally plots a CVaR efficient frontier:

```
CVaR.eff.frontier <- function(S,alpha,n.pf,plot=T)
{
  call(CVaR.globalmin.portfolio)
  call(CVaR.portfolio)
  Risk <- matrix(0, ncol=1, nrow=n.pf)
  Return <- matrix(0, ncol=1, nrow=n.pf)
  x <- CVaR.globalmin.portfolio(S,alpha)
  mu.min <- x$mu.min
  mu.max <- max(apply(S,2,mean))
  mu.range <- seq(mu.min, mu.max,
     (mu.max-mu.min)/(n.pf-1))
  x <- CVaR.portfolio(S, alpha, mu.target=mu.min)
  weight <- x$weight
  Risk[1] <- x$risk
  Return[1] <- mu.min

  for(i in 2:n.pf) {
     x <- CVaR.portfolio(S,alpha,
        mu.target=mu.range[i])
     Risk[i] <- x$risk
     Return[i] <- mu.range[i]
     weight <- cbind(weight,x$weight)
  }
  # Convert CVaR to a positive quantity
  Risk = - Risk
  if(plot) {
     par(mfrow=c(1,2))
     plot(Risk, Return, type="b",xlab = "RISK",
        ylab = "RETURN")
     title("MEAN vs. CVaR EFFICIENT FRONTIER")
     barplot(weight,legend = names(S))
     title("FRONTIER PORTFOLIOS")
  }
  list(Risk = Risk, Return = Return, Weights =
     weight)
}
```

Code 6.17 CVaR Efficient Frontier

Note that in order to use the function `CVaR.eff.frontier`, as in the examples below, you need to have already created the functions `CVaR.model` (Code 5.9) and `CVaR.portfolio` (Code 5.10).

Figure 6.49 shows the time series of five years of monthly returns for four stocks, with tickers BKE, GG, GYMB, and KRON, for which we will compute a CVaR efficient frontier.

The plots shown in Figure 6.49 were made by extracting the stocks from the `smallcap.ts` time series object and using the `seriesPlot` (see Code 6.7) function as follows:

```
tickers <- c("BKE","GG","GYMB","KRON")
returns.ts <- smallcap.ts[,tickers]
seriesPlot(returns.ts,strip.text =
  colIds(returns.ts),
  trellis.args = list(as.table = T,type = "l"),
  one.plot = F)
```

Note the positive and negative returns outliers and that depending upon the investor's knowledge he may wish to take any one of the three views we have proposed. For example, the investor may know that the large outlier in the GG returns was associated with a singular event that is not expected to recur in the next year or two and may feel that the two positive outlier returns in GYMB present an overly optimistic view of future performance. Consequently, he will want a CVaR optimal portfolio constructed with a pessimistic view. On the other hand, the investor may feel that most of the negative outliers are sufficiently far in the past, or, as in the case of KRON, are left in the dust by a strong positive trend, leading him to construct a CVaR portfolio based on a positive view. Finally, the investor may feel that the positive and negative outliers tend to have cancelling effects and have no good reason to believe they will occur in the next year. Consequently, he will compute a CVaR portfolio based on a robust view that reflects the behavior of the bulk of the returns.

Assuming we have created the `returns.ts` object as above, we can compute the standard CVaR efficient frontier and barplot of weights in Figure 6.50 with the commands (the computation takes noticeably longer than a mean-variance optimal frontier):

```
returns = returns.ts@data
CVaR.eff.frontier(returns, alpha = .05, n.pf = 10)
```

You can now use `trimmed.returns` to compute the CVaR efficient frontier based on one of the three possible manager views. For the robust view, use:

```
returns.tr <-
     trimmed.returns(returns)$returns.trimmed
```

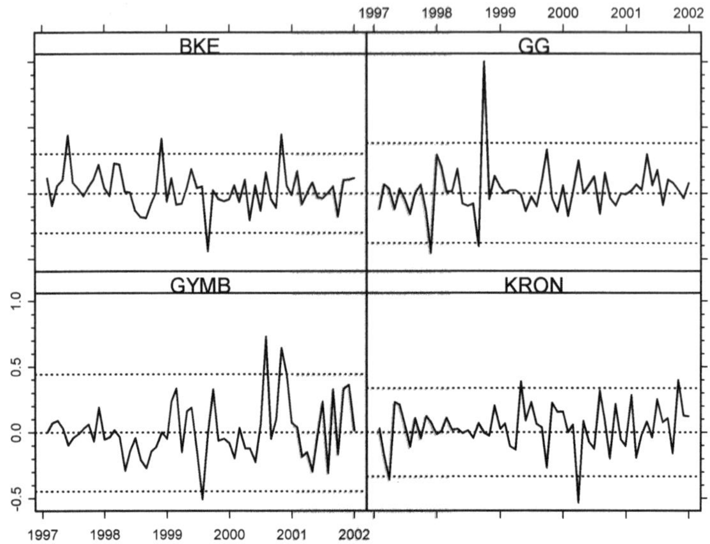

Figure 6.49 Time Series of Returns for Four Stocks

```
CVaR.eff.frontier(returns.tr,alpha = .05,n.pf = 10)
```

The results are the CVaR efficient frontier and portfolio weights in Figure 6.51, where upon careful inspection you notice the change in efficient frontier location and the change in weights relative to those in Figure 6.50.

Of course, what we really want to have is overlaid efficient frontiers and displayed values of the mean return and CVaR for each stock as in Section 6.8, where standard deviations were used as the risk measure. We can easily do this by modifying Code 6.11. The only additional function specific to the CVaR context that we need is a little function to compute the CVaR of each set of returns, rather than the standard deviation, so that we can display each stock in the mean return versus CVaR coordinates. This simple function is given in Code 6.18.

```
CVaR.simple <- function(x, alpha = .05) {
  k = floor(length(x)*alpha)
  #convert CVaR to a positive quantity
  -mean(sort(x)[1:k])
}
```

Code 6.18 CVaR Computation Function

Now we show a few examples before providing the code needed to produce them. Figure 6.52 provides an overlaid version of the CVaR efficient frontiers of

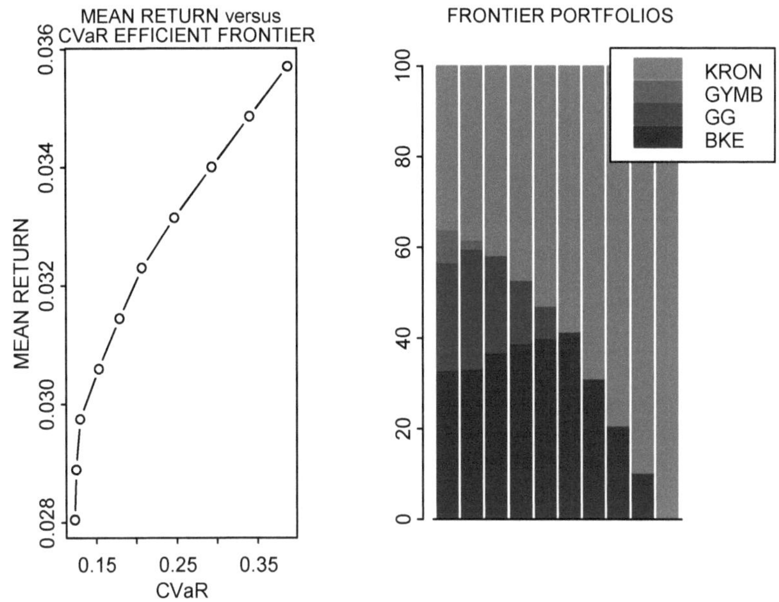

Figure 6.50 CVaR Efficient Frontier and Weights for Stock Returns of Figure 6.48

Figure 6.50 and Figure 6.51 along with the individual stocks' sample means and sample CVaRs.

In Figure 6.52 one sees that the robust CVaR efficient frontier yields larger returns than the standard CVaR efficient frontier, with the difference increasing with increasing CVaR risk. Figure 6.53 and Figure 6.54 show the results for pessimistic and optimistic views, respectively.

Figure 6.53 shows that the manager with a pessimistic view gets lower returns than the standard CVaR manager at all levels of CVaR below that of KRON, with the largest difference at the global minimum CVaR values. Finally, Figure 6.54 shows that the manager with an optimistic view gets mean returns that are uniformly higher than the classic CVaR returns at all levels of CVaR. Note also that the gain of the optimistic CVaR portfolio in Figure 6.53 relative to the robust view CVaR portfolio in Figure 6.54 is most substantial for the smaller values of CVaR. This is quite understandable based on the differences in trimming for these two views and the fact that both negative and positive outlier returns are evident in Figure 6.49.

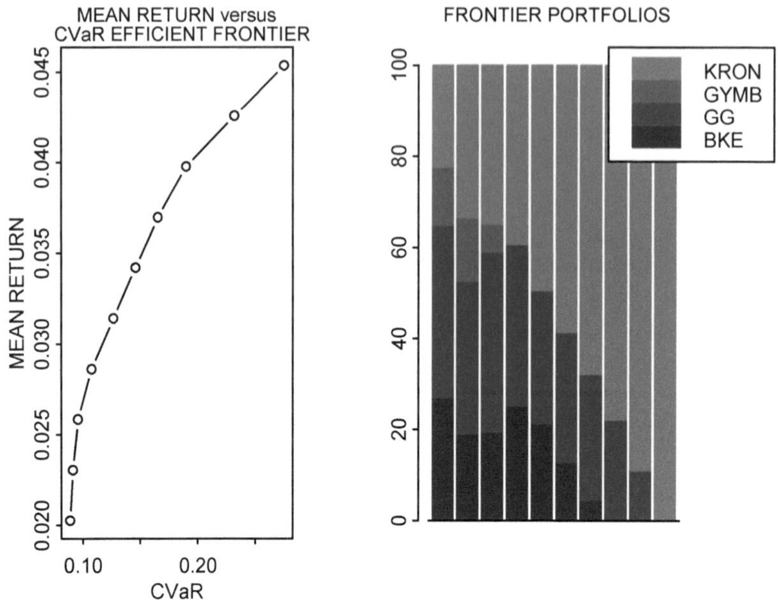

Figure 6.51 CVaR Efficient Frontier and Weights with Robust View Trimming of Returns

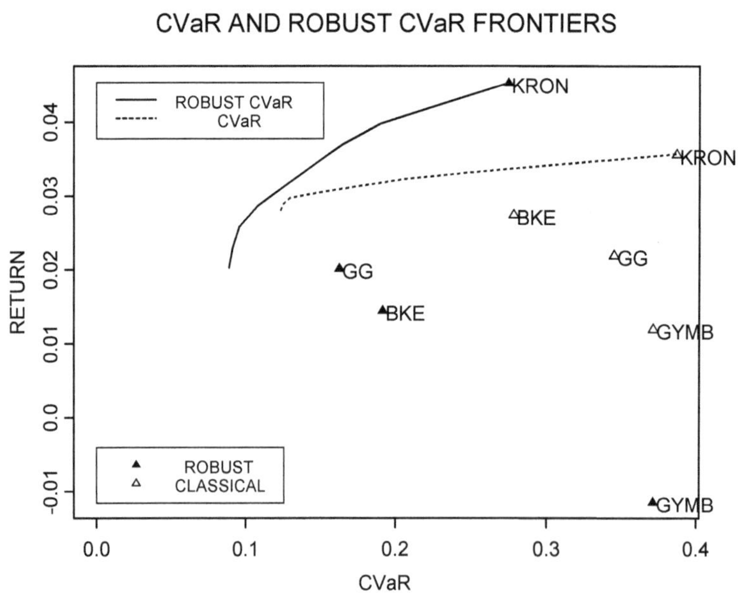

Figure 6.52 CVaR and Robust View CVaR Efficient Frontiers

Code 6.19 gives the code for making the plots above (just change `view = "robust"` to `view = "pessimistic"` and `view = "optimistic"` to get Figure 6.53 and Figure 6.54, respectively):

```
tickers <- c("BKE","GG","GYMB","KRON")
returns.ts <- smallcap.ts[,tickers]
returns <- returns.ts@data
p <- ncol(returns)
# Parameters
alpha <- .05
view <- "robust"
display.letters <- F
display.points <- T
display.names <- T
n.pf <- 10
plot.weights <- F
series.plots <- F
# Time Series Plots
if(series.plots)
  seriesPlot(returns.ts,
     strip.text = colIds(returns.ts),
     trellis.args = list(as.table = T,type = "l"),
     one.plot = F)
# Compute Standard CVaR Efficient Frontier
ef.cvar <- CVaR.eff.frontier(returns, alpha, n.pf,
  plot = F)
ef.cvar$Weights
# Compute Robust CVaR Efficient Frontier
ret.trimmed <-
  trimmed.returns(returns,view)$returns.trimmed
ef.cvar.robust <- CVaR.eff.frontier(ret.trimmed,
  alpha, n.pf, plot = F)
ef.cvar.robust$Weights

# Plot Efficient Frontiers
if(display.letters || display.points) {
  mu1 <- apply(returns,2,mean)
  mu2 <- apply(ret.trimmed,2,mean)
  cvar1 <- apply(returns,2,CVaR.simple,alpha=alpha)
  cvar2 <- apply(ret.trimmed,2,CVaR.simple,
     alpha=alpha)
  xlim <- range(ef.cvar$Risk,ef.cvar.robust$Risk,
     cvar1,cvar2,0)
```

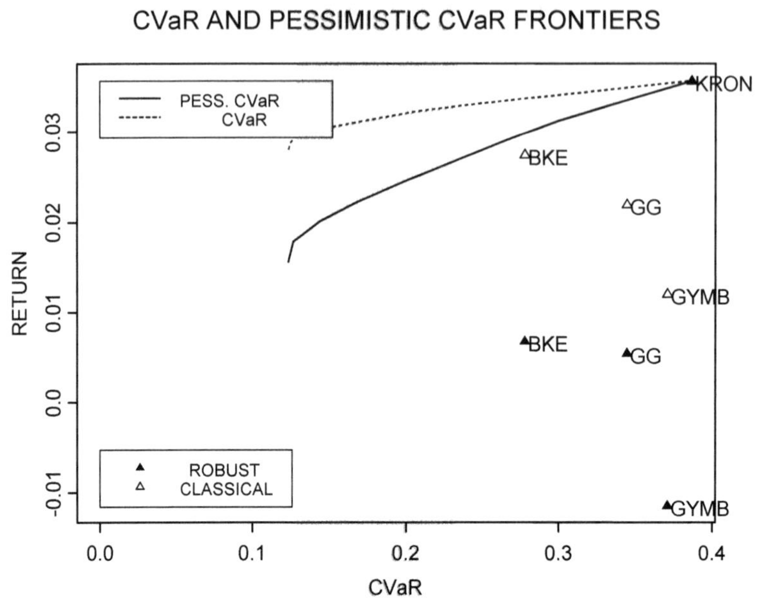

Figure 6.53 CVaR and Pessimistic View CVaR Efficient Frontiers

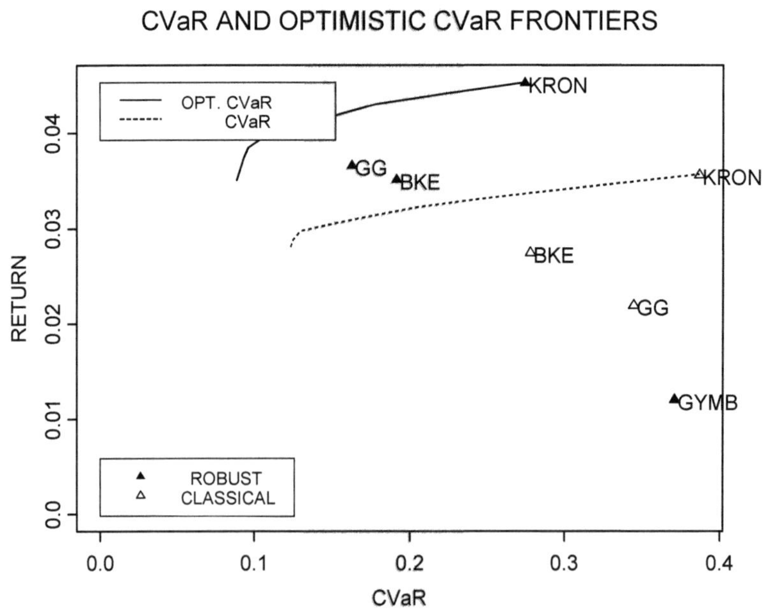

Figure 6.54 CVaR and Optimistic View CVaR Efficient Frontiers

```
  ylim <-
     range(ef.cvar$Return,ef.cvar.robust$Return,
     mu1,mu2,0)
}
else {
  xlim <- range(ef.cvar$Risk,ef.cvar.robust$Risk,0)
  ylim <- range(ef.cvar$Return,
     ef.cvar.robust$Return,0)
}
plot(ef.cvar$Risk,ef.cvar$Return,xlim=xlim,
  ylim=ylim,  type = "n",xlab="CVaR",ylab="RETURN")
lines(ef.cvar$Risk,ef.cvar$Return,lty = 8,lwd = 2)
lines(ef.cvar.robust$Risk,ef.cvar.robust$Return,
  lwd=2)
if(view == "robust") {
  title(main="CVaR AND ROBUST CVaR FRONTIERS")
}
else if(view == "pessimistic") {
  title(main="CVaR AND PESSIMISTIC CVaR FRONTIERS")
}
else if(view == "optimistic") {
  title(main="CVaR AND OPTIMISTIC CVaR FRONTIERS")
}

# Add Frontiers Legend
x <- xlim[1]+.0*(xlim[2]-xlim[1])
y <- ylim[2]-.0*(ylim[2]-ylim[1])
if(view == "robust") {
  leg.names <- c("ROBUST CVaR","         CVaR")
}
else if(view == "pessimistic") {
  leg.names <- c("PESS. CVaR","         CVaR")
}
else if(view == "optimistic") {
  leg.names <- c("OPT. CVaR","         CVaR")
}
legend(x,y,leg.names,lty=c(1,8), lwd = 2)

# Plot Stock Mu's and CVaR's and Add Legend
if(display.letters){
  for(i in 1:p) {
     points(cvar1[i],mu1[i],pch = letters[i])
     points(cvar2[i],mu2[i],pch = LETTERS[i])
  }
```

```
  if(display.names) {
     text(cvar1 + 0.002, mu1, names(mu1), adj= 0)
     text(cvar2 + 0.002, mu2, names(mu2), adj= 0)
  }
  x <- xlim[1]+.15*(xlim[2]-xlim[1])
  y <- ylim[1]+.10*(ylim[2]-ylim[1])
  leg.names <- c("A,B,. Robust Mu's,CVaR's",
     "a,b,. Classic Mu's,CVaR's")
  text(x,y,leg.names[1])
  text(x,y-.002,leg.names[2])
}
if(display.points) {
  points(cvar1,mu1,pch = 2)
  points(cvar2,mu2,pch = 17)
  if(display.names) {
     text(cvar1 + 0.002, mu1, names(mu1), adj= 0)
     text(cvar2 + 0.002, mu2, names(mu2), adj= 0)
  }
  x <- xlim[1]+.00*(xlim[2]-xlim[1])
  y <- ylim[1]+.13*(ylim[2]-ylim[1])
  leg.names <- c("  ROBUST","CLASSICAL")
  legend(x,y,leg.names, marks = c(17,2))
}
# Plot Portfolio Weights for Both Efficient
# Frontiers
if(plot.weights) {
  par(mfrow = c(1,2))
  barplot(ef.cvar$Weights,legend = names(returns))
  title(main = "CVaR FRONTIER WEIGHTS")
  barplot(ef.cvar.robust$Weights,
     legend = names(returns))
  if(view == "robust") {
     title(main="ROBUST CVaR FRONTIER WEIGHTS")
  }
  else if(view == "pessimistic") {
     title(main="PESSIMISTIC CVaR FRONTIER
        WEIGHTS")
  }
  else if(view == "optimistic") {
     title(main="OPTIMISTIC CVaR FRONTIER WEIGHTS")
  }
  par(mfrow = c(1,1))
}
```

Code 6.19 CVaR Efficient Frontier Plots

6.10.1 Manager Views and What-If Predictive Diagnostics

We want to stress the predictive and diagnostic nature of the optimal CVaR efficient frontiers above. First of all, any kind of efficient frontier calculation is an "in-sample" predictive model in the sense that when one selects a particular portfolio to use based on such an efficient frontier, one is predicting that the efficient frontier is a reasonable predictor of future mean return-risk trade-off. Second, it may often be the case that an asset manager is at first unwilling to take any one of the views proposed above (robust, pessimistic, optimistic) or even a "standard" view that there should be no outlier treatment. In such cases, the manager may derive considerable diagnostic benefit from a "what-if" analysis based on computing CVaR efficient frontiers for each of the views. If the results are all in reasonable agreement, there is little cause to worry about influential outliers. But if there are substantial differences between two or more of the views, such analysis can act as a catalyst for the asset manager to investigate any unusual positive or negative returns that may be influencing the results. This may lead the manager to adopt a particular view based on a deeper knowledge of what has caused the events and his belief about whether they are likely to occur during the investment horizon.

We also remark that the simple return trimming method used here for CVaR portfolio optimization could also be used as a preprocessor for Markowitz mean-variance optimization. The additional computational burden of the robust covariance matrix calculation (in the case of a large number of assets) could then be avoided by instead using the classical mean and covariance matrix estimates based on univariate trimmed returns.

6.10.2 Choice of Alpha

Section 5.6.1 shows that estimates of CVaR are much more variable than estimates of VaR, which are in turn much more variable than estimates of standard error. This is natural in that CVaR is based on the mean value of the smallest $\alpha\%$ of the returns. Consequently, one can expect that CVaR efficient frontiers will be more variable than Markowitz mean-variance frontiers. (This could be checked with bootstrap experiments.) One way to mitigate this problem is to increase the value of α, say, to .1 or .2. Note that when $\alpha = .5$, CVaR is the mean of the returns below the median, which differs from lower semi-variance only by using the median in place of the overall mean. This choice may be interesting to asset managers in view of its very simple interpretation as the average losses below the median loss. Of course, the resulting portfolio weights do not pay as much relative attention to the downside returns when using larger values of alpha as when using smaller values of alpha. Examples of CVaR frontiers for $\alpha = .05$, .2, and .5 are provided in Figure 6.55, which is produced with Code 6.20.

```
tickers <- c("BKE","GG","GYMB","KRON")
returns <- smallcap.ts[,tickers]@data
cvar05 <- CVaR.eff.frontier(returns, alpha = .05,
  n.pf = 10, plot = F)
cvar2 <- CVaR.eff.frontier(returns, alpha = .2,
  n.pf = 10, plot = F)
cvar5 <- CVaR.eff.frontier(returns, alpha = .5,
  n.pf = 10, plot = F)
xlim <- range(cvar05$Risk,cvar2$Risk,cvar5$Risk)
ylim <-
     range(cvar05$Return,cvar2$Return,cvar5$Return)
plot(cvar05$Risk,cvar05$Return, type = "l",
  xlim = xlim, ylim = ylim,
  xlab = "CVaR", ylab = "MEAN RETURNS")
title(main = "CVaR EFFICIENT FRONTIERS FOR VARIOUS
  ALPHAS")
lines(cvar05$Risk,cvar05$Return, lty = 1, lwd =2)
lines(cvar2$Risk,cvar2$Return, lty = 4, lwd = 2)
lines(cvar5$Risk,cvar5$Return, lty = 8, lwd = 2)
leg.names = c("ALPHA = .05","ALPHA = .1",
  "ALPHA = .5")
legend(.26,.029,legend = leg.names, lty = c(1,4,8),
  lwd = 2)
```

Code 6.20 CVaR Efficient Frontiers for Different Values of Alpha

From Figure 6.55 it is clear that the efficient frontiers are not obtained from one another simply by a uniform scaling with respect to CVaR values. It will be useful to compare the portfolio weight profile for each of the frontiers (Exercise 11).

6.11 Influence Functions for Portfolios

Influence functions are powerful statistical tools for characterizing three key aspects of an estimator, namely: (a) the influence of individual data values on the estimator, particularly the influence of outliers; (b) the maximum bias of the estimator caused by small fractions of outliers; and (c) the asymptotic variance of the estimators.[17] While influence functions have been widely applied in statistics (see, for example, Hampel et al., 1986), there has been almost no use of them in finance and in portfolio construction in particular. This section

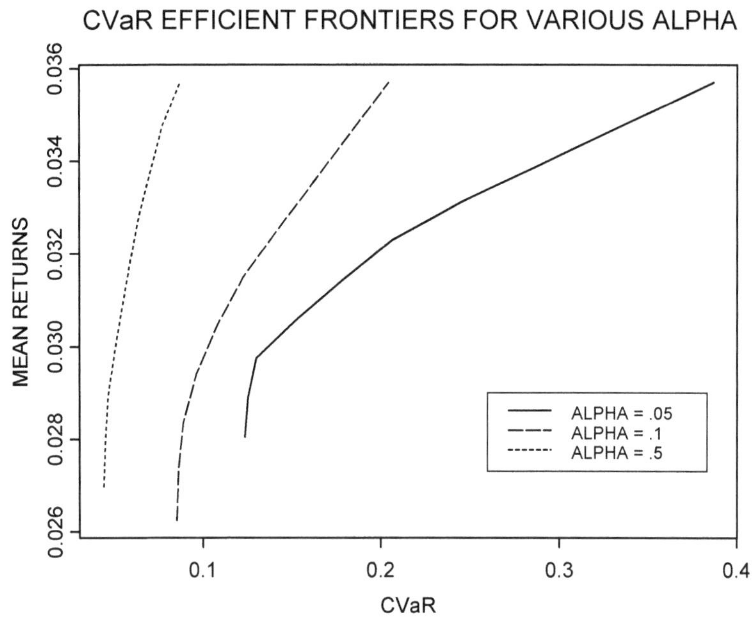

Figure 6.55 CVaR Efficient Frontiers for Three Values of Alpha

introduces the basic definitions and uses of influence functions and provides some initial applications to portfolio construction for the purpose of sensitivity analysis.

6.11.1 Introduction to Influence Functions

Influence functions have both finite sample and asymptotic versions. The intuitive motivation for the asymptotic influence function (**influence function** for short) comes from a finite sample form, one version of which is as follows. Let $\hat{\theta}_n = \hat{\theta}_n(x)$ be an estimator of a parameter θ based on a sample of data $\boldsymbol{x} = (x_1, x_2, \cdots, x_n)$ of size n, and let x be an additional data point. To fix ideas you can think of $\hat{\theta}_n$ as a sample mean of returns or a sample standard deviation estimate of volatility for a particular stock. Then an empirical influence function (EIF) of $\hat{\theta}_n$ at $\boldsymbol{x}$ is the function of x given by

$$\mathrm{EIF}(x; \hat{\theta}_n, \boldsymbol{x}) = (n+1) \cdot \left(\hat{\theta}_n(x, \boldsymbol{x}) - \hat{\theta}_n(\boldsymbol{x}) \right), \tag{6.23}$$

where the factor $n+1$ is used to normalize the result across sample sizes.[18] For a few simple estimators (such as the sample mean and sample median), it is possible to compute an analytical expression for the EIF, but for most robust estimators this is not possible. However, one can numerically compute the EIF for a "typical" sample of returns $\boldsymbol{x}$ and plot the results as a function of the value of the additional data point x. Thinking of a normally distributed sample of returns, one can choose $\boldsymbol{x}$ to be normal random numbers whose mean and standard deviation are those of typical returns. A better approach that eliminates the variability of the random sample and gives a good rendition for small as well as large sample sizes is to let the sample $\boldsymbol{x}$ be the quantiles of a normal distribution.

Figure 6.56 displays the EIFs of the sample mean, the sample median, a 10% trimmed mean, and the optimal location M-estimate obtained from `lmRob`, as described in Section 6.3, using twenty quantiles of a standard normal distribution for the prototype data sample $\boldsymbol{x}$. The main messages from these EIFs are:

(a) The unbounded character of the EIF for the sample mean reflects the fact that a single outlier can cause an arbitrarily large influence on the sample mean.
(b) All the other estimates have bounded EIFs, reflecting the fact that a single outlier can only influence the estimate by a limited amount.
(c) The median has a nearly discontinuous EIF, which reflects the fact that the median has a certain "roughness" character.
(d) Very large outliers have zero EIF values for the optimal M-estimate, reflecting the fact that this estimate "rejects" sufficiently large outliers.

Note that the trimmed mean, which at first glance appears to discard large outliers, does not in fact accomplish this goal in the same effective way as the optimal M-estimate. The computations and plots of Figure 6.56 are produced by Code 6.21.

```
n <- 20
probs <- (1:n - .5)/n
xn <- qnorm(probs)
x <- seq(-5,5,.1)
k <- length(x)
eif <- rep(0,k)
par(mfrow = c(2,2))
par(pty = "s")
for(i in 1:k) {
  eif[i] <- (n+1)*(mean(c(x[i],xn))-mean(xn))
}
```

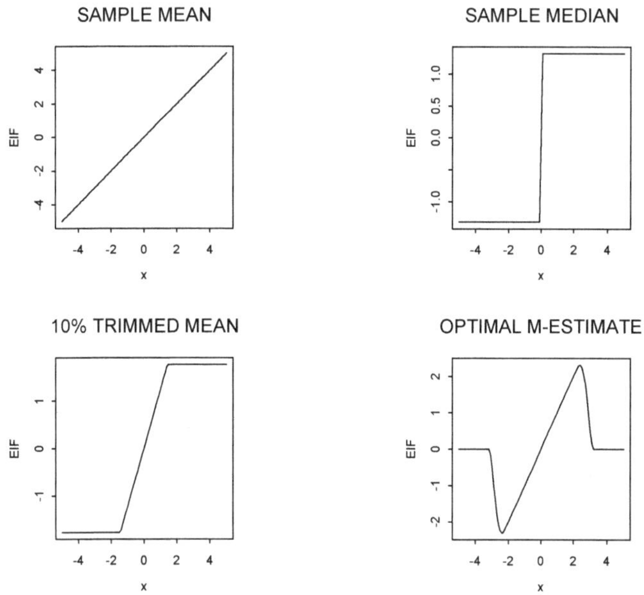

Figure 6.56 EIFs for Sample Mean, Median, Trimmed Mean, and Huber M-Estimate

```
plot(x,eif,type = "l",ylab = "EIF",
  main = "SAMPLE MEAN")
for(i in 1:k) {
  eif[i] <- (n+1)*(median(c(x[i],xn))-median(xn))
}
plot(x,eif,type = "l",ylab = "EIF",
  main = "SAMPLE MEDIAN")
for(i in 1:k) {
  eif[i] <- (n+1)*(mean(c(x[i],xn),trim=.1)-
     mean(xn,trim=.1))
}
plot(x,eif,type = "l",ylab = "EIF",
  main = "10% TRIMMED MEAN")
for(i in 1:k) {
  eif[i]=(n+1)*(coef(lmRob(c(x[i],xn)~1))-
     coef(lmRob(y~1)))
}
plot(x,eif,type = "l",ylab = "EIF",
  main = "OPTIMAL M-ESTIMATE")
```

Code 6.21 EIFs for Mean Returns Estimates

You can easily compute EIFs for the sample standard deviation volatility estimate by replacing the estimator function (e.g., `mean` in one of the "for" loops in Code 6.21 with the function `stdev`. You can do likewise for a robust scale estimate of volatility, for example, by replacing the function `mean` with the function `scale.tau`. The results, shown in Figure 6.57, show that (a) a single outlier has rapidly unbounded influence on the conventional standard deviation volatility estimate, and (b) outliers have only a bounded influence on the tau-scale estimate. With respect to the sample standard deviation EIF, note that when the additional data point is close to zero, which is the value of the sample mean for the prototype $\boldsymbol{x}$, the additional data point is an **inlier** that results in a negative value of the EIF because such an inlier decreases the value of the standard deviation. Note that the tau-scale volatility estimate EIF has a shape similar to that of the standard deviation in the central region from –2 to 2, except that small values of the added data point do not have so much negative influence except right at zero.

As the sample size tends towards infinity, the empirical influence function will (under regularity conditions) converge to the influence function defined as follows. It is assumed that the data are generated by a parametric distribution F_θ, where θ is the true parameter value. Let $\theta(F)$ be the asymptotic value of the parameter estimate $\hat{\theta}_n = \hat{\theta}_n(\boldsymbol{x})$ when the data have an arbitrary distribution function F, and note that typically $\theta(F) \neq \theta$ for an arbitrary F. It is also assumed that the parameter estimate is **consistent** (i.e., $\hat{\theta}_n$ converges to θ_o in probability), and that $\theta = \theta(F_\theta)$.[19] We represent the asymptotic version of the prototype sample $\boldsymbol{x}$ for an arbitrary distribution F and additional data point x by the mixture distribution

$$F_\gamma = (1-\gamma)\cdot F + \gamma \cdot \delta_x, \tag{6.24}$$

where γ is the mixture probability and δ_x is a point mass probability distribution located at x. The influence function $\mathrm{IF}(x) = \mathrm{IF}(x;\theta(F),F)$ is defined as

$$\mathrm{IF}(x) = \lim_{\gamma \downarrow 0} \frac{\theta(F_\gamma) - \theta(F_o)}{\gamma}. \tag{6.25}$$

Equivalently, $\mathrm{IF}(x)$ is the derivative of $\theta(F_\gamma)$ evaluated at $\gamma = 0$[20]:

$$\begin{aligned}\mathrm{IF}(x) &= \lim_{\gamma \downarrow 0} \frac{\theta(F_\gamma) - \theta(F_o)}{\gamma} \\ &= \frac{d}{d\gamma}\theta(F_\gamma)\Big|_{\gamma=0}.\end{aligned} \tag{6.26}$$

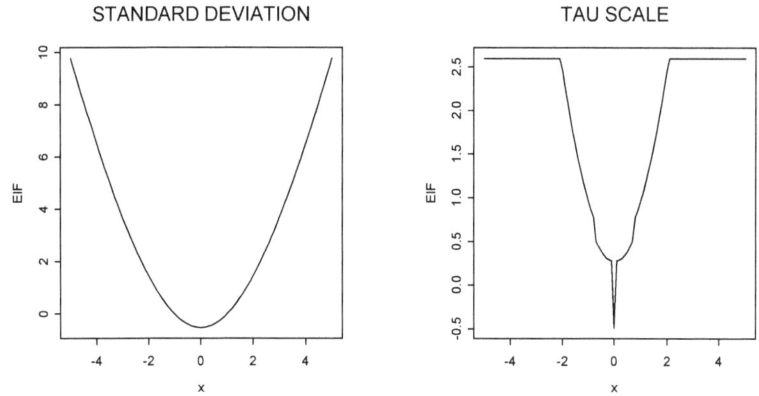

Figure 6.57 EIFs of Standard Deviation and Robust Tau-Scale Volatility Estimates

A very important property of the influence function is that it provides an approximate expression for the large sample bias

$$BIAS(x;\gamma) \triangleq \theta(F_\gamma) - \theta(F_\theta) = \theta(F_\gamma) - \theta$$

due to a small fraction γ of data located at x,

$$BIAS(x;\gamma) \approx \gamma \cdot \mathrm{IF}(x), \tag{6.27}$$

where the influence function $\mathrm{IF}(x)$ is evaluated at F_θ. There is evidence that for good robust estimators this local linear approximation of the bias is reasonably good for fractions γ as large as 5% to 10%, which covers many situations of importance in finance.[21]

An easy computation shows that the influence function of the sample mean estimate $\bar{x}$ is

$$IF(x;\bar{x}) = x - \mu, \tag{6.28}$$

where $\theta = \mu$ is the true mean value (Exercise 13). A slightly more involved computation shows that the influence function of an M-estimate $\hat{\mu}$ of location (see Equation 6.2) is

$$IF(x;\hat{\mu}) = \frac{\psi_{\mu,s}(x)}{E_F\left[\psi_{\mu,s}(x)\right]}, \tag{6.29}$$

where

$$\psi_{\mu,s}(x) = \psi\left(\frac{x-\mu}{s}\right) \tag{6.30}$$

and μ and s are the true location and scale parameters of the returns (Exercise 14). This influence function is the same as $\psi_{\mu,s}(x)$ except for the scale factor in the denominator. So, for the optimal bias-robust location M-estimate ψ function shown in the right-hand plot of Figure 6.6, the corresponding approximating EIF (shown in the lower-right-hand plot of Figure 6.56) differs from Figure 6.6 only by a scale factor and a small finite-sample approximation error.

6.11.2 Influence Functions for Sample Mean and Covariance Estimates of Returns

The definition of influence function extends in the obvious way to the case of multivariate data sets of returns and multidimensional parameter estimates. The functional representation of the sample mean estimate $\bar{\boldsymbol{x}} = (\bar{x}_1, \bar{x}_2, \cdots, \bar{x}_k)'$ is $\boldsymbol{\mu}(F) = \int \boldsymbol{x} dF(\boldsymbol{x})$. A straightforward calculation leads to (Exercise 15)

$$\text{IF}(\boldsymbol{x}; \bar{\boldsymbol{x}}) = \boldsymbol{x} - \boldsymbol{\mu}. \tag{6.31}$$

For the sample covariance matrix estimate

$$\hat{\boldsymbol{\Omega}} = \frac{1}{n}\sum_{i=1}^{n}(\boldsymbol{x}_i - \bar{\boldsymbol{x}})(\boldsymbol{x}_i - \bar{\boldsymbol{x}})', \tag{6.32}$$

one finds that the influence function is (Exercise 16)

$$\text{IF}(\boldsymbol{x}; \hat{\boldsymbol{\Omega}}) = (\boldsymbol{x} - \boldsymbol{\mu})(\boldsymbol{x} - \boldsymbol{\mu})' - \boldsymbol{\Omega}. \tag{6.33}$$

For the case of a single asset where $k = 1$, the covariance matrix estimate becomes a sample variance, $\boldsymbol{\Omega} = \sigma^2$, and we get the influence function of the sample variance:

$$\text{IF}(x; \hat{\sigma}^2) = (x - \mu)^2 - \sigma^2. \tag{6.34}$$

This shows that the influence of an outlier on the sample variance is quadratically unbounded, while that of an "inlier" at $x = \mu$ is $-\sigma^2$. It is easy to see from its definition that the influence function of the sample standard deviation $\hat{\sigma} = \sqrt{\hat{\sigma}^2}$ is

$$\mathrm{IF}(x;\hat{\sigma}) = \frac{1}{2\sigma}\left((x-\mu)^2 - \sigma^2\right),$$

which for a standard normal distribution is $.5 \cdot (x^2 - 1)$. In the left-hand plot of Figure 6.57, you see that the EIF for the sample standard deviation is a good approximation to $\mathrm{IF}(x;\hat{\sigma})$ for the standard normal distribution.

6.11.3 Influence Functions for Mean-Variance Optimal Tangency Portfolios

The Markowitz mean-variance efficient frontier for unconstrained portfolios is completely determined by the mean vector and covariance matrix of the returns. In order to estimate quantities on the efficient frontier, such as the mean return and risk of the global minimum variance and tangency portfolios and the maximum Sharpe ratio, one substitutes estimates of the mean vector and covariance matrix for their true values in the corresponding formulas. For the case of the tangency portfolio with return vector $\boldsymbol{\mu}$ and an excess return vector $\boldsymbol{\mu}_e = \boldsymbol{\mu} - \mathbf{1} \cdot r_f$, the weights vector, mean return, and variance estimates, respectively, are

$$\hat{\mathbf{w}}_T = \frac{\hat{\boldsymbol{\Omega}}^{-1}\hat{\boldsymbol{\mu}}_e}{\mathbf{1}'\hat{\boldsymbol{\Omega}}^{-1}\hat{\boldsymbol{\mu}}_e}, \tag{6.35}$$

$$\hat{\mu}_T = \hat{\boldsymbol{\mu}}_e'\hat{\mathbf{w}}_T = \frac{\hat{\boldsymbol{\mu}}_e'\hat{\boldsymbol{\Omega}}^{-1}\hat{\boldsymbol{\mu}}_e}{\mathbf{1}'\hat{\boldsymbol{\Omega}}^{-1}\hat{\boldsymbol{\mu}}_e}, \tag{6.36}$$

$$\hat{\sigma}_T^2 = \hat{\mathbf{w}}_T'\hat{\boldsymbol{\Omega}}\hat{\mathbf{w}}_T = \frac{\hat{\boldsymbol{\mu}}_e'\hat{\boldsymbol{\Omega}}^{-1}\hat{\boldsymbol{\mu}}_e}{\left(\mathbf{1}'\hat{\boldsymbol{\Omega}}^{-1}\hat{\boldsymbol{\mu}}_e\right)^2}. \tag{6.37}$$

The functional representation of the quantities above, needed for computing their influence functions, is obtained by replacing the mean and covariance matrix estimates by their functional representations $\boldsymbol{\mu}_e(\gamma) = \int \boldsymbol{x}\, dF_\gamma(\boldsymbol{x}) - \mathbf{1} \cdot r_f$ and $\boldsymbol{\Omega}(\gamma) = \int (\boldsymbol{x} - \boldsymbol{\mu}(\gamma))(\boldsymbol{x} - \boldsymbol{\mu}(\gamma))'\, dF_\gamma(\boldsymbol{x})$. This results in corresponding tangency portfolio functional representations for $\mathbf{w}_T(\gamma)$, $\mu_T(\gamma)$, and $\sigma_T^2(\gamma)$, from which one can compute influence functions. The influence function for the tangency portfolio weights is

$$\mathrm{IF}_{W_T}(\boldsymbol{x}) = \frac{d}{d\gamma}\frac{\boldsymbol{\Omega}^{-1}(\gamma)\boldsymbol{\mu}_e(\gamma)}{\mathbf{1}'\boldsymbol{\Omega}^{-1}(\gamma)\boldsymbol{\mu}_e(\gamma)}\bigg|_{\gamma=0} \tag{6.38}$$

$$= \frac{\mathbf{1}'\boldsymbol{\Omega}^{-1}\boldsymbol{\mu}_e - \boldsymbol{\Omega}^{-1}\boldsymbol{\mu}_e\mathbf{1}'}{(\mathbf{1}'\boldsymbol{\Omega}^{-1}\boldsymbol{\mu}_e)^2}\cdot\left[-\boldsymbol{\Omega}^{-1}\mathrm{IF}_{\Omega}(\boldsymbol{x})\boldsymbol{\Omega}^{-1}\boldsymbol{\mu}_e + \boldsymbol{\Omega}^{-1}\mathrm{IF}_{\mu}(\boldsymbol{x})\right],$$

as can be verified by careful calculation (Exercise 17). Since the influence function $\mathrm{IF}_{\Omega}(\boldsymbol{x})$ for the covariance matrix is quadratically unbounded in $\boldsymbol{x}$ (i.e., the value increases quadratically with size $\boldsymbol{x}$), the same is true of the influence function for the weights given above. Since $\mu_T(\gamma) = \boldsymbol{\mu}_e'(\gamma)\cdot\mathbf{w}_T(\gamma)$, one gets (Exercise 18)

$$\mathrm{IF}_{\mu_T}(\boldsymbol{x}) = \boldsymbol{\mu}_e'\cdot\mathrm{IF}_{W_T}(\boldsymbol{x}) + (\boldsymbol{x}-\boldsymbol{\mu})'\cdot\mathbf{w}_T. \tag{6.39}$$

One can also show that (Martin and Zhang, 2004)

$$\mathrm{IF}_{\sigma_T}(\boldsymbol{x}) = \frac{1}{2\sigma_T}\mathrm{IF}_{\sigma_T^2}(\boldsymbol{x}).$$

$$= \frac{1}{2\sigma_T}\left[2\cdot\mathrm{IF}_{W_T}'(\boldsymbol{x})\cdot\boldsymbol{\Omega}\cdot\mathbf{w}_T + \mathbf{w}_T'\cdot\mathrm{IF}_{\Omega}(\boldsymbol{x})\cdot\mathbf{w}_T\right] \tag{6.40}$$

Notice that when the additional data value $\boldsymbol{x}$ is located at the mean return vector $\boldsymbol{\mu}$, (i.e., $\boldsymbol{x} = \boldsymbol{\mu}$), the influence function for the weight vector is

$$\mathrm{IF}_{W_T}(\boldsymbol{\mu}) = \frac{\mathbf{1}'\boldsymbol{\Omega}^{-1}\boldsymbol{\mu}_e - \boldsymbol{\Omega}^{-1}\boldsymbol{\mu}_e\mathbf{1}'}{(\mathbf{1}'\boldsymbol{\Omega}^{-1}\boldsymbol{\mu}_e)^2}\cdot\left[-\boldsymbol{\Omega}^{-1}(-\Omega)\boldsymbol{\Omega}^{-1}\boldsymbol{\mu}_e + \mathbf{0}\right]$$

$$= \frac{\mathbf{1}'\boldsymbol{\Omega}^{-1}\boldsymbol{\mu}_e\boldsymbol{\Omega}^{-1}\boldsymbol{\mu}_e - \boldsymbol{\Omega}^{-1}\boldsymbol{\mu}_e\mathbf{1}'\boldsymbol{\Omega}^{-1}\boldsymbol{\mu}_e}{(\mathbf{1}'\boldsymbol{\Omega}^{-1}\boldsymbol{\mu}_e)^2} \tag{6.41}$$

$$= \mathbf{0}.$$

One can also show that $\mathrm{IF}_{\mu_T}(\boldsymbol{x}) = 0$ and $\mathrm{IF}_{\sigma_T}(\boldsymbol{x}) = -\frac{1}{2}\sigma_T$. These results are intuitively appealing in that when a data value $\boldsymbol{x}$ is located at the mean returns vector $\boldsymbol{\mu}$, it is reasonable that it have no perturbing influence on the portfolio weights or mean return, and the negative influence on the portfolio risk is consistent with that of the influence of an inlier on a simple standard deviation. Some other interesting results on portfolio influence functions may be found in Martin and Zhang (2004).

We test the use of the influence formulas above on a very simple case where we know the tangency portfolio solution immediately and can interpret the influence function results most easily. Suppose we have 60 monthly observations of returns on two stocks with equal mean returns, equal volatilities, and a diagonal covariance matrix constructed as follows:

```
> stock.names <- c("STOCK.1", "STOCK.2")
> mutest <- c(0.01, 0.01)
> names(mu.test) <- stock.names
> Vtest <- diag(c(0.003, 0.003))
> dimnames(Vtest) <- list(stock.names, stock.names)
> mutest
[1] 0.01 0.01
> Vtest
        STOCK.1 STOCK.2
STOCK.1   0.003   0.000
STOCK.2   0.000   0.003
```

Our influence function code (Code 6.22) incorporates a function `port.tan` for computing the tangency portfolio:

```
port.tan <- function(V, mu, rf)
{
  p <- length(mu)
  one <- rep(1, p)
  mue <- mu - rf
  a <- solve(V, mue)
  Vinv <- solve(V)
  d <- as.numeric(inprod(one, a))
  wts <- a/d
  n <- as.numeric(qform(mue, Vinv))
  muep <- n/d
  sigma <- n^0.5/abs(d)
  sr <- sign(d) * n^0.5
  list(Weights = wts, Mu.e = muep, Sigma = sigma,
     "Sharpe Ratio" = sr)
}
```

Code 6.22 Unconstrained Tangency Portfolio

Code 6.23 gives three simple functions for computing an inner product and quadratic form in `port.tan`, as well as an outer product function needed for our influence function calculations:

```
inprod <- function(x, y) {
  as.numeric(t(matrix(x)) %*% matrix(y))
}

qform <- function(x, A) {
  x = matrix(x)
```

```
  as.numeric(t(x) %*% A %*% x)
}

outprod = function(x, y) {
  matrix(x) %*% t(matrix(y))
}
```

Code 6.23 Inner Product, Outer Product, and Quadratic Form Functions

Now we compute the influence function value for a data point $\boldsymbol{x} = (.065, -.065)$ on the tangency portfolio weights, mean return, risk (standard deviation), and Sharpe ratio, assuming a risk-free rate of .004 and using $\gamma = 1/61$ in the bias approximation $BIAS(x;\gamma) \approx \gamma \cdot \text{IF}(x)$:

```
rf <- .004
Gamma <- 1/61
x <- c(.0648,-.0448)

if.tan(x, Vtest, mutest, rf, print.results = T,
  Gamma, IF.relative = T)

* TANGENCY PORTFOLIO WEIGHTS AND PERFORMANCE *

     WT1      WT2   MUE    SIGMA SHARPE
     0.5      0.5 0.006 0.0387 0.1549

* INFLUENCE FUNCTION OF TANGENCY PORTFOLIO *

[1] "GAMMA = 0.016"

         X1      X2  WT1   WT2 MUE  SIGMA SHARPE
[1,] 1.001 -1.001 0.15 -0.15   0 -0.008  0.008
```

The tangency portfolio weights `(WT1, WT2)`, excess mean return `(MUE)`, risk `(SIGMA)`, and Sharpe ratio `(SHARPE)` provided as the first output above are exactly as one expects. In the last line of output the `X1` and `X2` are standardized versions of $\boldsymbol{x} = (.065, -.065)$ (i.e., they represent one standard deviations of the added data from the mean returns). Because we used the optional argument `IF.relative = T`, the influence function values `WT1`, `WT2`, `MUE`, `SIGMA`, `SHARPE` are all relative to the true tangency portfolio values. For example, the `WT1` value of `.15` represents an increase in the weight of 15%, etc.

Code 6.24 gives the S-PLUS code for the calculation above.

```
if.tan = function(x, V, mu, rf, print.results = T,
  Gamma = 1, IF.relative = F)
{
  # Compute tangency portfolio for IF.relative
  # calculations
  tanport <- port.tan(V, mu, rf)
  wts.tan <- tanport$Weights
  mu.tan <- tanport$Mu
  sigma.tan <- tanport$Sigma
  sr.tan <- tanport$"Sharpe Ratio"
  opt.tan <- c(wts.tan, mu.tan, sigma.tan, sr.tan)

  p <- length(mu)
  # Optionally print tangency portfolio
  if(print.results) {
     names(opt.tan) = c(paste("WT", 1:p, sep = ""),
        "MUE","SIGMA", "SHARPE")
        cat("\n* TANGENCY PORTFOLIO WEIGHTS AND
        PERFORMANCE *\n\n")
     print(round(opt.tan, 4)); cat("\n")
  }
  # Compute excess returns, vector of 1's,
  # Vinverse*mu.e, Vinverse
  mu.e <- mu - rf
  one <- rep(1, p)
  xcent <- x - mu
  a <- solve(V, mu.e)
  Vinv <- solve(V)
  # Compute influence functions
  IF.cov <- outprod(xcent, xcent) - V
  A <- Vinv %*% matrix(xcent) -
     Vinv %*% IF.cov %*% a
  B <- 1/inprod(one, a)
  IF.wts <- B * A - (B^2) * a * inprod(one, A)
  IF.mu <- inprod(mu.e, IF.wts) +
     inprod(xcent,wts.tan) #IF mu.e
  IF.sigma <- (2*t(wts.tan) %*% V %*% IF.wts +
     qform(wts.tan,IF.cov))/(2*sigma.tan)
  IF.sr <- (sr.tan^2 - 1 +
     (inprod(a,xcent)-1)^2)/(2*sr.tan)
  IF <- Gamma * c(t(IF.wts), IF.mu, IF.sigma,
     IF.sr)
  if(IF.relative) {
     IF <- IF/opt.tan
```

```
        digits <- 3
        x <- xcent/as.vector(diag(V)^0.5)
    }
    else {
        digits <- 4
    }
    IF <- t(as.matrix(c(x, IF)))
    dimnames(IF)[[2]] <- c(paste("X", 1:p, sep = ""),
        paste("WT", 1:p, sep = ""), "MUE", "SIGMA",
        "SHARPE")

    if(print.results) {
        cat("\n* INFLUENCE FUNCTION OF TANGENCY
            PORTFOLIO *\n\n")
        cat(paste("GAMMA =", round(Gamma,3)))
        cat("\n\n")
    }
    round(IF, digits)
}
```

Code 6.24 Influence Function of Tangency Portfolio

Now we can use the function `if.tan` in a function `if.comp2`, given in Code 6.25 to compute contour plots of the tangency portfolio influence function for weights, mean return, and risk for a range of values of $\boldsymbol{x} = (x_1, x_2)$. The resulting influence function contours shown in Figure 6.58 for the tangency portfolio weight w_1 show that when X1 = X2 there is no influence bias on the weights, while X1 > X2 results in positive bias and X2 > X1 results in negative bias.

Figure 6.59 and Figure 6.60 show corresponding results for the tangency portfolio mean return and standard deviation.

Figure 6.59 shows that when X1+X2 > 0, the influence bias in mean return is positive, and when X1 + X2 < 0, this bias is negative. Figure 6.60 shows that the standard deviation of the tangency portfolio increases with increasing values of |X1+X2|.

The results in Figure 6.58, Figure 6.59, and Figure 6.60 may seem intuitively reasonable based on the very simple structure of the assumed mean vector and covariance matrix. In any event, by careful examination of the expressions for the influence functions for the weight vectors, mean return, and standard deviation, one can easily check that the results are qualitatively correct (Exercise 19).

Here is a short bit of code that uses the function `if.comp2` in Code 6.25 to carry out the computation above and produce the plot of Figure 6.58:

```
rf <- .004
Gamma <- 1/61
if.comp2(Vtest, mutest, rf, plotchoice = "if.wts",
  k=3, nsteps=9, Gamma,IF.relative = T)
```

To get the plots of Figure 6.59 and Figure 6.60, use the optional arguments `plotchoice = "if.mu"` and `plotchoice = "if.sigma"`, respectively.

```
if.comp2 <- function(V, mu, rf, plotchoice =
      "if.wts", k=5, nsteps=3, Gamma = 1,
      IF.relative = T)
{
  sigma <- diag(V)^0.5
  rng1 <- c(mu[1]-k*sigma[1], mu[1]+k*sigma[1])
  rng2 <- c(mu[2]-k*sigma[2], mu[2]+k*sigma[2])
  x1 <- rep(seq(rng1[1], rng1[2], length = nsteps),
     times = nsteps)
  x2 <- rep(seq(rng2[1], rng2[2], length = nsteps),
     times = rep(nsteps, times = nsteps))
```

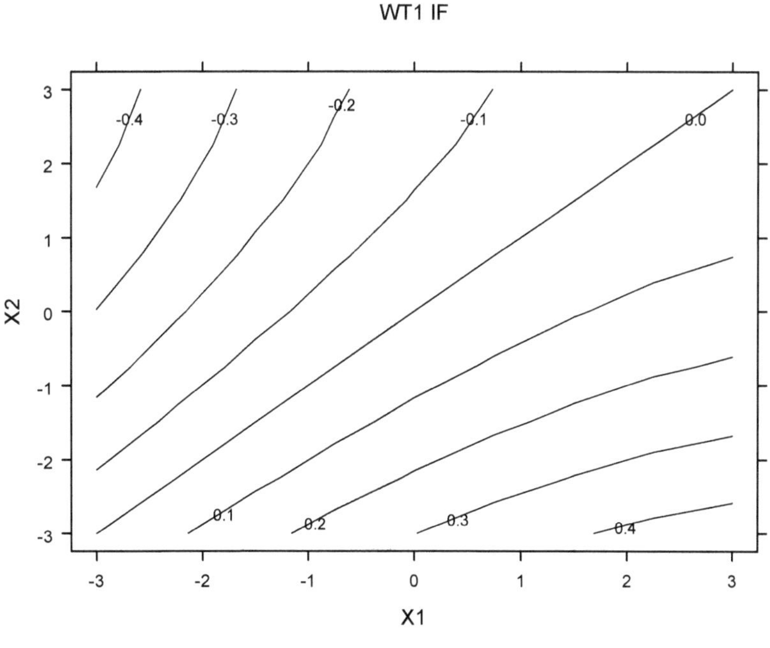

Figure 6.58 Influence Function Contours for Tangency Portfolio Weights

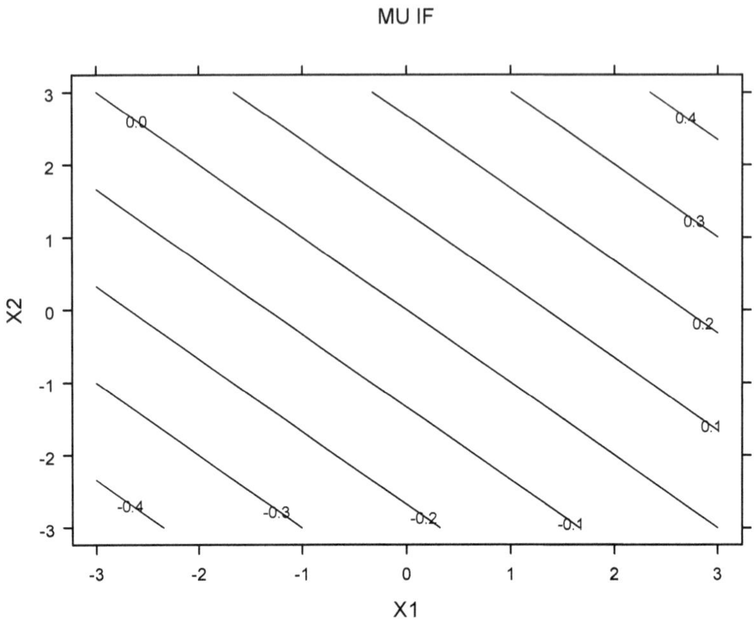

Figure 6.59 Influence Function Contours for Tangency Portfolio Mean Return

```
x <- cbind(x1, x2)
IF <- data.frame(matrix(rep(0, 7 * nsteps^2),
   ncol = 7))

for(i in 1:(nsteps^2)) {
   IF[i, ] <- if.tan(x[i, ], V, mu, rf,
      print.results = F, Gamma, IF.relative)
}

names(IF) <- dimnames(if.tan(x[1, ],V,mu,rf,
   print.results = F))[[2]]

if(plotchoice == "if.wts") {
   contourplot(WT1 ~ X1*X2, data = IF,
      main = "WT1 IF")
}
else if(plotchoice == "if.mu") {
   contourplot(MUE ~ X1*X2, data = IF,
      main = "MU IF")
}
else if(plotchoice == "if.sigma") [
```

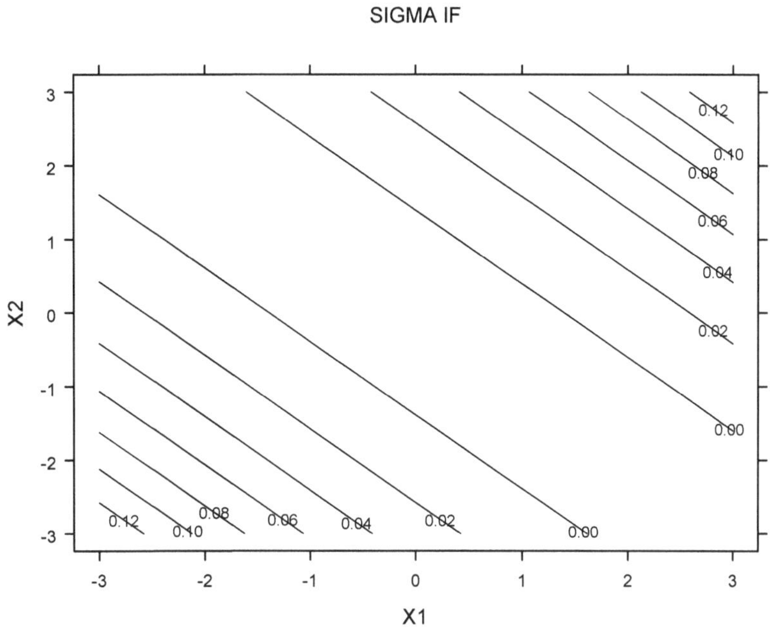

Figure 6.60 Influence Function Contours for Tangency Portfolio Risk

```
        contourplot(SIGMA ~ X1*X2, data = IF,
           main = "SIGMA IF")
    }
}
```

Code 6.25 IF Plots for Tangency Portfolio

We recommend that the reader experiment with the tangency portfolio influence functions above using a more realistic set of mean returns and covariance matrix that may arise in practice. For example, you might use the following monthly mean returns, covariance matrix, and volatilities (Exercise 20)[22]:

```
> mu3
  SP500 GOV.BOND SMALL.CAP
 0.0101   0.0043    0.0137

> V3
                SP500 GOV.BOND SMALL.CAP
    SP500 0.00325  0.00023   0.00420
 GOV.BOND 0.00023  0.00050   0.00019
SMALL.CAP 0.00420  0.00019   0.00764
```

```
> round(sqrt(diag(V3)),4)
[1] 0.0570 0.0224 0.0874
```

You can get the correlation matrix from the covariance matrix with the function `cov.to.corr` in Code 6.26.

```
cov.to.corr <- function(v) {
   dimnames <- dimnames(v)
   sigma <- diag(v)^0.5
   s.inv <- diag(1/sigma)
   rho <- s.inv %*% v %*% s.inv
   dimnames(rho) <- dimnames
   rho
}
```

Code 6.26 Convert Covariance Matrix to Correlation Matrix

```
> round(cov.to.corr(V3),4)
               SP500 GOV.BOND SMALL.CAP
     SP500 1.0000    0.1804    0.8429
  GOV.BOND 0.1804    1.0000    0.0972
 SMALL.CAP 0.8429    0.0972    1.0000
```

You can compute the tangency influence function for all three of these assets with `if.tan`, but when using `if.comp` you need to work with two at a time. Note that if you use SP500 and SMALL.CAP, you have a high correlation, but if you use GOV.BOND and SMALL.CAP, you have a small correlation. Do not be surprised if you find that the influence of additional outlier data is greater in the case of high correlation between assets.

6.11.4 Influence of Outliers on the Sharpe Ratio

It turns out that the influence function for the Sharpe ratio of the tangency portfolio has the simple form (see Martin and Zhang, 2004)

$$\mathrm{IF}_{SR}(x) = \frac{1}{2SR}\left(SR^2 + 1 - (y-1)^2 \right), \tag{6.42}$$

where

$$y = \boldsymbol{\mu}_e \boldsymbol{\Omega}^{-1}(\boldsymbol{x} - \boldsymbol{\mu}) \tag{6.43}$$

Thus, somewhat surprisingly, the value of $\mathrm{IF}_{SR}(\boldsymbol{x})$ is bounded above by $\left(SR^2+1\right)/2SR$ (i.e., an outlier can cause at most a bounded positive bias in the maximum Sharpe ratio). But since y can be made arbitrarily large by making the size of $\boldsymbol{x}$ arbitrarily large, an outlier can cause an arbitrarily large negative bias of the Sharpe ratio. This represents a fundamental kind of asymmetry in the potential influence of outliers on the maximum Sharpe ratio. The reader is encouraged to explore the influence of outliers on the maximum Sharpe ratio by appropriately modifying the function `if.comp2`, first in such a way as to explore the influence of outliers in one coordinate direction at a time (i.e., in one set of returns at a time).

6.11.5 Empirical Influence Functions for Unconstrained and Constrained Portfolios

It is a straightforward matter to compute influence functions for weights, mean return, and risk of other unconstrained portfolios (such as the global minimum variance portfolio). Since the method is an infinitesimal one, providing a valid approximation for small fractions of influential data, it could in principle be applied to a portfolio optimized under constraints, provided none of the constraints are binding and the influence of an outlier does not cause some of the constraints to become binding. The obvious first approach to computing influence functions for constrained portfolios (e.g., long-only portfolios, sector constraints, etc.) is to compute empirical influence functions (EIFs) along the lines of the calculations leading to Figure 6.56. While this will be computationally burdensome since one has to solve a QP or LP problem for each outlier data position, it can no doubt be done. A first step would be to compute the empirical influence function for the tangency portfolio quantities in the unconstrained case as a check on its accuracy. Our earlier results with EIFs for simple location estimates were quite encouraging, but since the tangency portfolio estimates are much more complicated, this initial check will be useful. A deeper study of the QP optimization structure might lead to some efficient methods of computing influence functions or EIFs for optimal portfolios under constraints. This is a topic for further research.

Exercises

1. Use the `plot` function in a `for` loop to plot time series of returns for each microcap stock in `microcap.ts` so that you can get an overall visual grasp of the behavior of each stock in this microcap group with respect to outliers and time-varying volatility. Use the `plot` function with arguments as in Code 6.1, except use a generic first argument `returns`, setting `returns` equal to one of the stock's returns series for each cycle of the `for` loop. Use the graphics command `par(mfrow(n1,n2))` to automate plotting a number of time series of returns per page for each of the market cap groups. Use the S-PLUS Windows toolbar menu choice "Options > Graph Options" and select "Every Graph" from the Auto Pages drop-down list in the "Traditional Graphics" region of the dialog, as this will result in each new page of plots appearing as a separate page of the Graph Sheet. Make similar plots for each of the groups `smallcap.ts`, `midcap.ts`, and `largecap.ts`.

2. Make Q-Q plots of returns for a few time series from each of the four market cap groups, both with and without 95% simulation envelopes. Automate making Q-Q plots for all stock returns in the four market cap groups in a manner similar to the way you plotted all the time series in Problem 1.

3. Compute classical and robust means and standard deviations of returns for all the stocks in the microcap group, and plot the means versus standard deviations for these estimates. Use one plotting symbol for the classical estimates and another plotting symbol for the robust estimates. Add the ticker symbols as text labels. (Warning: with a lot of stocks there may be confusing overlap of these text labels.)

4. Use the classical and robust EWMA volatility estimate and UMT functions (Code 6.4 and Code 6.5) on a few stocks from each of the four market cap categories. What do you conclude about the prevalence of outliers and overestimation of volatility following isolated outliers? What do you conclude about the potential usefulness of a robust UMT?

5. Modify the UMT Code 6.4 to use $\hat{\sigma}_{t-1}$ in place of $\hat{\sigma}_t$ in the test statistic and evaluate the improvement in detecting initial outlier returns (unusual price movements). Do you think the modified method is adequate for detecting a returns outlier that occurs shortly after another returns outlier?

6. Convert Code 6.6 into an S-PLUS function that makes an array of plots (like that of Figure 6.17) of the input returns series, overlays the LS and robust regression lines, and places the legend and annotations automatically. Then run the function on a few of the stock returns in each of the time series data sets `microcap.ts`, `smallcap.ts`, `midcap.ts`, `largecap.ts`. You might want to do this with a `for` loop as in Problem 1, in which case you can do it for all twenty stock returns in each of the market-cap groups. (Alternatively, the Trellis graphics functions in S-Plus, if you are familiar with them, provide a clean way to do this.) For which stock returns do the least squares and robust betas differ significantly because of the presence of outliers?

7. Alphas are of considerable interest to investors because they represent excess returns obtainable from investing in a given stock over and above what is predicted by the Capital Asset Pricing Model (CAPM). Modify the code in Exercise 6 so that you obtain the least squares and robust alphas and their standard errors. For what firms do the least squares and robust alphas differ significantly? What do you conclude about the usefulness of robust alphas?

8. Explore small subsets of four to six multivariate stock returns in one of the time series data sets `microcap.ts`, `smallcap.ts`, `midcap.ts`, `largecap.ts`, and `midcapD.ts` by making pairwise scatterplots and classical versus robust correlations to find a subset that exhibits one or more substantial differences between the classic and robust correlations. Explain why the substantial difference or differences between classical and robust correlations are reasonable given the nature of the data. For such a subset, compute and display classical and robust Mahalanobis distances, and comment on any interesting aspects of the multidimensional outliers found.

9. Use the multivariate returns data set of Exercise 8 for this exercise. Compute and display classical and robust mean-variance efficient frontiers, and discuss how you would use the results to guide a portfolio selection investment decision.

10. Use the multivariate returns data of Exercises 8 and 9 or other similarly interesting multivariate returns data for this problem. Compute and display bootstrapped classical and robust efficient frontiers and boxplots of paired differences in classical and robust Sharpe ratios. What is the effect of changing the number of bootstrap samples? How many bootstrap samples appear to be adequate to you?

11. Compute and display the portfolio weights for the three efficient frontiers in Figure 6.55. What do you find? Do the results make sense?

12. Use the multivariate returns data of Exercises 8 and 9 or other similarly interesting multivariate returns data for this problem. For alpha equal to .05, compute robust, optimistic, and pessimistic CVaR optimal portfolios and display the results of each along with the classical CVaR optimal portfolio (also using an alpha value of .05). Discuss how you would use the results to guide a portfolio selection investment decision. Do likewise for alpha values of .1, .2, and .5.

13. Derive the expression for the influence function of the sample mean of a single-asset return.

14. Derive the expression for the influence function of a location M-estimate.

15. As a slight extension to Exercise 13, derive the expression for the influence function of a sample mean of multivariate returns.

16. Derive the expression for the influence function of the sample covariance matrix.

17. Verify the expression for $\mathrm{IF}_{W_T}(\boldsymbol{x})$.

18. Verify the expression for $\mathrm{IF}_{\mu_T}(\boldsymbol{x})$.

19. Verify that the results in Figure 6.58, Figure 6.59, and Figure 6.60 are qualitatively correct.

20. Compute and display tangency portfolio influence functions assuming that the mean returns vector is `mu3` and the covariance matrix is `V3`. Explain why the results are reasonable.

Endnotes

[1] This is true even when adjusting for time-varying volatility with a GARCH model. See, for example, calculations by Menn (2003).
[2] If you wanted to get a data frame object with a single variable rather than a vector object (the default simple data object in S-PLUS), you would drop the subscript part "`[,1]`" in the following two commands. Some functions in S-PLUS will work on S-PLUS V4 time series objects directly (e.g., `mean` and `var` will do so, though with slightly different output formats, but `stdev` will not). Many functions that do not work on S-PLUS V4 time series objects will work on the data frame component of a time series object that you would obtain as described above. Unfortunately, some functions, such as `stdev`, will not even work on a data frame but will work on a vector object, which is why we elected to extract the data from `returns.ts` as a vector in this example.
[3] The functions `qqnorm` and `qqline` are other examples of functions that do not work on data frames.
[4] See the online manual for the Robust Library in S-PLUS 6 for further details on the Q-Q plot simulation envelopes. See also Atkinson (1985).
[5] We remark that when there is a lot of data (e.g., a few hundred observations), one may be able to fit a heavy-tailed distribution to asset returns with a reasonably high degree of accuracy (see, for example, Rachev and Mittnik, 2000). Then although one cannot predict just when a future outlier will occur, one can be certain that a certain number will occur on average over a certain time interval, and this is can be very useful in the context of risk management.
[6] M-estimators are generalizations of maximum likelihood estimators introduced by Huber (1964) for estimates of location and by Huber (1973) for regression. See also Huber (2004) and Hampel et al. (1986)
[7] It should be noted that the weights in this case are data-dependent, which means that this weighted least squares equation is nonlinear.
[8] This is similar to the fact that the classical *t*-test lacks robustness of power toward heavy-tailed deviations from normality.
[9] Capital Asset Pricing Model.
[10] This general form of shrinkage estimator was justified using a Bayesian argument by Vasicek (1973) and by Blume (1971) using an argument of regression toward the mean.
[11] See the "Current Commercial Practice" section of Martin and Simin (2003) for further details.
[12] The horizontal dashed lines are located at ± 2.5 times the scaled median absolute deviation about the median (MADM) robust scale estimate, which is an approximately unbiased estimate of the standard deviation when the returns are normally distributed.
[13] The modified function `seriesPlot`, and a modified function `panel.superpose.ts` that is called by `seriesPlot`, are provided in the code archive for this book (see the Preface).
[14] The script `multi.start.function.ssc`, written by Heiko Bailer and included in the code archive for this book (see the Preface), implements the classical and robust versions of the Stambaugh method.

[15] We note that there is nothing wrong with using a parametric bootstrap so long as the parametric model is adequate. The problem with the Jorion and Michaud approach is the evaluation of the performance of their resampled portfolios using the original sample mean and covariance rather than the resampled means and covariances.
[16] See also Bradley and Taqqu (2003).
[17] For a thorough introduction to influence functions in the context of robust statistics, see Hampel et al. (1986), which discusses all the key properties of influence functions. Here we focus primarily on the influence of outliers and the approximate bias caused by outliers.
[18] This is one of several possible definitions of a finite sample influence function. See, for example, Mallows (1975).
[19] The latter condition is called Fisher consistency in the statistical literature. See, for example, Huber (2004) or Hampel et al. (1986).
[20] This is a directional or Gateaux derivative of the functional $\theta(F)$ at F_o in the "direction" F_γ.
[21] See, for example, Hampel et al. (1986) and the maximum bias curves in Martin, Yohai, and Zamar (1989).
[22] These are the values used in Rockafellar and Uryasev (2000) but have been slightly rounded.

7 Bayes Methods

7.1 The Bayesian Modeling Paradigm

Let $\boldsymbol{\theta} = (\theta_1, \theta_2, \cdots, \theta_p)$ be a vector of model parameters for a financial model (e.g., $\boldsymbol{\theta} = (\theta_1, \theta_2) = (\mu, \sigma^2)$ represents the mean and variance for the returns of a single asset, $\boldsymbol{\theta} = (\theta_1, \theta_2) = (\alpha, \beta)$ represents the intercept and slope in the single-factor market model, or $\boldsymbol{\theta} = (\theta_1, \theta_2, \cdots, \theta_K) = (\mu_1, \mu_2, \cdots, \mu_K)$ represents a set of mean returns in a mean-variance portfolio optimization problem). For $t = 1, 2, \cdots, T$, let $\mathbf{r}_t = (r_{t1}, r_{t2}, \cdots, r_{tK})$ be the row vector of returns on p assets, and let $\mathbf{R}$ be the $T \times K$ matrix of such row vectors. Bayes' Theorem for probability densities states that the **posterior** density $p(\boldsymbol{\theta}|\mathbf{R})$ for $\boldsymbol{\theta}$ given $\mathbf{R}$ is given by the Bayes formula

$$p(\boldsymbol{\theta} \mid \mathbf{R}) = \frac{p(\mathbf{R} \mid \boldsymbol{\theta}) p(\boldsymbol{\theta})}{p(\mathbf{R})} = \frac{L(\boldsymbol{\theta}) p(\boldsymbol{\theta})}{p(\mathbf{R})}, \tag{7.1}$$

where $L(\boldsymbol{\theta}) = p(\mathbf{R} \mid \boldsymbol{\theta})$ is the **likelihood function** (viewed as a function of $\boldsymbol{\theta}$ with $\mathbf{R}$ fixed), $p(\boldsymbol{\theta})$ is the prior probability density, and $p(\mathbf{R})$ is the **marginal** density of $\mathbf{R}$:

$$p(\mathbf{R}) = \int p(\mathbf{R} \mid \boldsymbol{\theta}) p(\boldsymbol{\theta}) d\boldsymbol{\theta} = \int L(\boldsymbol{\theta}) p(\boldsymbol{\theta}) d\boldsymbol{\theta}. \tag{7.2}$$

Note that because $\mathbf{R}$ is fixed in the posterior density for $\boldsymbol{\theta}$ given $\mathbf{R}$, we have

$$p(\boldsymbol{\theta} \mid \mathbf{R}) \propto L(\boldsymbol{\theta}) p(\boldsymbol{\theta}). \tag{7.3}$$

Let $\tilde{\mathbf{R}}$ be a new data matrix whose rows are $\tilde{\mathbf{r}}_t = (\tilde{r}_{t1}, \tilde{r}_{t2}, \cdots, \tilde{r}_{tK})$, $t = 1, 2, \cdots, \tilde{T}$, and assume that $\tilde{\mathbf{R}}$ and $\mathbf{R}$ are conditionally independent given $\boldsymbol{\theta}$, which is typically the case. Then the Bayesian **predictive density** of $\tilde{\mathbf{R}}$ given $\mathbf{R}$ is

$$\begin{aligned} p(\tilde{\mathbf{R}} \mid \mathbf{R}) &= \int p(\tilde{\mathbf{R}}, \boldsymbol{\theta} \mid \mathbf{R})\, d\boldsymbol{\theta} \\ &= \int p(\tilde{\mathbf{R}} \mid \boldsymbol{\theta}, \mathbf{R}) p(\boldsymbol{\theta} \mid \mathbf{R})\, d\boldsymbol{\theta} \qquad (7.4) \\ &= \int p(\tilde{\mathbf{R}} \mid \boldsymbol{\theta}) p(\boldsymbol{\theta} \mid \mathbf{R})\, d\boldsymbol{\theta}. \end{aligned}$$

A major advantage of using a Bayesian model in portfolio management and investment decisions is that it allows the manager to combine prior information from one or more sources with sample returns information to arrive at a decision. This is particularly important in the commonly occurring situation where the manager has sample information in the form of a limited amount of returns data (e.g., monthly returns for a year or two in some cases). These are situations in which the prior information is not dominated by the sample information in the form of the likelihood, and consequently the prior information is likely to add value in the investment decision process. The kinds of information the investor may be able to use in constructing a prior include outputs of returns forecasting models, cross-section information, fundamentals research, and views or bets made based on one or more of these kinds of information.

In order to construct a Bayesian model for a particular investment problem, all one has to do "in principle" is specify the form of the likelihood and the form of the prior, compute the marginal density of $\mathbf{R}$, and plug these results into the Bayes formula. Then one can compute the posterior density $p(\boldsymbol{\theta}|\mathbf{R})$ and the posterior predictive density $p(\tilde{\mathbf{R}} \mid \mathbf{R})$ and make whatever posterior probability statement one wishes (e.g., a posterior mean or median, a posterior standard deviation, or a posterior confidence interval).

7.1.1 Portfolio Construction via Posterior Expected Utility Maximizaton

Suppose we have returns $\tilde{\mathbf{r}} = (\tilde{r}_1, \tilde{r}_2, \cdots, \tilde{r}_K)$ for a time period spanning the present to the next time of interest (e.g., from the end of this month to the end of the next month), and we wish to choose portfolio weights $\boldsymbol{w}' = (w_1, w_2, \cdots, w_K)$ to optimize the portfolio return $\tilde{r}_p = \boldsymbol{w}'\tilde{\boldsymbol{r}}$ in some way. If we have a probability density function $p(\tilde{\boldsymbol{r}} \mid \boldsymbol{\theta})$ with parameter vector $\boldsymbol{\theta}$ and a utility function U, then we choose the weights vector $\boldsymbol{w}$ to maximize the expected utility $E[U(\boldsymbol{w}'\tilde{\boldsymbol{r}})] = \int U(\boldsymbol{w}'\tilde{\boldsymbol{r}})p(\tilde{\boldsymbol{r}} \mid \boldsymbol{\theta})d\boldsymbol{r}$; that is:

$$\boldsymbol{w}_{opt} = \underset{\boldsymbol{w}}{\text{argmax}} \int U(\boldsymbol{w}'\tilde{\boldsymbol{r}})p(\tilde{\boldsymbol{r}} \mid \boldsymbol{\theta})d\boldsymbol{r}. \tag{7.5}$$

It is common practice to replace the unknown value of $\boldsymbol{\theta}$ in $p(\tilde{\boldsymbol{r}} \mid \boldsymbol{\theta})$ with an estimate $\hat{\boldsymbol{\theta}}$ based on prior returns history. But then the maximized utility $E\left[U(\boldsymbol{w}'_{opt}\tilde{\boldsymbol{r}})\right] = E\left[U(\boldsymbol{w}'_{opt}\tilde{\boldsymbol{r}}) \mid \hat{\boldsymbol{\theta}}\right]$ contains **estimation risk** since it does not account for the uncertainty in the estimate $\hat{\boldsymbol{\theta}}$.

The Bayesian approach provides a natural way to account for this estimation risk. Let $p(\tilde{r}_p \mid \mathbf{R}) = p(\boldsymbol{w}'\tilde{\boldsymbol{r}} \mid \mathbf{R})$ be the posterior predictive density of the future portfolio return $\tilde{r}_p$:

$$p(\tilde{r}_p \mid \mathbf{R}) = \int p(\tilde{r}_p \mid \boldsymbol{\theta})p(\boldsymbol{\theta} \mid \mathbf{R})\, d\boldsymbol{\theta}. \tag{7.6}$$

Then we compute optimal portfolio weights by maximizing the expected utility under this posterior predictive density:

$$\begin{aligned} \boldsymbol{w}_{opt} &= \underset{\boldsymbol{w}}{\text{argmin}}\, E[U(\boldsymbol{w}'\tilde{\boldsymbol{r}}) \mid \mathbf{R}] \\ &= \underset{\boldsymbol{w}}{\text{argmin}} \int\int U(\boldsymbol{w}'\tilde{\boldsymbol{r}})p(\boldsymbol{w}'\tilde{\boldsymbol{r}} \mid \boldsymbol{\theta})p(\boldsymbol{\theta} \mid \mathbf{R})\, d\boldsymbol{\theta}\, d\boldsymbol{r}. \end{aligned} \tag{7.7}$$

The resulting maximized expected utility $E\left[U(\boldsymbol{w}'_{opt}\tilde{\boldsymbol{r}}) \mid \mathbf{R}\right]$ reflects the uncertainty in the parameter $\boldsymbol{\theta}$ by virtue of averaging with respect to $p(\boldsymbol{\theta} \mid \mathbf{R})$ in the integral above.

7.1.2 The MCMC Method

The use of Bayesian models in finance (and other areas of study) has been quite limited until recent years because of a major impediment: except for the special case of conjugate priors where the posterior and prior densities are in the same parametric family of densities, it is seldom possible to obtain a tractable analytic expression for the posterior density of the parameters and posterior predictive density. Consequently, much of the early work on Bayesian models in econometrics and finance concentrated on linear models with normal errors and multivariate normal models for mean returns, where conjugate priors were available, namely a normal inverse chi-squared (inverse Gamma) distribution for the normal linear model and a normal inverse Wishart distribution for multivariate normal mean returns.[1]

Unfortunately, normal distributions are often quite inadequate models for asset returns, as we have seen in Chapters 5 and 6, and this leads to the need for non-normal likelihoods. Furthermore, the cross-sectional behavior of parameters such as mean return, volatility, and betas are often non-Gaussian, which means that we need non-normal prior distributions. Use of realistic distributions for returns and priors such as normal mixtures and *t* distributions takes one outside the conjugate priors framework and forces one to use a computationally intense method of computing posterior distributions. The most viable and popular approach these days is one of several versions of the **Markov Chain Monte Carlo** (MCMC) method of approximating posterior distributions.[2]

The basic idea behind the MCMC method is to create a Markov chain whose stationary distribution is $p(\boldsymbol{\theta}|\mathbf{R})$. Under reasonable conditions, one can start the chain at an arbitrary point and the distribution of the values of the chain will converge to the stationary distribution $p(\boldsymbol{\theta}|\mathbf{R})$ exponentially fast.[3] Based on this assumption, the MCMC method proceeds in the following three steps: (1) First generate a long sequence of samples $\tilde{\boldsymbol{\theta}}_i,\ i=1,2,\cdots,n$ from the Markov chain with n large enough, (2) discard the first n_o "burn-in" values, and (3) use the remaining values $\tilde{\boldsymbol{\theta}}_i,\ i=n_o+1,n_o+2,\cdots,n$ to form an estimate $\hat{p}(\boldsymbol{\theta}|\mathbf{R})$ of the posterior density $p(\boldsymbol{\theta}|\mathbf{R})$, or more simply use these values in the obvious way to compute estimates of quantities such as the posterior mean, posterior standard deviation, posterior percentage points, and so on.

A powerful aspect of this approach is that the posterior distribution of complicated nonlinear functions $\mathbf{g}(\boldsymbol{\theta})$ may be easily estimated using the MCMC sequence $\mathbf{g}(\tilde{\boldsymbol{\theta}}_i)$. For example, if $\boldsymbol{\theta}=\sigma^2$ is the variance of returns and $g(\theta)=C(\sigma^2)$ is the Black-Scholes call price of a European option, then the posterior density $p\left(C(\sigma^2)\,|\,\mathbf{R}\right)$ is easily estimated using the MCMC method. Or if $\boldsymbol{\theta}=(\mu,\sigma^2)$ and we are interested in the Sharpe ratio $SR=\mu/\sigma$, the posterior density $p(\mu/\sigma\,|\,\mathbf{R})$ is easily estimated by MCMC. And if we want to optimize a portfolio by maximizing the expected utility with respect to the posterior predictive distribution, we can use post-burn-in samples $\tilde{\boldsymbol{\theta}}_i$ to generate samples $\tilde{r}_{p,i}=\boldsymbol{w}'\tilde{\boldsymbol{r}}_i$ from $p(\tilde{r}_p\,|\,\boldsymbol{\theta}_i)$ and choose $\boldsymbol{w}$ to maximize the posterior estimate of expected utility

$$\frac{1}{n}\sum_{i=1}^{n}U(\tilde{r}_{p,i})=\frac{1}{n}\sum_{i=1}^{n}U(\boldsymbol{w}'\tilde{\boldsymbol{r}}_i)\,. \tag{7.8}$$

The devil is in the details, and there are a variety of special forms of MCMC such as the Gibbs sampler, the Metropolis algorithm, and the Metropolis-Hastings algorithm, with the Gibbs sampler being the workhorse for many common types of models, such as linear regression. See Gilks et al. (1996) for descriptions of these MCMC variants. The S+Bayes module used for many of the calculations in this chapter uses one of these three types of MCMC methods, depending on the model at hand.

In order to construct a good Bayesian model for investment decisions, one needs to combine exploratory data analysis (EDA) with experimentation with Bayesian model fits based on alternative prior and likelihood specifications. This in turn requires that one have good, flexible Bayesian modeling and analysis software that provides convenient specification of priors and likelihoods and the MCMC computation of posteriors. The S+Bayes module provided with this book provides good support for such efforts.

In Section 7.2, we provide an extensive discussion of Bayes models for mean returns and volatility, with a substantial number of code examples, in order to provide a solid foundation for dealing with more complex Bayes models. Section 7.3 discusses Bayes linear regression models as applied to simple factor models such as the single-factor market model and cross-section regression factor models. Section 7.4 discusses applications of the Bayes linear regression model to Black-Litterman models of mean returns and extensions to non-normal priors and likelihoods. Finally, Section 7.5 discusses in detail one form of Bayes-Stein estimate of mean returns.

7.2 Bayes Models for the Mean and Volatility of Returns

7.2.1 Normal Bayes Model of Mean Returns with Volatility Known

Let's assume that we have returns $\mathbf{r} = (r_1, r_2, \cdots, r_T)$ for a single asset that are independent and identically distributed, each with a normal distribution $N(r \mid \mu, \sigma^2)$ with mean μ and variance σ^2. For now, we assume that the portoflio manager knows the value of σ^2. The resulting "normal" likelihood for μ is

$$L(\mu) = \prod_{t=1}^{T} N(r_t \mid \mu, \sigma^2) \propto \exp\left\{ -\frac{1}{2\sigma^2} \sum_{t=1}^{T} (r_t - \mu)^2 \right\}. \tag{7.9}$$

We also assume a normal prior $p(\mu) = N(\mu \mid \mu_o, \sigma_o^2)$. This is a simple example of a **conjugate prior** distribution—one that has the same general form as the likelihood and consequently results in a posterior distribution of the same form as the prior. The latter part of this statement will be clear from the calculations below. The parameters μ_o and σ_o^2 are called **hyperparameters**, and they need to be specified by the portfolio manager. For the moment, let's assume that, in addition to specifying the values of these hyperparameters, the manager knows a value for the returns volatility σ.

Since the posterior density $p(\mu \mid \boldsymbol{r})$ is proportional to the product of the likelihood and the prior, we have

$$p(\mu \mid \boldsymbol{r}) \propto \exp\left\{ -\frac{(\mu - \mu_o)^2}{2\sigma_o^2} - \frac{\sum_{t=1}^{T} (r_t - \mu)^2}{2\sigma^2} \right\}. \tag{7.10}$$

It is an easy exercise (Exercise 7.1) to show that

$$\sum_{t=1}^{T} (r_t - \mu)^2 = \sum_{t=1}^{T} (r_t - \overline{r})^2 + n(\overline{r} - \mu)^2,$$

where $\overline{r}$ is the sample mean of the returns, and plugging this in above gives

$$p(\mu \mid \mathbf{r}) \propto \exp\left\{ -\frac{(\mu - \mu_o)^2}{2\sigma_o^2} - \frac{(\overline{r} - \mu)^2}{2\sigma^2 / T} \right\}. \tag{7.11}$$

Expanding the exponent and completing the square leads to (Exercise 7.2):

$$p(\mu \mid \mathbf{r}) = N(\mu \mid \hat{\mu}_T, \sigma_T^2), \tag{7.12}$$

where the posterior mean is

$$\mu_T = \frac{\frac{1}{\sigma_o^2}\mu_o + \frac{T}{\sigma^2}\overline{r}}{\frac{1}{\sigma_o^2} + \frac{T}{\sigma^2}} \tag{7.13}$$

and the posterior variance is

$$\sigma_T^2 = \left(\frac{1}{\sigma_o^2} + \frac{T}{\sigma^2} \right)^{-1}. \tag{7.14}$$

The posterior mean is the **precision-weighted average** of the prior mean μ_o and the observed return r, where $1/\sigma_o^2$ is the **precision of the prior** and T/σ^2 is the **precision of the sample mean** $\bar{r}$. We sometimes refer to the latter as the **likelihood precision** or **data precision**. It is easy to see that the **posterior precision** $1/\sigma_T^2$ is just the sum of the prior precision and the likelihood precision:

$$\frac{1}{\sigma_T^2} = \frac{1}{\sigma_o^2} + \frac{T}{\sigma^2}. \tag{7.15}$$

This example reflects a fundamental **shrinkage** property that appears in Bayes posterior means for many models: the posterior mean μ_T is shrunk away from the maximum likelihood estimate (the sample mean in this case) and toward the prior mean by an amount depending upon the relative precisions of the prior and likelihood. Notice that when the likelihood precision T/σ^2 is large relative to $1/\sigma_o^2$, the majority of the weight in μ_T is given to $\bar{r}$, the likelihood precision is said to dominate the prior precision, and in the limit μ_T becomes the maximum likelihood estimate $\bar{r}$ and $\sigma_T^2 \to \sigma^2/T$. Conversely, if $1/\sigma_o^2$ is large relative to T/σ^2, the majority of the weight in μ_T is given to μ_o, the prior precision is said to dominate the likelihood precision, and in the limit $\mu_T \to \mu_o$ and $\sigma_T^2 \to \sigma_o^2$.

If $\tilde{r}$ is a single new return, then the posterior predictive density for $\tilde{r}$ is

$$\begin{aligned} p(\tilde{r} \mid \boldsymbol{r}) &= \int p(\tilde{r}, \mu \mid \boldsymbol{r})\, d\mu \\ &= \int p(\tilde{r} \mid \mu, \boldsymbol{r})\, p(\mu \mid \boldsymbol{r})\, d\mu \\ &= \int p(\tilde{r} \mid \mu)\, p(\mu \mid \boldsymbol{r})\, d\mu \\ &\propto \int \exp\left\{ -\frac{(\tilde{r} - \mu)^2}{2\sigma^2} - \frac{(\mu - \mu_T)^2}{2\sigma_T^2} \right\} d\mu. \end{aligned} \tag{7.16}$$

The exponent in the integral above is a quadratic form in the variables μ, $\tilde{r}$, and the integral is an integral of a bivariate normal distribution with respect to μ. Thus the posterior predictive density is the marginal density of a bivariate normal density and so must itself be a normal density,

$$p(\tilde{r} \mid \boldsymbol{r}) = N(\tilde{r} \mid \mu_P, \sigma_P^2), \tag{7.17}$$

for some values of the mean μ_P and variance σ_P^2. It may be shown (Exercise 3) that

$$\begin{aligned} \mu_P &= \mu_T \\ \sigma_P^2 &= \sigma^2 + \sigma_T^2. \end{aligned} \tag{7.18}$$

We see that the mean of the posterior predictive distribution is the same as the posterior mean for μ. However, the variance of the posterior predictive distribution is the variance of the returns plus the posterior variance, reflecting the additional uncertainty in the posterior prediction due to the uncertainty in the posterior estimate of mean returns. This uncertainty tends towards zero as T tends towards infinity.

We illustrate application of the simple Bayes model above for the data of Figure 7.1, which shows monthly returns for the stock with ticker KRON for five years from January 1997 to December 2001. The investor sees that the returns have exhibited substantial time-varying volatility and perhaps a time-varying mean value, and she decides to use a Bayesian model for mean returns using returns from the last twelve months of 2001 as the sample information. For the previous four-year period 1997 through 2000, the mean monthly return is .023 and the monthly volatility is .165. Based on this information, the investor makes the choices $\mu_o = .02$ and $\sigma_o = .04$, the latter because she has relatively little confidence in the prior mean value of .02, and she specifies $\sigma = .17$ as a "known" value.[4]

The resulting likelihood and prior and posterior densities are displayed in Figure 7.2, where it is apparent that neither the prior nor the likelihood is dominant. The posterior mean value .047 is the result of substantial shrinkage from the sample mean value .087 toward the prior mean value .02. The posterior standard deviation is .03 as compared with the prior standard deviation of .04. Both of these changes are due to the additional information provided in the sample of twelve monthly returns.

The posterior prediction density is shown in Figure 7.3 along with the prediction density one would have if there were no uncertainty about the mean of the prediction density (i.e., the latter is the *certainty equivalent* prediction density). In this case there is little difference between the two because the returns volatility is so much larger than the prior and likelihood uncertainties. This will not always be the case in Bayesian modeling of asset returns and Bayesian portfolio construction.

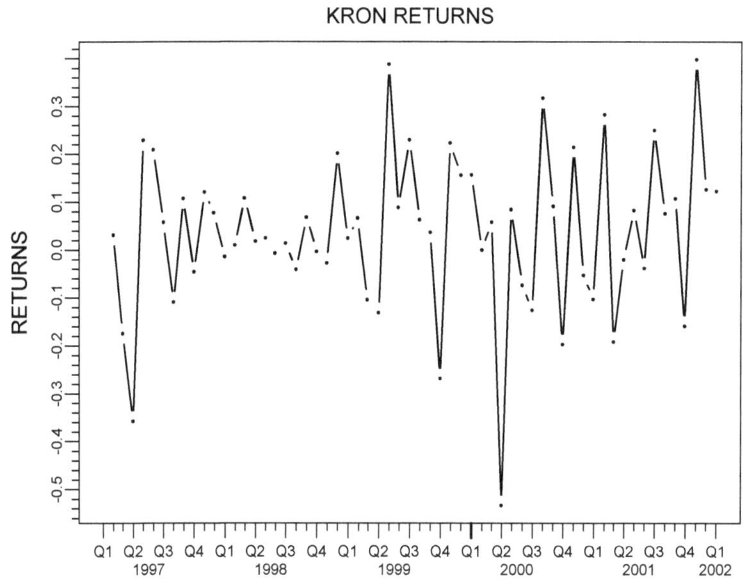

Figure 7.1 KRON Monthly Returns

It is important to keep in mind that the choice of values of hyperparameters μ_o and σ_o^2 and returns volatility σ^2 (and sample size T) have considerable influence on whether or not the prior or likelihood dominate in the Bayesian analysis. For example, suppose that the investor decides that he or she is more certain about the prior value $\mu_o = .02$ by choosing $\sigma_o = .015$ instead of $\sigma_o = .04$, still using as data the returns for the twelve months of 2001 and the value $\sigma = .17$.

Figure 7.4 shows that, in this case, the prior strongly dominates the likelihood. On the other hand, if the investor is very unsure about the prior value $\mu_o = .02$ and chooses $\sigma_o = .15$ to reflect this uncertainty, then the prior is relatively noninformative and the likelihood dominates the prior as shown in Figure 7.5. In the limit as $\sigma_o \to \infty$ the prior is said to be **noninformative**. In the case of a noninformative prior for the mean, the posterior distribution is identical to the likelihood, the Bayes posterior mean is equal to the sample mean, and the posterior variance is the variance of the sample mean.[5]

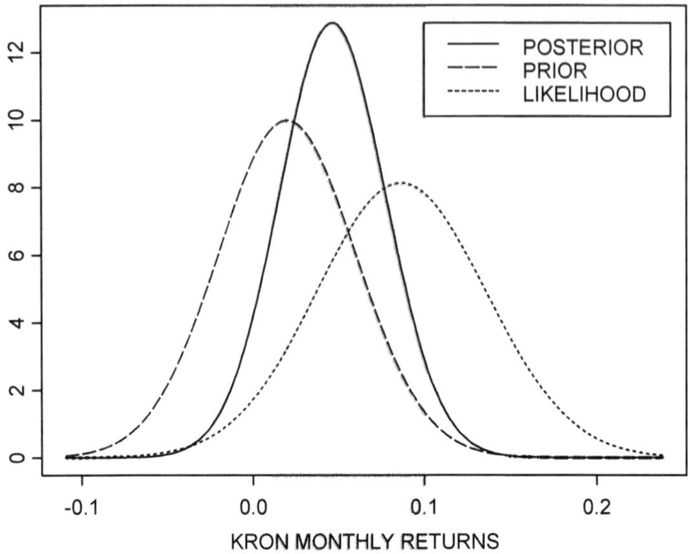

Figure 7.2 Bayes Model for KRON Returns ($\mu_0 = .02, \sigma_0 = .04, \sigma = .17$)

Code 7.1 gives the S-PLUS code for making the normal Bayesian model computations and plots shown in Figure 7.1 through Figure 7.5:

```
ticker <- "KRON"
returns.ts <- smallcap.ts[,ticker]
ret.name <- colIds(returns.ts)
plot(returns.ts, plot.args = list(type = "b",
  pch = "."),reference.grid= F, ylab = "RETURNS",
  main = paste(ticker,"RETURNS"))
returns <- seriesData(returns.ts[49:60,1])[,1]
length(returns)
sigma <- .17; mu0 <- .02; sigma0 <- .04
n <- length(returns )
mu.min <- min(returns); mu.max <- max(returns)
# Plot Posterior Predictive Density
x <- seq(1.7*mu.min, 1.2*mu.max, 0.001)
dpost.pred <- dnorm(x, mu.n,
prec.prior <- sigma0^-2
prec.like <- n * sigma^-2
```

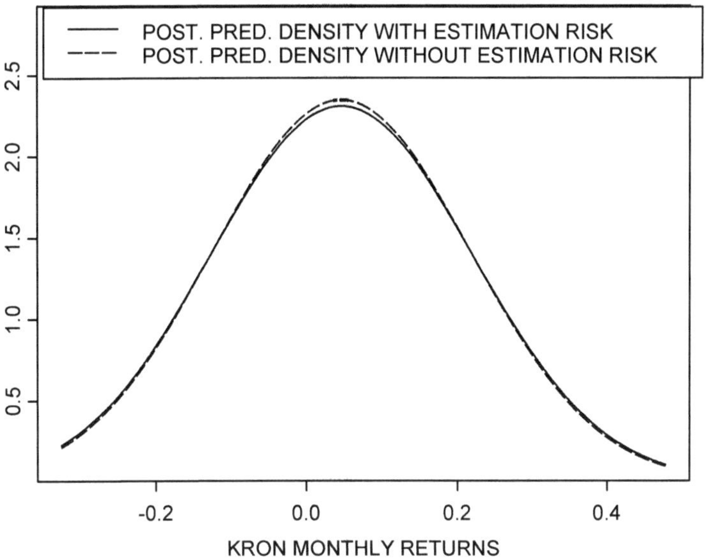

Figure 7.3 Posterior Predictive Densities With and Without Estimation Risk

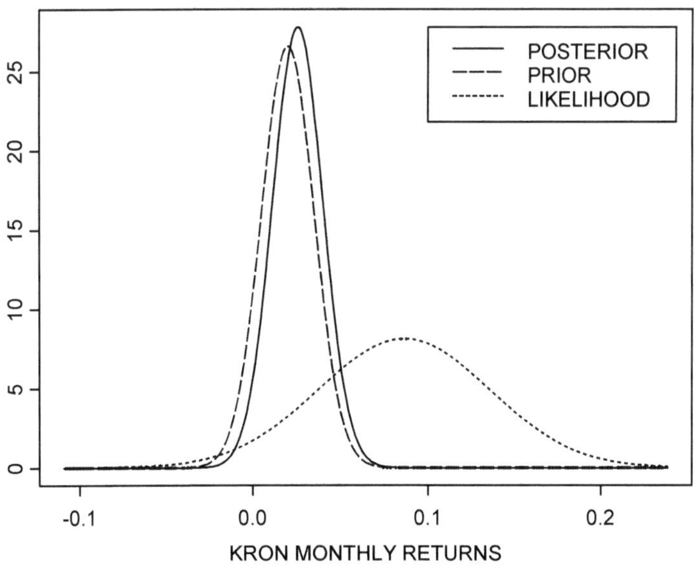

Figure 7.4 Bayes Model for KRON Returns ($\mu_0 = .02, \sigma_0 = .015, \sigma = .17$)

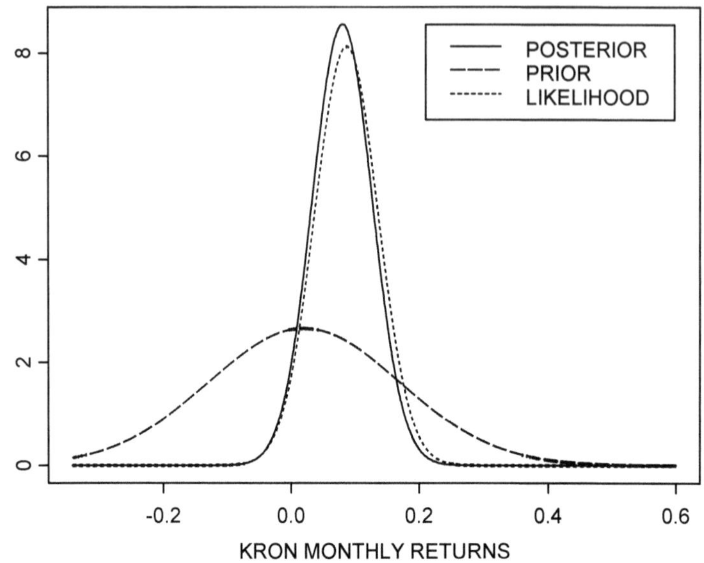

Figure 7.5 Bayes Model for KRON Returns (μ_0 = .02, σ_0 = .15, σ = .17)

```
prec.post <- prec.prior + prec.like
returns.mean <- mean(returns )
mu.n <- (prec.prior * mu0 +
  prec.like * returns.mean)/prec.post
x <- seq(.57*mu.min, .6*mu.max, 0.001)
dprior <- dnorm(x, mu0, sigma0)
dlike <- dnorm(x, returns.mean, sigma/sqrt(n))
dpost <- dnorm(x, mu.n, 1/sqrt(prec.post))
ylim <- range(dprior,dlike,dpost)
# Plot Prior, Likelihood and Posterior
plot(x, dpost, type = "l", ylim=ylim, ylab="",
  xlab = paste(ticker,"MONTHLY RETURNS"))
lines(x,dpost ,lwd = 2)    # Thicker posterior line
lines(x,dlike ,lty = 8,lwd = 2)    # Plot likelihood
lines(x,dprior,lty = 4,lwd = 2)    # Plot prior
xleg <- .12
yleg <- ylim[2]
legend(xleg,yleg,legend=c("POSTERIOR","PRIOR",
  "LIKELIHOOD"),lty=c(1,4,8),lwd = 2)
sqrt(sigma^2 + 1/prec.post))
dpost.pred.certain.mu <- dnorm(x, mu.n, sigma)

ylim <- 1.2*range(dpost.pred,dpost.pred.certain.mu)
xleg <- -.3
```

```
yleg <- ylim[2]
plot(x, dpost.pred, type = "l", ylim=ylim, ylab="",
  xlab = paste(ticker,"MONTHLY RETURNS"))
lines(x, dpost.pred,lwd = 2)
lines(x, dpost.pred.certain.mu, lty = 4,lwd = 2)
legend(xleg,yleg,legend=c("POST. PRED. DENSITY WITH
  ESTIMATION RISK","POST. PRED. DENSITY WITHOUT
  ESTIMATION RISK"),lty=c(1,4),lwd=2)
# Return Values
list("Prior Mean" = mu0,"Prior Std. Dev." =
  1/sqrt(prec.prior), "Sample Mean" =
  returns.mean, "Posterior Mean" = mu.n,
  "Posterior Std. Dev." = 1/sqrt(prec.post))
```

Code 7.1 Simple Normal Bayes Model for Mean Returns

The code shown produces Figure 7.1, Figure 7.2, and Figure 7.3. We obtained Figure 7.4 and Figure 7.5 by changing the values of σ_o in Code 7.1.

7.2.2 Normal Bayes Model for Variance of Returns with Mean Known

Suppose as before that you have T independent returns $\mathbf{r}' = (r_1, r_2, \cdots, r_n)$, each with an $N(r \mid \mu_o, \sigma^2)$ distribution but with σ^2 unknown and μ_o known.[6] Then the likelihood for σ^2 is

$$\begin{aligned} L(\sigma^2) &= \prod_{t=1}^{T} N(r_t \mid \mu_o, \sigma^2) \\ &= \frac{1}{(2\pi)^{T/2}(\sigma^2)^{T/2}} \exp\left\{ -\frac{1}{2\sigma^2} \sum_{t=1}^{T} (r_t - \mu_o)^2 \right\} \\ &= c \cdot \frac{1}{(\sigma^2)^{T/2}} \exp\left\{ -\frac{T \cdot \mathrm{v}}{2\sigma^2} \right\}, \end{aligned} \tag{7.19}$$

where

$$\mathrm{v} = \frac{1}{T} \sum_{t=1}^{T} (r_t - \mu_o)^2 . \tag{7.20}$$

The conjugate prior in this case is a **scaled inverse chi-squared** distribution

$$p(\sigma^2) = Inv\chi^2(\sigma^2; \nu_o, \sigma_o^2)$$
$$= c \cdot \frac{1}{(\sigma^2)^{(\nu_o/2+1)}} \exp\left\{ -\frac{\nu_o \sigma_o^2}{2\sigma^2} \right\} \tag{7.21}$$

with degrees of freedom ν_o and scale parameter $\nu_o \sigma_o^2$.[7] For further details, see Appendix 7A.

Combining the likelihood and prior results in the inverse chi-squared posterior

$$p(\sigma^2 | \mathbf{r}) = c \cdot L(\sigma^2) p(\sigma^2)$$
$$= c \cdot \frac{1}{(\sigma^2)^{(\nu_o+T)/2+1}} \exp\left\{ -\frac{\nu_o \sigma_o^2 + T \cdot \mathrm{v}}{2\sigma^2} \right\} \tag{7.22}$$
$$= Inv\chi^2\left(\sigma^2; \nu_o + T, \frac{\nu_o \sigma_o^2 + T \cdot \mathrm{v}}{\nu_o + T} \right).$$

This posterior density has posterior degrees of freedom $\nu_{o,post} = \nu_o + T$ and posterior scale parameter $\sigma_{o,post}^2 = \nu_o \sigma_o^2 + T \cdot \mathrm{v}$.

7.2.2.1 Assuming $\mu_o = 0$

Because mean returns are typically small compared with their standard deviations, there is sometimes little accuracy lost if you assume $\mu_o = 0$ when computing $\mathrm{v} = \frac{1}{T} \sum_{t=1}^{T} (r_t - \mu_o)^2$. Code 7.2 gives a little script to check this claim for the case of five years of monthly returns for the four large caps stocks ORCL, MSFT, HON, and LLTC:

```
tickers <- c("ORCL","MSFT","HON","LLTC")
returns <- largecap.ts[,tickers]@data
n <- dim(returns)[1]
mu <- apply(returns, 2, mean)
sigmasq <- apply(returns, 2, var)
sigmasq0 <- sigmasq + mu^2
sigmasq0/sigmasq
```

Code 7.2 Zero Mean Effect

This script gives as output

```
 ORCL  MSFT   HON  LLTC
1.034 1.041 1.006 1.048.
```

The maximum error here is at most a little under five percent, so it may sometimes be safe to make the assumption that $\mu_o = 0$. The reader can check this out further by applying Code 7.2 to data sets for other returns (Exercise 7.4). Of course, the investor can input a best guess at a reasonable value μ_o if desired.

7.2.2.2 Determining ν_0 and σ_0^2

In order to use this Bayes model, you need to specify the parameters ν_o and σ_o^2 of the scaled inverse chi-squared prior. From Appendix 7A, we see that these parameters may be expressed in terms of $\mu_{\sigma^2} = E(\sigma^2)$ and $SD_{\sigma^2} = \sqrt{VAR(\sigma^2)}$ as

$$\nu_o = 2R^2 + 4, \tag{7.23}$$

$$\sigma_o^2 = \mu_{\sigma^2} \frac{R^2 + 1}{R^2 + 2}, \tag{7.24}$$

where $R = \mu_{\sigma^2} / SD_{\sigma^2}$. Suppose, for example, that a manager wants to carry out a Bayesian analysis of the volatility of the mid-cap stock with ticker KRON shown in using the last year of monthly returns as the data. The manager believes that $\mu_{\sigma^2} = .018$ and $SD_{\sigma^2} = .013$ are reasonable values for the mean and standard deviation of the variance σ^2, which gives $\nu_o = 7.83$ and $\sigma_o^2 = .013$. The resulting prior and posterior densities are shown in Figure 7.6. We get that figure by first running the function script in Code 7.3 and then running the script in Code 7.4.

```
density.invchisq <- function(x, dof, sigsq,
     scaled=T)
{
  # Set scaled = F for standard inverse
  # chi-squared density
  if(missing(sigsq)) sigsq <- 1
  dof2 <- dof/2

  if(scaled) {scale <- dof*sigsq} else {scale <- 1}

  d0 <- (1/Gamma(dof2)) * (scale/2)^dof2
```

```
    d1 <- 1/(x^(dof2 + 1))
    d2 <- exp( - scale/(2 * x)
    d0 * d1 * d2
}
```

Code 7.3 Scaled Inverse Chi-Squared Density

```
ticker <- "KRON"
returns <- smallcap.ts[,ticker]@data[,1]
n <- 12
returns <- returns[(60-n+1):60]
# Manager's mean and stdev. of variance
mu.var <- .018
sd.var <- .013
# Computation of prior d.o.f. and scale
rsq <- (mu.var/sd.var)^2
nu0.prior <- 2*rsq+4
sigmasq0.prior <- mu.var*(rsq+1)/(rsq+2)
scale.prior <- nu0.prior*sigmasq0.prior
# Computation of posterior d.o.f. and scale
v <- sum(returns^2)/n
nu0.posterior <- nu0.prior + n
sigmasq0.posterior <- (scale.prior +
  n*v)/nu0.posterior
# Plot prior and posterior densities
x <- seq(.001,.06,.0001)
dprior <-
  density.invchisq(x,nu0.prior,sigmasq0.prior)
dposterior <- density.invchisq(x,nu0.posterior,
  sigmasq0.posterior)
ylim <- c(0,max(dprior,dposterior))
plot(x,dposterior,type="l",ylim=ylim,xlab="",
  ylab="")
title(main = "PRIOR AND POSTERIOR VARIANCES FOR
  KRON RETURNS")
lines(x,dposterior,lwd=2)
lines(x,dprior,lty=8,lwd=2)
legend(.04,65,c("PRIOR","POSTERIOR"),lty=c(8,1),
  lwd = c(2,2))
#Same as mu.var
mean.prior <- scale.prior/(nu0.prior-2)
scale.posterior <- nu0.posterior*sigmasq0.posterior
mean.posterior <-
```

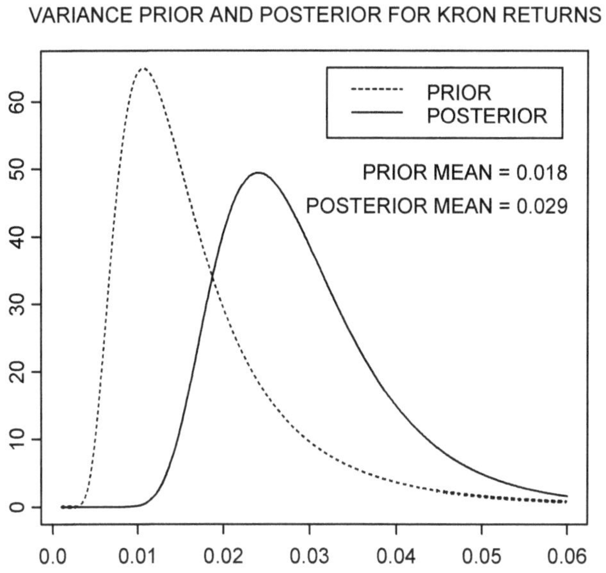

Figure 7.6 Bayes Model for Variance of KRON Returns

```
  round(scale.posterior/(nu0.posterior-2),3)
text(.05,45, paste("PRIOR MEAN =",mean.prior))
text(.048,40,
  paste("POSTERIOR MEAN =",mean.posterior))
```

Code 7.4 Posterior Variance of Returns with Mean Known

We leave it as Exercise 5 for the reader to modify Code 7.4 in order to plot the likelihood as well as the prior and posterior densities.

7.2.3 Normal Bayes Model for Volatility of Returns with Mean Known

The main reason we introduced the normal Bayes model of variance with mean known is to provide a convenient stepping stone to the normal Bayes model of mean returns with variance unknown, which we discuss next. However, in most finance contexts, one is interested in volatility (i.e., the standard deviation of returns) rather than variance of returns. Fortunately, it is quite easy to convert prior and posterior densities for variance to prior and posterior densities for volatility. One just uses the following standard conversion formula for densities under a change of variable:

$$p(\sigma) = p(\sigma^2) \cdot \frac{d\sigma^2}{d\sigma} \\ = 2\sigma \cdot p((\sigma)^2) . \tag{7.25}$$

We leave it as Exercise 6 for the reader to modify Code 7.4 so that it displays priors and posteriors for volatility rather than for variance.

7.2.4 Mean and Variance of Returns with Normal Likelihood and Conjugate Priors

We continue to assume a normal likelihood for independent and identically distributed returns $\mathbf{r}' = (r_1, r_2, \cdots, r_T)$:

$$L(\mu) = \prod_{t=1}^{T} N(r_t \mid \mu, \sigma^2), \tag{7.26}$$

except now we use the more realistic assumption that μ and σ^2 are both unknown, with joint prior density $p(\mu, \sigma^2)$. There are three main classes of joint prior distributions:

1. Noninformative priors
2. Conjugate priors
3. Independence priors.

7.2.4.1 Noninformative Priors

The joint noninformative prior combines the marginal noninformative prior $p(\mu) \propto$ constant for the mean with the marginal noninformative prior $p(\sigma^2) \propto \frac{1}{\sigma^2}$ for the variance by multiplying them together[8]:

$$p(\mu, \sigma^2) \propto \frac{1}{\sigma^2}. \tag{7.27}$$

This noninformative prior is a special case of both conjugate and independence priors.[9] It has been used extensively in the past primarily because it results in closed-form analytical expressions for densities of the marginal posterior distributions for μ and σ^2 [10]:

$$\mu \mid \mathbf{r} \sim ST_{T-1}\left(\bar{r}, \frac{s}{\sqrt{T}}\right)$$
$$\sigma^2 \mid \mathbf{r} \sim Inv\chi^2(T-1, s^2)\ . \tag{7.28}$$

Here $\bar{r}$ is the sample mean and $s^2 = \frac{1}{T-1}\sum_{t=1}^{T}(r_t - \bar{r})^2$ is the unbiased sample variance estimate of σ^2. The notation $\mu \mid \mathbf{r}$ means μ conditioned on $\mathbf{r}$, the notation "~" means "is distributed as," and $ST_\nu(\mu, \sigma)$ denotes a Student's t distribution with ν degrees of freedom, location parameter μ, and scale parameter σ.

7.2.4.2 Conjugate Priors

The joint conjugate prior for the normal likelihood is a **normal inverse chi-squared** distribution:[11]

$$p(\mu, \sigma^2) = N\left(\mu \mid \mu_o, \frac{\sigma^2}{k_o}\right) \cdot Inv\chi^2(\sigma^2 \mid, \nu_o, \sigma_o^2) \tag{7.29}$$

Like the noninformative prior, the conjugate prior has been used extensively in the past primarily because closed-form analytic expressions exist for the posterior densities.[12] The marginal posteriors for μ is a *t* distribution, and the marginal posterior for σ^2 is an inverse chi-squared distribution. The joint posterior is a normal inverse chi-squared distribution from which it is easy to draw random samples. See Appendix 7B for details.

Note that because σ^2 appears in the normal density factor of $p(\mu, \sigma^2)$, the random variables μ and σ^2 are *dependent*, with the variance of the prior for the mean depending upon the variance of the returns. There is little reason to expect this type of dependency in general applications of Bayesian modeling. In the particular context of asset returns, this type of dependency does not appear to capture the empirical evidence about the nature of dependency between mean returns and volatility.[13]

7.2.4.3 Independence Priors

Lacking a solid justification for the highly special dependence structure of the conjugate prior, one will often prefer to use an **independence prior** for which the joint density factors into the product of the marginal densities:

$$p(\mu,\sigma^2) = p(\mu)\cdot p(\sigma^2)\,. \tag{7.30}$$

The class of independence priors includes the so-called **semi-conjugate prior**, where the above conjugate prior is modified by the replacement $\sigma^2 / k_o \rightarrow \tau_o^2$ in the normal density factor. Other independence priors of interest include t distribution and normal mixture choices for $p(\mu)$.

In the case of independence priors, we do not have closed-form analytic expressions for even the marginal posterior densities let alone the joint posterior density. However, in the special case of an independence prior with an inverse chi-squared prior for $p(\sigma^2)$, one can show that the marginal posterior density for the mean is proportional to the product of a Student's t density and the prior density $p(\mu)$,[14]

$$p(\mu \mid \mathbf{r}) = c\cdot ST_{\nu_T}\left(\mu \mid \bar{r},\sigma_T^2\right)\cdot p(\mu), \tag{7.31}$$

where

$$\sigma_T^2 = \frac{1}{\nu_o + T}\left(\nu_o\sigma_o^2 + (T-1)s^2\right). \tag{7.32}$$

The normalizing constant c can be easily computed via numerical integration in this one-dimensional case.

7.2.5 Choice of Prior

The use of a joint noninformative prior for the mean and variance is relatively uninteresting in finance applications because it does not make use of a portfolio manager's prior information.[15] Use of conjugate priors in finance is at best somewhat suspect for at least two reasons: (a) the use of a normal prior for μ is not supported by the empirical behavior of asset returns, and (b) as we have already noted, the dependency structure of the conjugate prior does not reflect the empirical dependency behavior of μ and σ^2.[16]

There is another limitation of the conjugate prior: the posterior for the mean is unimodal and located at a fixed weighted average of the sample mean and the prior mean. To quote Leamer (1978, p.79):

> In this sense it never distinguishes sample information from prior information, no matter how strong their apparent conflict. This is so because a conjugate prior treats prior information as if it were a previous sample of the same process. It may be argued that most prior information is distinctly different from sample information, and when they are apparently in conflict, the posterior distribution ought to be

multimodal with modes at both the sample location and the prior location.

The posterior for the mean in an independence prior with an inverse chi-squared prior for the variance is proportional to the product of a *t* density and the prior density for the mean. This posterior for the mean can be bimodal when using a normal prior for the mean, as well as when using a non-normal *t* distribution or a normal mixture distribution prior for the mean. This is a strong argument in favor of using independence priors.

As for the claim that a normal prior μ is unsatisfactory, Figure 7.7 displays normal Q-Q plots of sample means of five years of monthly returns for each of the following market capitalization groups: microcaps, small caps, mid-caps, and large caps. The distributions of these sample means is clearly quite non-normal for all four market cap groups. For the microcaps and smallcaps, the piecewise linear segments suggest that a normal mixture with three components may provide a good model, while a two-component mixture may work for the mid-caps. Several normal mixture components or some other form of prior for the mean may be needed for the large caps.

The S-PLUS script given in Code 7.5 will compute plots similar to those in Figure 7.7 but on the smaller market cap data sets of size twenty provided with this book, where you will still see some evidence of non-normality of the means of returns[17]:

```
par(mfrow = c(2,2))
micro <- apply(seriesData(microcap.ts),2,mean)
small <- apply(seriesData(smallcap.ts),2,mean)
mid <- apply(seriesData(midcap.ts),2,mean)
large <- apply(seriesData(largecap.ts),2,mean)
qqnorm(micro,ylab = "RETURNS",
  main = "MICROCAPS",pch = ".")
qqline(micro)
qqnorm(small,ylab = "RETURNS",
  main = "SMALL CAPS",pch = ".")
qqline(small)
qqnorm(mid,ylab = "RETURNS",
  main = "MID-CAPS",pch = ".")
qqline(mid)
qqnorm(large,yylab = "RETURNS",
  main = "LARGE CAPS",pch = ".")
qqline(large)
par(mfrow = c(1,1))
```

Code 7.5 Normal Q-Q Plots of Sample Means of Returns

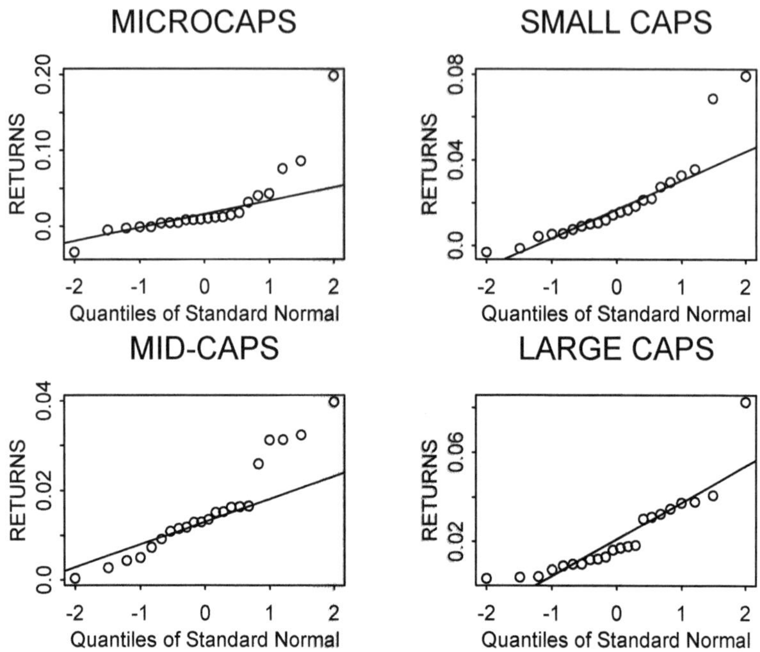

Figure 7.7 Non-normality of Sample Means of Stock Returns

7.2.6 Choice of Likelihood

The normal likelihood is typically as unrealistic a model for the likelihood for a given set of asset returns as the normal prior is for the mean returns. In support of this claim we compute normal Q-Q plots of returns for the twenty microcaps in `microcap.ts`, using Code 7.6.

```
returns <- microcap.ts@data[,-(1:2)]
n <- dim(returns)[[2]]
par(mfrow = c(n/5,5))
for(i in 1:20) {
  qqnorm(returns[,i]);qqline(returns[,i])
}
```

Code 7.6 Normal Q-Q Plots of Returns of Twenty Microcap Returns

The results shown in Figure 7.8 clearly indicate pervasive heavy-tailed non-normality with occasional extreme outliers. For small cap, mid-cap, and large cap stocks, you will find that each of these groups contains stocks with

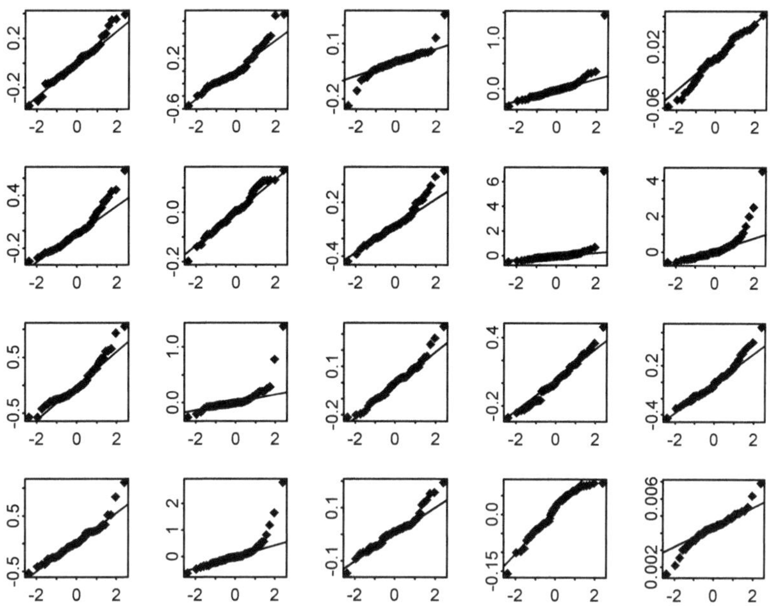

Figure 7.8 Normal Q-Q Plots of Returns of Twenty Microcaps

significant heavy-tailed non-normality and outliers (even the large cap stocks, for which the normal distribution approximation is often quite good).[18]

The results above provide strong motivation for using non-normal likelihoods such as t distribution and normal mixture distribution likelihoods. We remark that use of such likelihoods in effect provides a robust Bayesian model in which outliers in individual returns for a given asset will be down-weighted so as to have little influence on the posterior distribution obtained.

7.2.7 Composition Sampling From Posterior Distributions

Depending on the investment problem at hand, one needs to compute posterior probabilities or simple posterior summaries such as means, standard deviations, and quantiles for μ and σ^2 or nonlinear transformations thereof. In the case of nonlinear transformations, one is often faced with a numerical integration problem, even in the simplest cases of conjugate and noninformative priors. In such cases, we can obtain the desired result by sampling from a joint or marginal posterior distribution to estimate the quantities of interest. If the problem involves (μ, σ^2) jointly, we can use a particular form of sampling from the posterior distributions known as **composition sampling**.

7.2.7.1 Conjugate and Noninformative Priors

If one only wants a probability coverage, mean, standard deviation, or quantile of one of the marginal posteriors, these are easily computed using S-PLUS.[19] On the other hand, computing a probability or summary statistic for a nonlinear transformation of either μ or σ^2 separately (e.g., a Black-Scholes option price in the latter case) typically involves an unpleasant numerical integration. In such cases, a simple and effective approach is to sample from a Student's t distribution posterior for $p(\mu \mid \mathbf{r})$ or an inverse chi-squared distribution for $p(\sigma^2 \mid \mathbf{r})$.[20] When we need to compute a posterior probability coverage or summary statistic for a nonlinear transformation of (μ, σ^2) such as the Sharpe ratio μ / σ, we can use the following composition sampling approach. The key to composition sampling is having known simple forms for the conditional posterior density $p(\mu \mid \sigma^2, \mathbf{r})$ and the marginal posterior density $p(\sigma^2 \mid \mathbf{r})$ from which one can sample with relative ease. In the case of the conjugate prior, the former is a normal density $N(\mu \mid \mu_T, \sigma_T^2)$ and the latter is an inverse chi-squared density. Composition sampling consists of first sampling a value σ_1^2 from $p(\sigma^2 \mid \mathbf{r})$. Then, using this value, sample a value μ_1 from $p(\mu \mid \sigma_1^2, \mathbf{r})$. Repeat this M times to obtain the joint sample (μ_i, σ_i^2), $i = 1, 2, \cdots, M$ from the joint posterior $p(\mu, \sigma^2 \mid \mathbf{r})$, and use this quantity to estimate whatever probability or summary statistic one wants for μ, σ^2 or a nonlinear transformation thereof.

We illustrate the use of composition sampling by replicating a result in Scherer (2004) on optimal equity allocation at future time horizons. Assume we have sixty monthly returns $\mathbf{r} = (r_1, r_2, \cdots, r_{60})'$ for a single equity and that the returns follow a standard Brownian motion model of asset prices with mean μ and volatility σ. We want to determine the optimal equity allocation w^* in a simple cash and equity portfolio for different yearly time horizons in the future:

$$w^* = \frac{1}{\lambda} \cdot \frac{\mu_T}{\sigma_T^2}. \tag{7.33}$$

As in Section 4.1.1 of Scherer (2004), we assume $\lambda = 3$ and that the estimated mean and volatility based on the sixty months of returns are $\hat{\mu} = .3\%$ and $\hat{\sigma} = 6\%$. Assuming no estimation error, and noting that $\mu_T = T\mu$ and $\sigma_T^2 = T\sigma^2$, we get $w^* = 27.7\%$ independently of T. However, we have estimation error and need to get posterior predictive values for μ_T and σ_T^2 in order to properly compute a posterior estimate of w^*. This is easily done with composition sampling as follows.

The posterior for the variance σ^2 given $\mathbf{r}$ is the inverse chi-squared distribution

$$p(\sigma^2 \mid r) = Inv\chi^2(\sigma^2 \mid 59, (.06)^2), \qquad (7.34)$$

and the posterior for μ given $\mathbf{r}$ and σ^2 is the normal distribution

$$p(\mu \mid \mathbf{r}, \sigma^2) = N(\mu \mid .003, (.06)^2 / 60). \qquad (7.35)$$

The distribution of a future return $\tilde{r}_T$ at yearly time horizon T given μ and σ^2 under the assumed Brownian motion model is

$$p(\tilde{r}_T \mid \mu, \sigma^2) = N(\tilde{r}_T \mid T\mu, T\sigma^2), \qquad (7.36)$$

so we simulate N values of (μ, σ^2) by composition sampling and substitute these values into $p(\tilde{r}_T \mid \mu, \sigma^2)$ in order to simulate N values of $\tilde{r}_T$ from which we can compute estimates of μ_T and σ_T^2 under estimation error. For $\mathrm{N} = 100{,}000$ we get the results in Figure 7.9 using Code 7.7. The impact of estimation error is considerable, increasing with the length of the investment horizon.

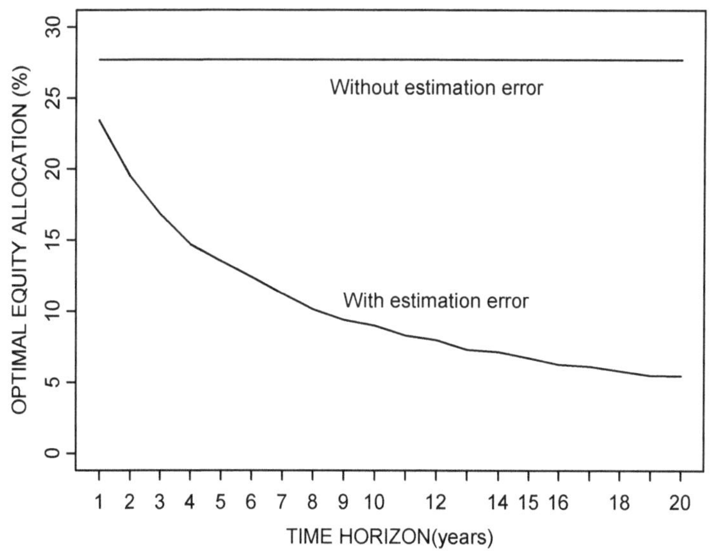

Figure 7.9 Equity Allocation versus Time with and without Estimation Error

```
# Set parameters
mu.hat <- 0.003
sigma.hat <- 0.06
```

```
months <- 60
lambda <- 3
horizon <- 1:20
w <- NULL
n <- 100000
for (k in horizon) {
  # composition sampling in for joint posterior
  # sample
  sigma2.samp <-
     (months-1)*(sigma.hat^2)/rchisq(n,months-1)
  mu.samp <-
     rnorm(n,mu.hat,sqrt(sigma2.samp/months))
  # posterior predictive sample
  r.tilde <-
     rnorm(n,12*k*mu.samp,sqrt(12*k*sigma2.samp))
  # compute the optimal equity allocation
  r.mu <- mean(r.tilde)
  r.sigma2 <- var(r.tilde)
  w[k] <- (r.mu/(lambda*r.sigma2))*100
}
plot(horizon, w, type="l", ylim=c(0,30),
  xlab="TIME HORIZON(years)",
  ylab="OPTIMAL EQUITY ALLOCATION (%)")
lines(1:20,rep(27.7,20))
text(12,26,"Without estimation error")
text(12,11,"With estimation error")
axis(1,at=horizon)
```

Code 7.7 Optimal Equity Allocation versus Time Horizon

7.2.7.2 Limitations of Composition Sampling

One might think that the composition sampling approach could be used for the case of independence priors, for one still has $p(\mu \mid \sigma^2, \boldsymbol{r}) = N(\mu \mid \mu_T, \sigma_T^2)$ as in the conjugate and noninformative cases. However, this is not the case because it is typically not easy to sample from $p(\sigma^2 \mid \mathbf{r})$. In the special case of a semi-conjugate prior, one can find an analytic expression for $p(\sigma^2 \mid \mathbf{r})$ up to a normalizing constant and determine the constant by numerical integration, but the expression is complex and does not lead to a simple method of sampling from $p(\sigma^2 \mid \mathbf{r})$.[21]

A further limitation of composition sampling occurs in the case of multi-dimensional parameters (e.g., as in multivariate returns and linear regression (factor) models for returns). In order to extend the method from the case of two parameters as above to the general case of k parameters $\theta_1,\theta_2,\cdots,\theta_k$, one needs to be able to sample from each of the conditional distributions $p(\theta_1 \mid \boldsymbol{r})$, $p(\theta_2 \mid \theta_1,\boldsymbol{r})$, ..., $p(\theta_k \mid \theta_1,\cdots\theta_{k-1},\boldsymbol{r})$. While it is often easy to sample from the so-called **full conditional posteriors** $p(\theta_k \mid \theta_1,\cdots\theta_{k-1},\boldsymbol{r})$, it is often difficult to sample from the **partial conditional posteriors** $p(\theta_j \mid \theta_1,\cdots\theta_{j-1},\boldsymbol{r})$, $j=2,\ldots,k-1$ and marginal posterior $p(\theta_1 \mid \boldsymbol{r})$.[22]

7.2.8 Gibbs Sampler Form of the Markov Chain Monte Carlo Method

The Gibbs sampler is a simple form of the Markov Chain Monte Carlo (MCMC) method that may be used to estimate multivariate posterior distributions through simulation of samples from the posterior.[23] It may be used for many common models in finance for which it is easy to sample from the set of *full* conditional densities as follows. Start with an arbitrary initial condition $\theta_2^{(0)}\cdots,\theta_k^{(0)}$, and then

Draw $\theta_1^{(1)} \sim p(\theta_1 \mid \theta_2^{(0)},\theta_3^{(0)},\cdots,\theta_k^{(0)},\boldsymbol{r})$
Draw $\theta_2^{(1)} \sim p(\theta_2 \mid \theta_1^{(1)},\theta_3^{(0)},\cdots,\theta_k^{(0)},\boldsymbol{r})$
Draw $\theta_3^{(1)} \sim p(\theta_3 \mid \theta_1^{(1)},\theta_2^{(1)},\theta_3^{(0)},\cdots,\theta_k^{(0)},\boldsymbol{r})$
....
....
....
Draw $\theta_k^{(1)} \sim p(\theta_k \mid \theta_1^{(1)},\theta_2^{(1)},\theta_3^{(1)},\cdots,\theta_{k-1}^{(1)},\boldsymbol{r})$

This yields $\boldsymbol{\theta}^{(1)} \mid \boldsymbol{r} = \left(\theta_1^{(1)},\theta_2^{(1)},\theta_3^{(1)},\cdots,\theta_{k-1}^{(1)},\theta_k^{(1)} \mid \boldsymbol{r}\right)$. Repeat the process above M times to obtain a vector sample of size M: $\boldsymbol{\theta}^{(t)} \mid \boldsymbol{r},\ t=1,2,\cdots,M$.

This sequence of vector random variables is not an independent sequence but instead is a first-order vector Markov process that is also called a Markov chain. Under regularity conditions, the distributions $p_t(\boldsymbol{\theta}^{(t)} \mid \boldsymbol{r})$ of this Markov chain converge to the **stationary distribution** of the chain, and this stationary distribution is the posterior distribution $p(\boldsymbol{\theta} \mid \boldsymbol{r})$.[24] In order to avoid distorting stochastic transient start-up effects when using the Gibbs sampler, one generates a long sequence (M large) and discards an initial "burn-in" segment of substantial size t_0. We can use the resulting Gibbs sampler sequence to estimate almost any statistical quantity of interest based on the full joint posterior or any finite-dimensional marginal, including the one-dimensional marginals, or any nonlinear transformation thereof.

As an example of how the Gibbs sampler works, consider the normal likelihood mean-variance model of returns with a semi-conjugate independence prior $N(\mu_o, \tau_o^2) \cdot Inv\chi^2(v_o, \sigma_o^2)$. The joint posterior for (μ, σ^2) is

$$p(\mu, \sigma^2 \mid \boldsymbol{r}) \propto \frac{1}{\left(\sigma^2\right)^{T/2}} \exp\left\{-\frac{TV}{2\sigma^2}\right\} \times \frac{1}{\left(\sigma^2\right)^{v_o/2+1}} \exp\left\{-\frac{v_o \sigma_o^2}{2\sigma^2}\right\} \times \exp\left\{-\frac{(\mu - \mu_o)^2}{2\tau_o^2}\right\},$$

where

$$V = \frac{1}{T} \sum_{t=1}^{T} (r_t - \mu)^2 . \tag{7.37}$$

From our earlier calculation for the normal likelihood with σ^2 known and a normal prior $N(\mu_o, \tau_o^2)$ for the mean, we know that $p(\mu \mid \sigma^2, \boldsymbol{r}) = N(\mu \mid \mu_T, \sigma_T^2)$, from which we can sample easily. By conditioning on μ and combining the first two factors in the representation above for the joint posterior, it follows that the conditional posterior for σ^2 given μ is

$$p(\sigma^2 \mid \mu, r) = Inv\chi^2\left(\sigma^2 \mid v_o + T, \frac{v_o \sigma_o^2 + TV}{v_o + T}\right), \tag{7.38}$$

from which we can also sample easily. So the Gibbs sampler in this case proceeds as follows. Choose an initial value $\sigma^{2(0)}$ and make alternate draws from the two conditional distributions above:

$$\begin{aligned}
\mu^{(1)} &\sim p(\mu \mid \sigma^{2(0)}, \boldsymbol{r}) \\
\sigma^{2(1)} &\sim p(\sigma^2 \mid \mu^{(1)}, \boldsymbol{r}) \\
\mu^{(2)} &\sim p(\mu \mid \sigma^{2(1)}, \boldsymbol{r}) \\
\sigma^{2(2)} &\sim p(\sigma^2 \mid \mu^{(2)}, \boldsymbol{r}) \\
&\cdots \\
&\cdots \\
\mu^{(M)} &\sim p(\mu \mid \sigma^{2(M-1)}, \boldsymbol{r}) \\
\sigma^{2(M)} &\sim p(\sigma^2 \mid \mu^{(M)}, \boldsymbol{r})
\end{aligned} \tag{7.39}$$

The sampler could equally well have been started with an initial $\mu^{(0)}$. Direct implementation of the simple Gibbs sampler above is left as Exercise 7, and use of the sampler to compute the posterior distribution of a Sharpe ratio is left to Exercise 8.

We note that the Gibbs sampler method may be easily extended to handle normal mixture and *t* distribution priors and likelihoods by using a **data augmentation** method initially introduced by Tanner and Wong (1987) and described in Tanner (1996).

7.2.9 Introduction to S+Bayes with Several Examples

We now introduce the S+Bayes (Insightful Corp., 2004) module in the context of several examples of computing the joint and marginal posterior distributions for the mean and variance of asset returns. We fit Bayes mean-variance models using the following S+Bayes functions for fitting a Bayes general linear model by specifying a linear model that contains only an intercept term.

- Specify priors: `bayes.normal`, `bayes.normal.mixture`, `bayes.t`, `bayes.t.mixture`, `bayes.invChisq`
- Fit Bayes model: `blm`, `blm.prior`, `blm.likelihood`, `blm.sampler`, `blm.control`
- Display model results: `print`, `summary`, `plot`
- Predict: `predict`
- MCMC diagnostics: `diagnostics`

We note that the above tasks will be essentially the same for fitting other types of models in S+Bayes and that the tasks will generally be carried out in the order shown.

The functions `bayes.normal`, `bayes.normal.mixture`, `bayes.t`, and `bayes.t.mixture` specify a prior for the mean as a special case of a multivariate prior for regression model coefficients. The function `bayes.invChisq` specifies a scaled inverse chi-squared prior for σ^2. The function `blm.prior` bundles the priors for the mean and variance, specifying whether you want a conjugate prior, an independence prior, or a noninformative prior (the default is independence priors), and produces an object to be used as an argument of the function `blm`. The function `blm.likelihood` specifies a normal or t distribution likelihood, and the functions `blm.sampler` and `blm.control` (an argument of `blm.sampler`) specify the characteristics of the Gibbs sampler. The generic functions `print`, `plot`, and `summary` call the Bayes linear model methods `blm.print`, `blm.summary`, and `blm.plot`. The generic function `diagnostics` calls the method `diagnostics.blm`, which allows you to check on the convergence of the Gibbs sampler. See S+Bayes (2004) for details on MCMC convergence diagnostics. We display these functions with their arguments as a partial guide to their use[25]:

```
bayes.normal(mean.vector, covmat, k0 = 1)

bayes.invChisq(df = 0, sigma0.sq = 1)

blm.prior(priorBeta = "noninformative",
  priorSigma = "non-informative", conjugate = F)

blm.likelihood(type = c("normal", "t"),
  errorCov=NULL, df = 3)

blm.sampler(control = blm.control(),
  sampler.type = c("Gibbs","Exact"),
  number.chains = 1,
  init.point = c("prior", "prior + likelihood",
  "likelihood","user's choice"), beta.init=NULL,
  sigma.init=NULL)

blm.control(bSize = 2000, simSize = 1000,
  freqSize = 1)

diagnostics.blm(x, type = c("geweke", "traces",
  "gelmanRubin","autoCorrelations"),
  by.chain = F, iterationUsed = 300,
  sampleInterval = 10, fractionFirstWindow = 0.1,
  fractionSecondWindow = 0.5)
```

7.2.9.1 Example 7.1: Noninformative Prior

The script in Code 7.8 fits a Bayes mean-variance model with a joint noninformative prior using as data the last twelve months of the KRON returns of Figure 7.1:

```
returns = smallcap.ts[49:60,"KRON"]@data
kron.fit = blm(KRON ~ 1, data = returns)
plot(kron.fit, which = "univariate")
kron.fit
```

Code 7.8 Bayes Model for KRON Returns with Noninformative Priors

By default, `blm` uses a normal likelihood. The resulting exact posterior density for the mean returns is plotted in Figure 7.10 along with a straight line to symbolically represent the noninformative prior. In the special case of a noninformative prior the posterior coincides with the likelihood, and comparison with the likelihood in Figure 7.2 confirms that `blm` is indeed computing the correct posterior.

The S+Bayes plots in Figure 7.11 show the Gibbs sampler result in the top two plots and the exact computation of the marginal posterior densities in the bottom two plots labeled "true posterior". Since we are computing the posterior mean using a linear model with intercept term only, the posterior mean is labeled "(Intercept)," and since S+Bayes displays posterior standard deviations rather than posterior variances the other plot is labeled "sigma".[26] These plots are a confidence builder with regard to the accuracy of the Gibbs sampler computations.

Here is the brief summary information produced by the last line of Code 7.8:

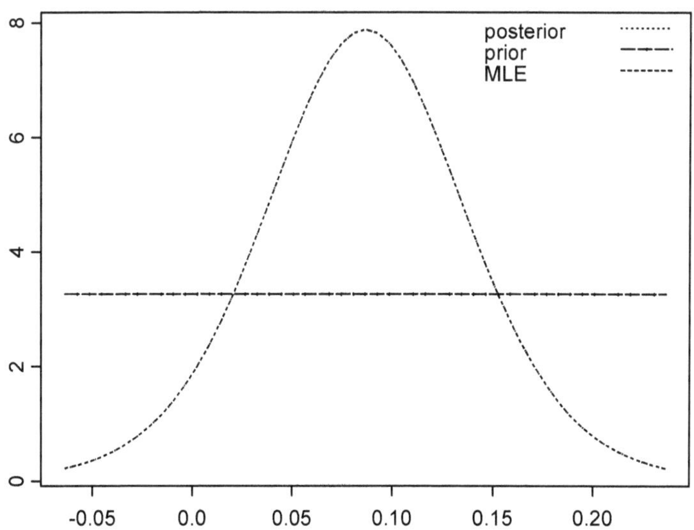

Figure 7.10 Noninformative Prior, Likelihood, and Posterior for KRON Returns

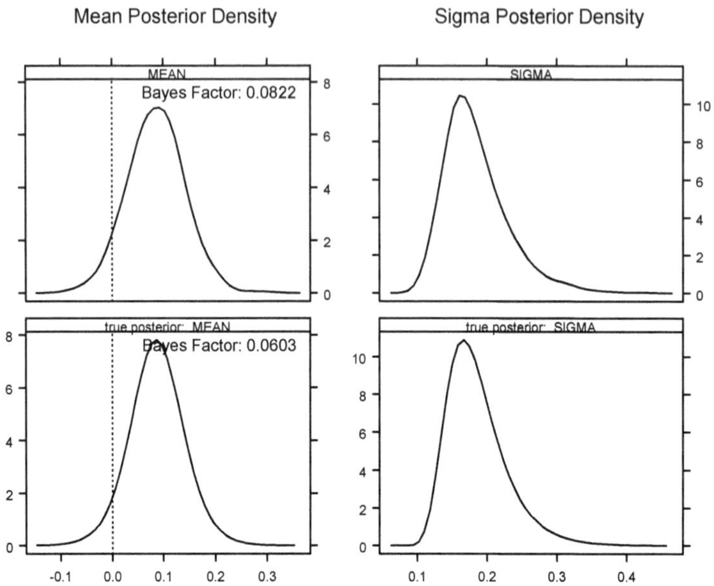

Figure 7.11 Gibbs Sampler and Exact Marginal Posteriors for Noninformative Prior

```
                   mean         stdev  Bayes factor
(Intercept) 0.08719671 0.05769628  0.070018

                 mean         stdev
sigma        0.1878148  0.04248314
```

The columns labeled "mean" and "stdev" are computed as the sample mean and sample standard deviation of the corresponding Gibbs sampler path outputs of S+Bayes. The **Bayes factor** is the ratio of the likelihoods for the nonzero mean versus a hypothesized zero mean model and can be viewed as the weight of evidence in favor of a nonzero mean versus evidence in favor of a zero mean. For intepretation of the values of Bayes factors, see Kass and Raftery (1995). For further discussion, see Section 2.3.3 of Carlin and Louis (2000).

7.2.9.2 Example 7.2: Normal Prior with Variance Known

Now we use the S+Bayes `blm` function based on the Gibbs sampler to fit the same normal Bayes model of mean returns with known variance as in Section 7.2.1. There we used a $N(.02,(.04)^2)$ prior for μ along a normal likelihood with noise variance $\sigma^2 = (.17)^2$ and used as data the last twelve months of the KRON returns shown in Figure 7.1. The expressions in Appendix 7A for the mean and variance of a scaled inverse chi-squared distribution show that you can approximate the case of a known fixed value of σ^2 by letting the degrees of freedom parameter ν_o be very large and setting $\sigma_o^2 = \sigma^2$. Code 7.9 is used to make the fit and plot the results:

```
returns <- smallcap.ts[49:60,"KRON"]@data
#mu0 = .02, tau0 = .04
prior.mean <- bayes.normal(.02,.04)
#nu0 = 100, sigma0 =.17
prior.var <- bayes.invChisq(100,(.17)^2)
my.prior <- blm.prior(priorBeta = prior.mean,
   priorSigma = prior.var)
#burn-in 500
my.sampler <- blm.sampler(blm.control(500,2000))
kron.fit <- blm(KRON ~ 1,prior = my.prior,
  sampler = my.sampler, data = returns)
plot(kron.fit, which = "univariate")
kron.fit
```

Code 7.9 Normal Model for Mean with Variance Known via Gibbs Sampler

In Code 7.9, we have illustrated how to use `blm.sampler` and `blm.control` to set the initial burn-in sample size at 500 and the final Gibbs sampler size at 2000 (rather than the default values of 2000 and 1000, respectively). The resulting prior, likelihood, and posterior are displayed in Figure 7.12, where the posterior has the largest mode, the prior has the second-largest mode, and the likelihood has the smallest mode. These results agree with those of Figure 7.2. We note that a plot of overlaid prior, posterior and likelihood such as those in Figure 7.10 and Figure 7.12 are only provided by S+Bayes for the simple mean-variance model.

The summary statistics produced by the last line of Code 7.9 are:

```
                 mean        stdev     Bayes factor
(Intercept)   0.0463251   0.0310182   0.08998547

                 mean        stdev
sigma         0.1711013   0.01175525
```

The posterior mean value of .046 is quite close to the exact value .047 obtained in Section 7.2.1.

7.2.9.3 Example 7.3: Semi-conjugate Prior

Now we fit a Bayes mean-variance model using a semi-conjugate prior with the same normal prior for the mean returns as above, but with the inverse chi-squared parameters $\nu_o = 7.83$ and $\sigma_o^2 = .013 = (.114)^2$ used in Section 7.2.2 to represent an unknown variance. We just replace the `prior.var = ...` line in Code 7.9 with `prior.var = bayes.invChisq(7.83, .013)`. Figure 7.13 shows that in this case the resulting prior, likelihood, and posterior are quite similar to those of Figure 7.12, with a slight difference in the likelihood.

Figure 7.14 displays the resulting posterior marginal densities for μ and σ^2 obtained with the Gibbs sampler along with "true" posterior marginal densities. In this case, "true" means using an analytic formula that is known up to a normalizing constant that is obtained by numerical integration. The Gibbs sampler results are in quite good agreement with the exact results, giving one further confidence in the Gibbs sampler.

In this case, the summary statistics yield values only slightly different from those in the previous example with known variance, giving a posterior mean value .050 instead of .046.

```
                   mean        stdev     Bayes factor
(Intercept)    0.05029134  0.03090387    0.06687123
```

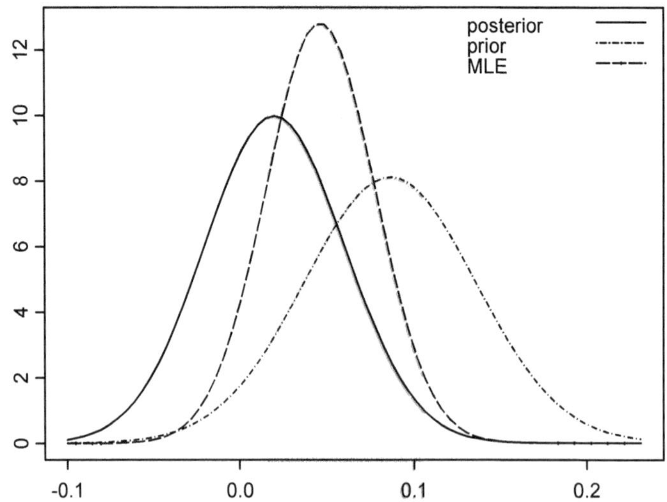

Figure 7.12 Normal Model for KRON Returns with Known Noise Variance

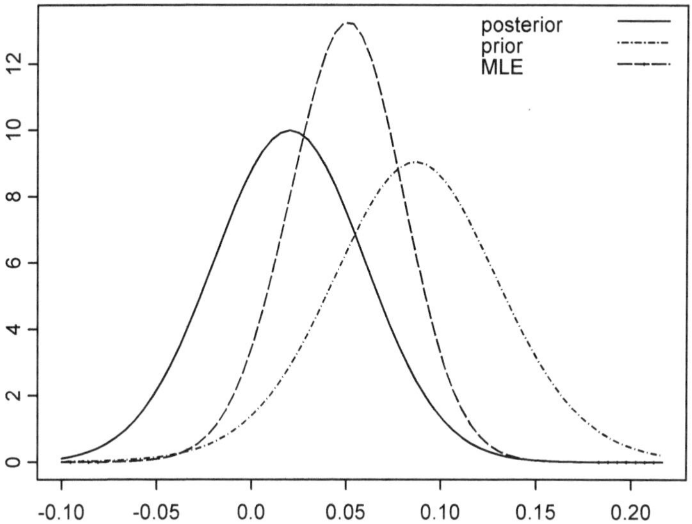

Figure 7.13 Bayes Model for KRON Mean Returns with Semi-conjugate Prior

```
                  mean        stdev
sigma        0.1578964   0.02713585
```

7.2.9.4 Example 7.4: Independent Normal Mixture Prior for the Mean

We continue to use the inverse chi-squared variance prior of Example 7.3 but now assume the investor wants to use a two-component normal mixture prior for the mean returns:

$$p(\mu \mid \gamma, \mu_{o1}, \mu_{o2}, \tau_{o1}^2, \tau_{o2}^2) = (1-\gamma)\cdot N(\mu \mid \mu_{o1}, \tau_{o1}^2) + \gamma \cdot N(\mu \mid \mu_{o2}, \tau_{o2}^2)\,.$$

Suppose the investor thinks there is a 65% chance that the mean return of the asset has a normal distribution with parameters $\mu_{o1} = .02$ and $\tau_{o1}^2 = (.02)^2$ and a 35% chance that it has a normal distribution with parameters $\mu_{o2} = .08$ and $\tau_{o1}^2 = (.02)^2$. Then he or she modifies Code 7.9, replacing the normal prior distribution function `bayes.normal` with the normal mixture function `bayes.normal.mixture` (and using the default Gibbs sampler), to get Code 7.10.

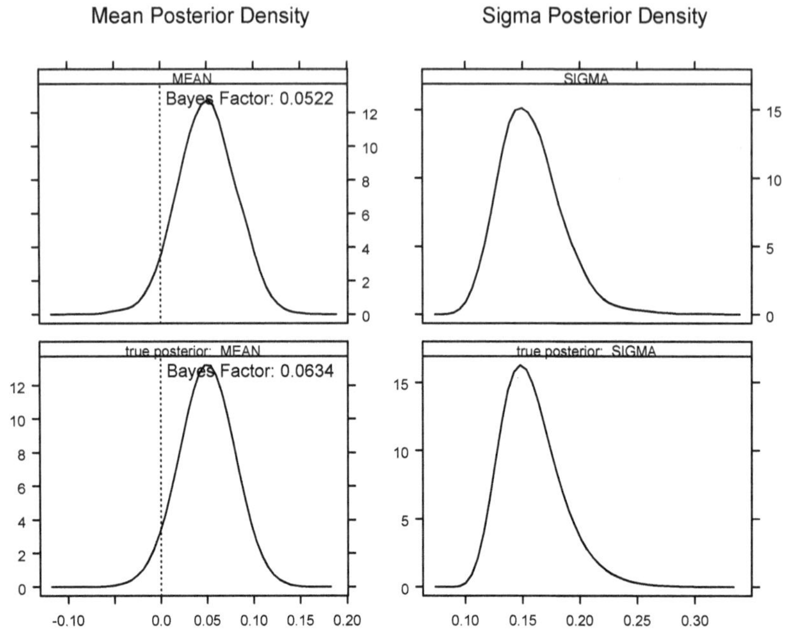

Figure 7.14 Gibbs Sampler Posteriors for KRON Returns with Semi-conjugate Prior

```
returns <- smallcap.ts[49:60,"KRON"]@data
props <- c(.65,.35)
mu0 <- c(.02,.08)
var <- .02          #Set tau01.sq
k <- 1              #Set tau02.sq = k*tau01.sq
prior.mean <- bayes.normal.mixture(mu0,var,k,props)
prior.var <- bayes.invChisq(7.83, .013)
my.prior <- blm.prior(priorBeta = prior.mean,
  priorSigma = prior.var)
kron.fit <- blm(KRON ~ 1,prior = my.prior,
  data = returns)
plot(kron.fit, which = "univariate")
```

Code 7.10 Bayes Model with Normal Mixture Prior for Mean Returns

The resulting prior, likelihood, and marginal posterior densities are shown in Figure 7.15.

The bimodal character of the posterior density is an interesting and intuitively appealing feature of this Bayesian analysis that is not achievable with a conjugate prior. In this case, the highest posterior mode favors a large mean return in the vicinity of 8%, even though the smallest prior mode was fairly well centered on the mean, $\mu_{o2} = .08$, of the second mixing component. This is because the data more strongly favor the mean of the second component.

The Gibbs sampler posteriors for the mean and variance in Figure 7.16 are as expected, comparing well with the true marginal posteriors computed as described for the case of the semi-conjugate prior in Example 7.3.

It should be noted that the investor's specification of the parameters of the normal mixture prior can have a considerable influence on whether or not bimodal posteriors are obtained and on the nature of the bimodality when it occurs. For example, if the investor chooses a 75%–25% mix rather than a 65%–35% mix, the result shown in Figure 7.17 is obtained. Now the situation is reversed from that of Figure 7.15, with the highest posterior mode favoring a relatively small return in the vicinity of 3%, while the lowest posterior mode favors the larger return of around 8%.

This example points out that posterior analysis based on normal mixture priors has a fair degree of sensitivity to the investor's specification of the normal mixture parameters. On the other hand, this type of prior gives the investor an opportunity to effectively use his prior information in a Bayesian model. The reader is encouraged to experiment further with the normal mixture parameters as well as the location of the marginal likelihood (Exercise 10).

7.2.9.5 Example 7.5: Normal Prior for the Mean and *t* Distribution Likelihood

Now we fit a Bayes model to the GAIT returns shown in Figure 7.18 using a *t* distribution likelihood and a normal likelihood along with a semi-conjugate prior. Note the outlier at the beginning of the series, which has the potential to adversely influence the normal likelihood.

Code 7.11 makes the plot in Figure 7.18, fits the Bayes model with both normal and *t* distribution likelihoods, and plots some results:

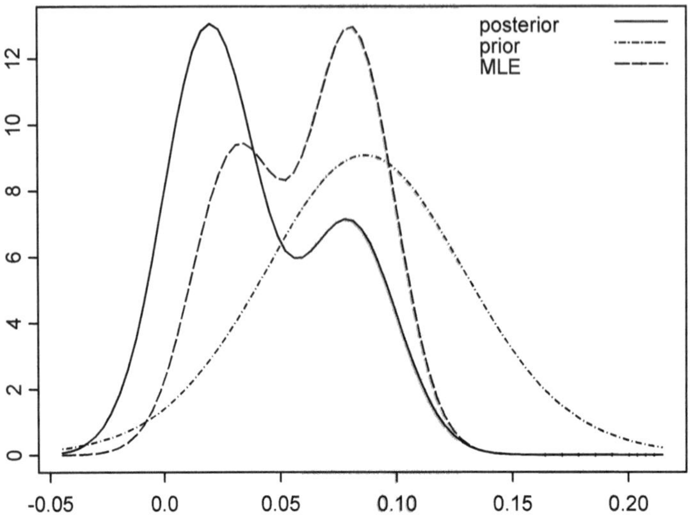

Figure 7.15 Bayes Model for KRON Mean Returns with 65%-35% Normal Mixture Prior

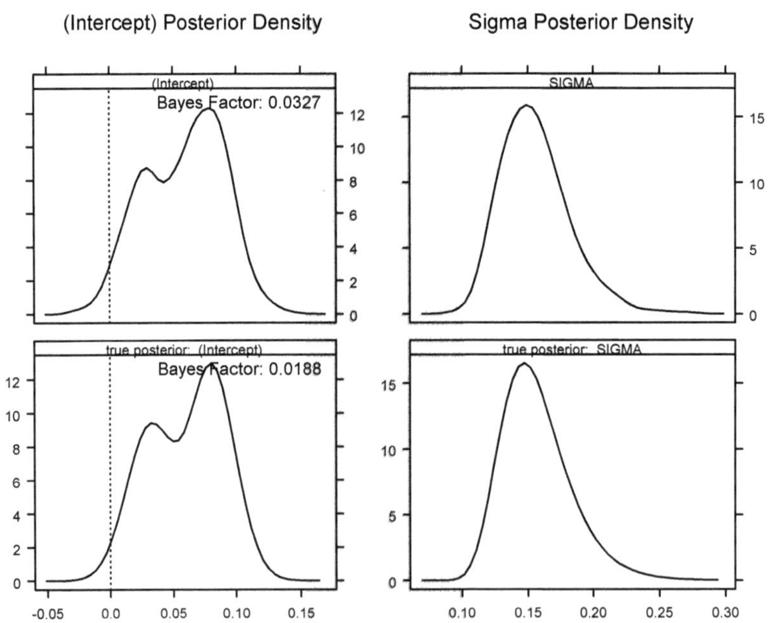

Figure 7.16 Gibbs Sampler Results for with 65%-35% Normal Mixture Prior

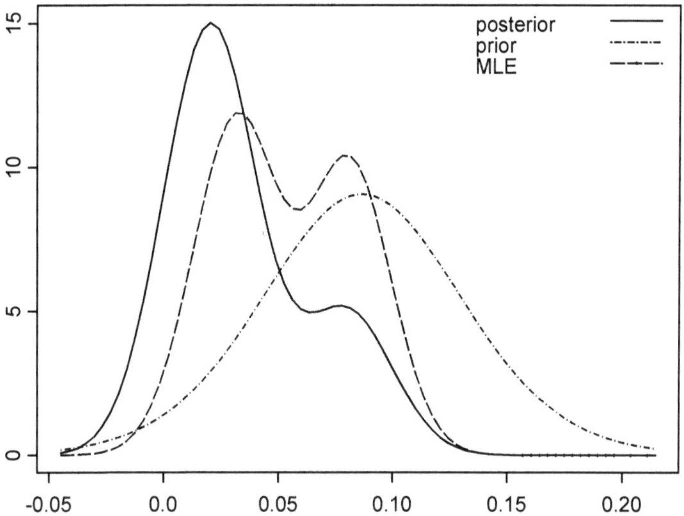

Figure 7.17 Bayes Model for KRON Mean Returns with 75%–25% Normal Mixture Prior

```
returns.ts <- microcap.ts[49:60,"GAIT"]
y.names <- colIds(returns.ts)
seriesPlot(returns.ts, one.plot = F,
  strip.text = y.names, type = "b",col = 1)
returns <- returns.ts@data
prior.mean <- bayes.normal(.05,.1)
prior.var <- bayes.invChisq(7.83, .013)
my.prior <- blm.prior(priorBeta = prior.mean,
  priorSigma = prior.var)
gait.fit <- blm(GAIT ~ 1,prior = my.prior,
  data = returns)
# Default d.o.f. = 3
likelihood <- blm.likelihood("t")
gait.fit.robust <- blm(GAIT ~ 1,prior = my.prior,
  likelihood = likelihood, data = returns)
plot(gait.fit,  which = "box", include.sigma = T)
plot(gait.fit.robust, which = "box",
  include.sigma = T)
gait.fit
gait.fit.robust
```

Code 7.11 Bayes Fit with Normal and *t* Likelihoods and Semi-conjugate Prior

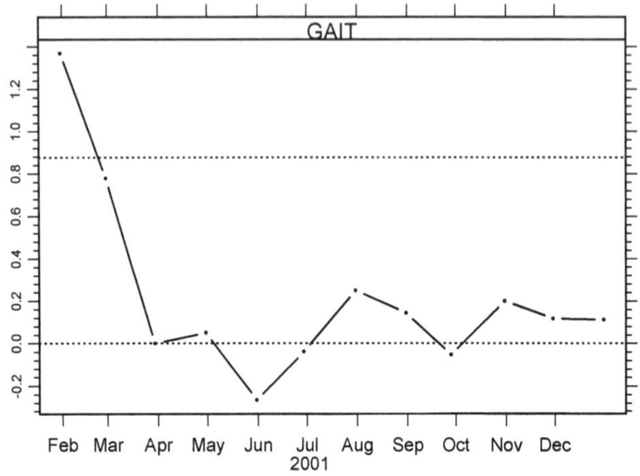

Figure 7.18 GAIT Monthly Returns

In Code 7.11, we used the same inverse chi-squared distribution hyperparameters as in Example 7.3 but used somewhat different hyperparameter values for the normal prior for the mean, reflecting anticipation of a higher return but with greater uncertainty.

Boxplots of the posterior mean and posterior standard deviation distributions for the normal and *t* likelihoods are shown in Figure 7.19 and Figure 7.20, respectively.

Comparison of the upper boxplots in Figure 7.19 and Figure 7.20 reveals that the initial outlier in the GAIT returns for the normal likelihood case causes the posterior distribution of the mean to be shifted to a substantially higher value, and to be more spread out, than with the *t* likelihood. Comparison of the lower boxplots in these figures reveals that for the normal likelihood the posterior distribution of the standard deviation is shifted to a considerably higher value, and has a greater spread, relative to the posterior distribution with the *t* likelihood. The reader should compare the posterior summary statistics for the normal likelihood and *t* likelihood computed by Code 7.11. A better display to facilitate comparisons of results would be to have the two alpha ("(Intercept)") posterior boxplots in the same plot with a common horizontal axis and likewise for the beta (market) posterior boxplots. In Section 7.3.2, we show how to do this by working directly with the S+Bayes Gibbs sampler output.

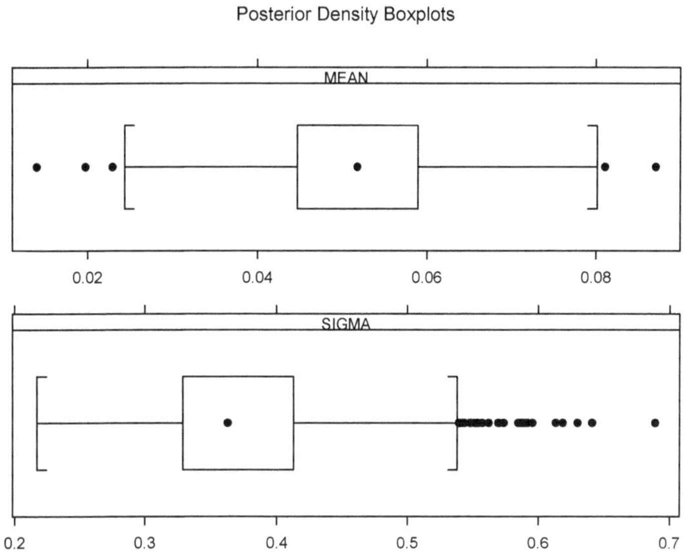

Figure 7.19 Bayes Model for GAIT with Normal Likelihood and Semi-conjugate Prior

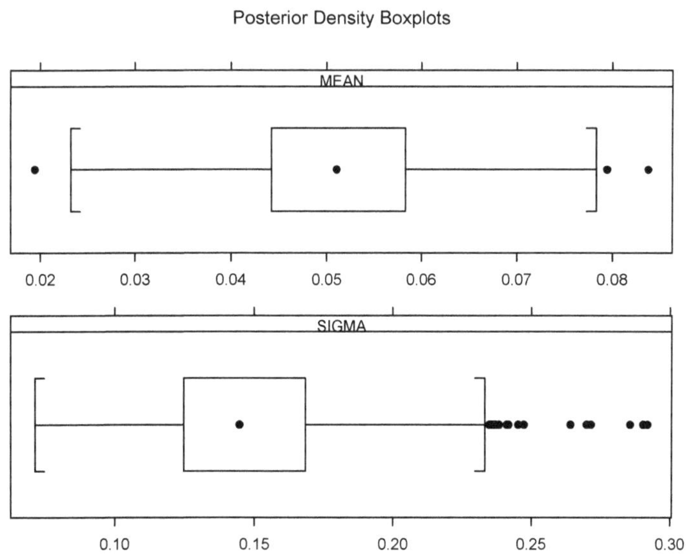

Figure 7.20 Bayes Model for GAIT with t-Likelihood and Semi-conjugate Prior

7.2.9.6 Example 7.6: A Few S+Bayes Plotting Options

We obtain **highest posterior density** (HPD) or **credible** regions for the default 95% level by the optional argument `region = "hpd"` or `region = "credible"` in the `plot` function. Non-default choices of level are specified with the `level =` optional argument to `plot`. For example, `plot(gait.fit, region = "hpd", which = "univariate", include.sigma = T)` gives Figure 7.21.

We get a bivariate contour plot of the joint posterior for (μ, σ) along with the marginal posteriors, as in Figure 7.22, by using `plot(gait.fit, which = "bivariate", include.sigma = T)`.

7.2.10 Empirical Bayes Estimates of Hyperparameters

Whether we use a conjugate, semi-conjugate, or other independence prior, we need to specify the hyperparameters for the joint prior for μ and σ^2. For example, we need to specify the parameters $\mu_o, k_o, \nu_o, \sigma_o^2$ in the case of the conjugate normal inverse-chi-squared prior, the parameters $\mu_o, \tau_o^2, \nu_o, \sigma_o^2$ in the case of the semi-conjugate prior, or other hyperparameters in the case of other independence priors such as a normal mixture or t distribution prior. In the case of a t likelihood, we also need to specify the degrees of freedom. These hyperparameter values might be specified by the portfolio manager in a more or less *subjective* Bayesian manner (i.e., they might be values that the manager believes to be reasonable based on his or her extensive experience combined with ad hoc data analysis). This is the spirit in which the examples of Section 7.2.9 were presented.

On the other hand, cross sections of returns provide substantial data to support the use of **empirical Bayes** methods in which unknown hyperparameter values are obtained as point estimates based on cross-section returns data. A natural statistical methodology for doing so is to compute such point estimates by maximizing the marginal likelihood of the hyperparameters. Let $\boldsymbol{\eta}$ denote the hyperparameters and let $p(\boldsymbol{\theta} \mid \boldsymbol{\eta})$ be the prior for $\boldsymbol{\theta}$ given the hyperparameters, for example, in the semi-conjugate model for mean returns $\boldsymbol{\theta} = \mu$ and $\boldsymbol{\eta} = (\mu_o, \tau_o^2)$. Then Bayes' Theorem, taking into account conditioning on the hyperparameters, is given by

$$p(\boldsymbol{\theta} \mid \mathbf{R}, \boldsymbol{\eta}) = \frac{p(\mathbf{R} \mid \boldsymbol{\theta}) p(\boldsymbol{\theta} \mid \boldsymbol{\eta})}{p(\mathbf{R} \mid \boldsymbol{\eta})}, \tag{7.40}$$

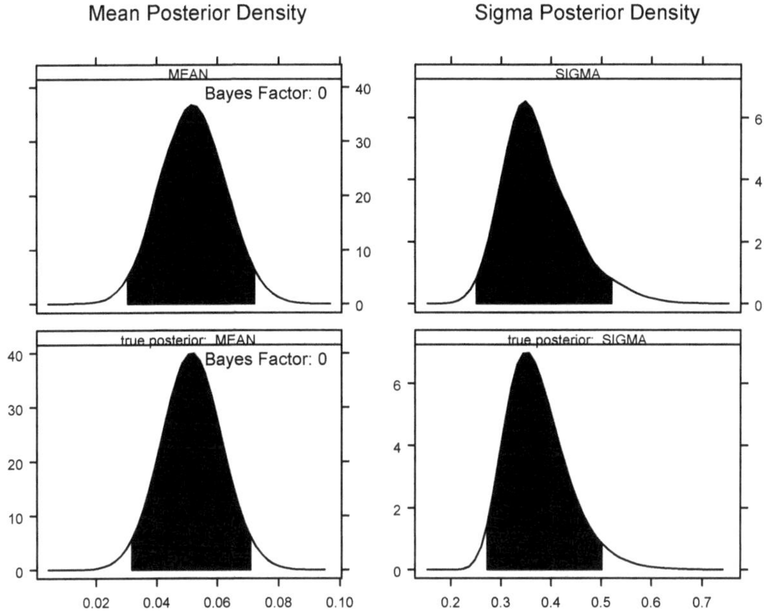

Figure 7.21 Posterior Densities with 95% Highest Posterior Density Regions

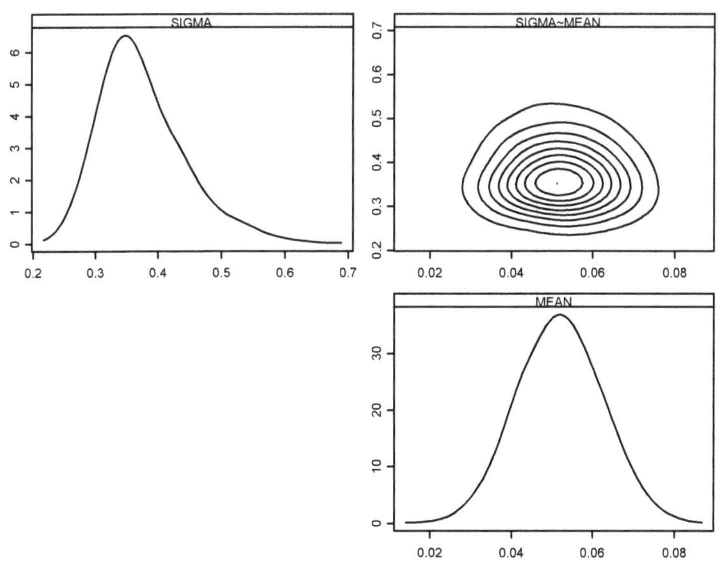

Figure 7.22 Bivariate Posterior Contour Plot and Marginal Posterior Densities

where the conditional marginal distribution for $\boldsymbol{R}$ in the denominator is computed as

$$p(\mathbf{R} \mid \boldsymbol{\eta}) = \int p(\mathbf{R} \mid \boldsymbol{\theta}) p(\boldsymbol{\theta} \mid \boldsymbol{\eta}) d\boldsymbol{\theta} . \tag{7.41}$$

Given an analytical form for $p(\mathbf{R} \mid \boldsymbol{\eta})$, we can in principle compute $\hat{\boldsymbol{\eta}}$ by maximizing the marginal likelihood $m(\boldsymbol{\eta}) = p(\mathbf{R} \mid \boldsymbol{\eta})$. Unfortunately it is often difficult or impossible to obtain a manageable expression for $p(\mathbf{R} \mid \boldsymbol{\eta})$, and consequently $\hat{\boldsymbol{\eta}}$ is often computed by more or less intuitively appealing ad hoc methods. We illustrate some such methods for estimating the hyperparameters in a mean-variance returns model.

7.2.10.1 Empirical Bayes Estimates of Hyperparameters in the Normal Prior

In the case of either the conjugate or semi-conjugate prior, a natural empirical Bayes estimate of the hyperparameter μ_o is obtained by computing the overall sample mean of a representative set of asset returns over a common time interval. For example, if the 20 small cap stocks in `smallcap.ts` were considered to be a representative set of returns, we would compute $\hat{\mu}_o$ quite simply:

```
> x <- microcap.ts@data[,-(1:2)]
> mean(apply(x, 2, mean))
[1] 0.02623051
```

In the case of a conjugate prior, the parameter k_o represents the strength of the dependency between the the variance σ^2 / k_o of the prior for mean returns and the squared volatility of returns σ^2. In this case, there is no simple method of computing a point estimate of k_o.[27]

In the case of a semi-conjugate independence prior we can compute a point estimate of the hyperparameter τ_o^2 based on the generalization of the standard one-way random effects analysis-of-variance statistical model,

$$r_{tk} = \mu_k + \varepsilon_{tk}, \quad t = 1,2,\cdots,T, \quad k = 1,2,\cdots,K \tag{7.42}$$

for a set of K asset returns. In this model, the μ_k are independent and identically distributed random variables with common distribution, $N(\mu_o, \tau_o^2)$, and the ε_{tk} are independent random variables with distribution $N(0, \sigma_k^2)$. The standard one-way random effects model assumes a common error variance for all assets, which is clearly an unrealistic assumption, so our generalization allows for different squared volatilities $\sigma_k^2 = \text{var}(\varepsilon_{tk})$ for each asset.

Let $\hat{\mu}_k$ and $\hat{\sigma}_k^2$ be the sample mean and sample variance of returns for the k-th asset, and let $\hat{\mu}$ be the overall sample mean of all K assets. Then our estimate of τ_o^2 is $\hat{\tau}_o^2 = \min\left[\hat{\sigma}_{\hat{\mu}_k}^2 - \frac{1}{T}\bar{\sigma}^2,\ 0\right]$, where $\hat{\sigma}_{\hat{\mu}_k}^2$ is the cross-sectional sample variance of the sample means $\hat{\mu}_k$ and $\bar{\sigma}^2$ is the cross-sectional average of the sample variances of each set of asset returns. It can be shown that $\hat{\tau}_o^2$ is an unbiased estimate of τ_o^2 under the model assumptions above.[28]

We can apply the method above to the small cap stocks with Code 7.12.

```
x <- microcap.ts@data[,-(1:2)]
N <- dim(x)[1]   #Use N in place of T
K <- dim(x)[2]
N1 <- N/(N-1)
K1 <- K/(K - 1)
mean.stocks <- apply(x, 2, mean)
mean.stocks
var.stocks <- N1 * apply(x, 2, var)
var.stocks
var.means <- K1 * var(mean.stocks)
mean.vars <- mean(var.stocks)/N
if(var.means > mean.vars)
  tau0sq <- var.means - mean.vars
else
  tau0sq <- 0
list(TAU0 = sqrt(tau0sq), TAU0SQ = tau0sq,
  VAR.MEANS = var.means, MEAN.VARS = mean.vars)
```

Code 7.12 Estimate of τ_o^2

This results in $\hat{\tau}_o^2 = .00055$ and correspondingly $\hat{\tau}_o = .023$.

7.2.10.2 Estimates of Hyperparameters of the Inverse Chi-Squared Prior

We also need estimates of the hyperparameters ν_o and σ_o^2 of the inverse chi-squared prior for σ^2. One way to do this is to compute estimates $\hat{\mu}_{\sigma^2}$ and $\hat{V}_{\sigma^2}$ of the mean and variance of σ^2 and substitute these into the expressions for ν_o and σ_o^2 given in Appendix 7A. The mean and variance estimates needed can be obtained by computing the sample mean and sample variance of the sample variances $\hat{\sigma}_k^2$ of each stock.[29] As a check on the resulting scaled inverse chi-squared model, we plot the fitted inverse chi-squared density along with a histogram of the sample variances to confirm whether or not the fitted density and histogram are in reasonable agreement. We use Code 7.13 to carry out the computations and plot the results in Figure 7.23.

```
# Set parameters
x1.mult <- 4
step <- .001
trim.at <- 100
# Compute invchisq parameters
returns <- microcap.ts@data[,-(1:2)]
dim(returns)
var.stocks <- apply(returns, 2, var)
mu.var <- mean(var.stocks [var.stocks < trim.at])
sd.var <- stdev(var.stocks [var.stocks < trim.at])
rsq <- (mu.var/sd.var)^2
nu0 <- 2*rsq+4
sigmasq0 <- mu.var*(rsq+1)/(rsq+2)
scale <- nu0*sigmasq0
# Plot histogram and density fit
x1 <- seq(.001,mu.var+x1.mult*sd.var,step)
d1 <- density.invchisq(x1,nu0,sigmasq0)
hist(var.stocks [var.stocks <trim.at],nclass="fd",
  probability = T, ylim = c(0,15),
  xlab = "VARIANCE", col = 0)
lines(x1, d1)
list(nu0 = nu0, sigmasq0 = sigmasq0, scale = scale)
```

Code 7.13 Fitted Inverse Chi-Squared Density Variances and Histogram Check

We see immediately that the density fit is poor and that this is likely caused by two outlying variances with values just above .6 and .8, respectively. For this reason, Code 7.13 has a `trim.at` parameter that was initially set at the very large value of 100 so that no trimming was done. If we change this value to `trim.at = .4` and change `ylim = c(0,15)` to `ylim = c(0,30)`, we get the result in Figure 7.24, which is a reasonably satisfactory fit.

This example points out the need to trim outliers in one way or another in order to get a robust fit to the bulk of the data for a non-normal parametric density.[30] The reader is encouraged to experiment with Code 7.13 on the stock returns in `smallcap.ts`, `midcap.ts`, and `largecap.ts` to see if reasonably good fits can be obtained, trimming outliers if needed (Exercise 12).

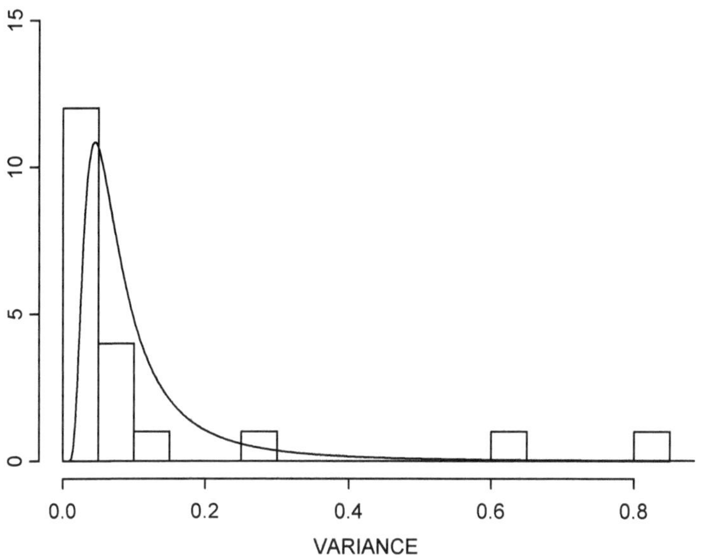

Figure 7.23 Fitted Inverse Chi-Squared Density and Histogram for Microcap Variances

7.2.10.3 Limitations of Empirical Bayes Modeling

While empirical Bayes modeling is an intuitively appealing simple method, there are several difficulties with its use. Focusing first on the estimate $\hat{\tau}_o^2$, we note that it is truncated at zero in order to avoid negative values.[31] However, the resulting truncated value $\hat{\tau}_o^2 = 0$ leads to the unnatural assumption that the unknown means of the returns of each asset have the same value. The second problem is that asset returns often have a nonzero correlation with one another and the estimate $\hat{\tau}_o^2$ does not correct for this.[32] The third problem is that the normality assumption is seldom justified.

As for the inverse chi-squared parameter estimates, it remains to be seen through further study whether or not the inverse chi-squared model is fully adequate for modeling the variances of asset returns. At the very least, we have seen that variance estimates for a collection of asset returns can have outliers that need to be trimmed to obtain a good fit to the bulk of the data. The proper way to deal with this is to use an appropriate heavy-tailed alternative to the inverse chi-squared distribution. Finally, an overarching difficulty with empirical Bayes methods is that when we plug in point estimates for the hyperparameters, the posterior distribution does not reflect the uncertainty in these point estimates.

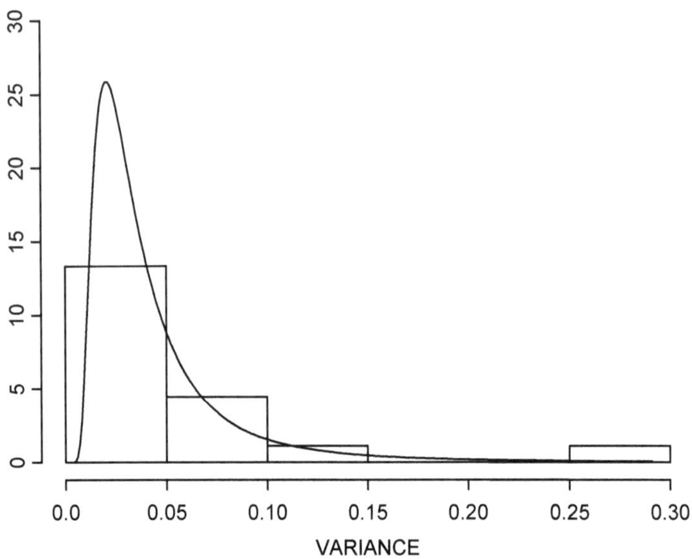

Figure 7.24 Fitted Inverse Chi-Squared Density with Outliers Trimmed

7.2.10.4 Hiearchical Model Alternatives to Empirical Bayes Methods

A good way to incorporate uncertainty about hyperparameters as well as circumvent the problem of negative variance estimates $\hat{\tau}_o^2$ is to fit a full Bayes hiearchical model by introducing a hyperprior density $p(\boldsymbol{\eta})$ and average the conditional posterior $p(\boldsymbol{\theta} \mid \mathbf{R}, \boldsymbol{\eta})$ with respect to $p(\boldsymbol{\eta})$:

$$p(\boldsymbol{\theta} \mid \mathbf{R}) = \int p(\boldsymbol{\theta} \mid \mathbf{R}, \boldsymbol{\eta}) p(\boldsymbol{\eta}) d\boldsymbol{\eta}. \tag{7.43}$$

Until the advent of MCMC methods, a full hierarchical Bayes modeling approach was generally not feasible. For further details on hierarchical Bayes modeling as a basis for potentially fruitful applications in finance and portfolio construction, see Carlin and Louis (2000) and Gelman et al. (2004).

7.3 Bayes Linear Regression Models

The methodology of Section 7.2 for single-asset mean-variance Bayes modeling may be extended to fit a Bayes general linear regression model

$$\boldsymbol{r} = \mathbf{X}\boldsymbol{\beta} + \boldsymbol{\varepsilon}, \tag{7.44}$$

where $\boldsymbol{r}' = (r_1, r_2, \cdots, r_K)$ is a vector of returns for K assets, $\mathbf{X}$ is a $K \times p$ matrix of risk factors (also known as factor exposures, factor loadings, or factor sensitivities), $\boldsymbol{\beta}$ is a $p \times 1$ vector of coefficients (also known as factor returns) to be estimated, and $\boldsymbol{\varepsilon}$ is a $K \times 1$ zero mean error vector with covariance matrix $\boldsymbol{\Sigma}$. This model includes as special cases cross-section factor models at fixed times t, Black-Litterman models, and time series macroeconomic models. It is usually assumed that a cross-section model has sufficient independent variable structure that the errors are essentially uncorrelated in the cross section, in which case a diagonal error covariance $\boldsymbol{\Sigma} = \mathbf{D}$ will suffice.

Assume for the moment that $\boldsymbol{\varepsilon}$ has a normal distribution with mean zero and a known covariance matrix $\boldsymbol{\Sigma}$ that has full rank. In this special and rather unrealistic case the conjugate prior for $\boldsymbol{\beta}$ is $p(\boldsymbol{\beta}) = MVN(\boldsymbol{\beta} \mid \boldsymbol{\beta}_{o,} \mathbf{V}_o)$, a multivariate normal density with mean $\boldsymbol{\beta}_o$ and covariance matrix $\mathbf{V}_o$. It can be shown that for $\mathbf{X}$ and $\mathbf{V}_o$, the posterior marginal joint distribution for the regression coefficient vector $\boldsymbol{\beta}$ is a multivariate t distribution with posterior mean given by the shrinkage formula

$$\hat{\boldsymbol{\beta}} = \left(\mathbf{V}_o^{-1} + \mathbf{X}'\boldsymbol{\Sigma}^{-1}\mathbf{X}\right)^{-1} \cdot \left(\mathbf{V}_o^{-1} \cdot \boldsymbol{\beta}_o + \mathbf{X}'\boldsymbol{\Sigma}^{-1}\mathbf{X} \cdot \hat{\boldsymbol{\beta}}_{GLS}\right), \tag{7.45}$$

where $\hat{\boldsymbol{\beta}}_{GLS} = (\mathbf{X}'\boldsymbol{\Sigma}^{-1}\mathbf{X})^{-1}\mathbf{X}'\boldsymbol{\Sigma}^{-1}\boldsymbol{r}$ is the generalized least squares estimate. When $\mathbf{X}$ has less than full rank, as is often the case in the Black-Litterman applications discussed in Section 7.4, the posterior mean is still well-defined and is given by

$$\hat{\boldsymbol{\beta}} = \left(\mathbf{V}_o^{-1} + \mathbf{X}'\boldsymbol{\Sigma}_o^{-1}\mathbf{X}\right)^{-1} \cdot \left(\mathbf{V}_o^{-1} \cdot \boldsymbol{\beta}_o + \mathbf{X}'\boldsymbol{\Sigma}_o^{-1}\boldsymbol{r}\right). \tag{7.46}$$

The conjugate model above with known σ^2 is easily extended to the "full" conjugate model, where $\boldsymbol{\Sigma} = \sigma^2\boldsymbol{\Sigma}_o$ with $\boldsymbol{\Sigma}_o$ known and σ^2 is unknown, by using a multivariate normal inverse chi-squared prior (see Leamer, 1978, who uses an inverse Gamma equivalent of the inverse chi-squared distribution). However, this prior is often quite unrealistic, as in the case of the simple mean-variance model of Section 7.2, which again motivates use of alternative semi-conjugate priors and independence priors, with non-normal priors for $\boldsymbol{\beta}$ and/or t distribution likelihoods. As in the mean-variance model, we do not have closed-form analytic expressions for the posterior distributions of interest, but instead can use the S+Bayes Gibbs sampler to compute posterior distributions for the linear regression model.

7.3.1 Bayes Alphas and Betas

A simple important application of Bayes linear regression is provided by the single-factor market model

$$r_t = \alpha + r_{m,t} \cdot \beta + \varepsilon_t, \quad t = 1, 2, \cdots, T, \tag{7.47}$$

where r_t and r_{mt} are risk-adjusted returns for an individual stock and the overall market. We note that for the single-factor market model it is often reasonable to assume that the ε_t are serially uncorrelated.[33] If in addition we assume a constant error variance σ^2, we only need to specify a joint prior distribution for the two-dimensional coefficient vector (α, β) and the error variance σ^2 in order to compute posterior distributions with S+Bayes.[34] We now show how to do so for several choices of prior.

7.3.1.1 Example 7.7: Noninformative Prior

In order to check on the accuracy of the S+Bayes Gibbs sampler, we first use a noninformative prior with a normal likelihood and compare the result with the LS estimate using Code 7.14.

```
returns <- largecap.ts@data
msft.fit <- blm(MSFT-t90 ~ market-t90, data =
     returns)
msft.fit@model[[1]]@betaMatrix.names <-
  c("ALPHA","BETA")
plot(msft.fit, which = "bivariate")
msft.lsfit <- lm(MSFT-t90 ~ market-t90,
  data = returns)
names(msft.lsfit$coefficients) = c("ALPHA","BETA")
msft.fit
msft.lsfit
```

Code 7.14 Bayes Alpha and Beta for Microsoft

This results in Figure 7.25 and the posterior summary statistics

```
                mean          stdev       Bayes factor
ALPHA       0.009989242  0.01562003     0.4035703
BETA        1.692417577  0.27536771     0.0000000

             mean        stdev
sigma   0.1199513 0.01160441
```

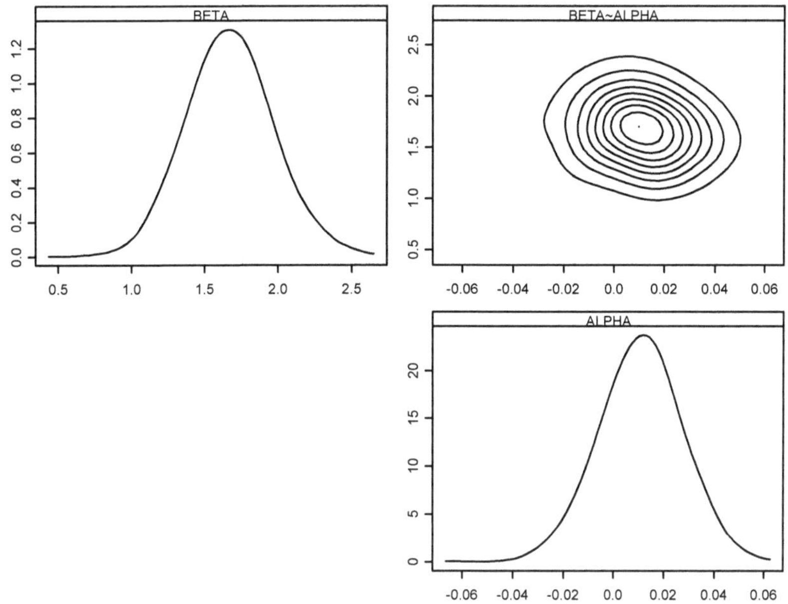

Figure 7.25 Marginal and Joint Posteriors for Microsoft Alpha and Beta

The least-squares-fitted model `msft.lsfit` gives the following estimates of alpha and beta:

```
     ALPHA       BETA
0.01047983    1.69152
```

So we have good agreement between the S+Bayes Gibbs sampler posterior means of alpha and beta and the exact posterior means given by the least squares estimates.

We see in Figure 7.25 that the marginal posterior densities are consistent with the posterior summary statistics above. We also see that the contour plot for the joint posterior indicates that there is a small amount of negative correlation in the joint posterior for alpha ("ALPHA") and beta ("BETA"). We use the function `summary(msft.lsfit)` to calculate a correlation of –.17 for the least squares estimate of alpha and beta, which is reasonably consistent with the contour plot.

7.3.1.2 Example 7.8: Informative Priors for Alpha and Beta

The idea to use informative priors in computing Bayesian betas was proposed a long time ago by Vasicek (1973), who used a semi-conjugate model with a normal prior for beta and a joint noninformative prior for the intercept and σ^2.[35] Naturally this led to a shrinkage estimator of beta, and the result provided a Bayesian motivation for computing so-called "adjusted" betas as is done by some financial data services providers. Let's see what happens when we compute Bayesian alphas and betas using a semi-conjugate model with informative priors for both α and β and a noninformative prior for σ^2.

We assume a bivariate normal prior with zero correlation between α and β. Under the CAPM, α is zero and in any event one does not expect large deviations from zero, so it is natural to assume that $\alpha \sim N(\mu_\alpha, \sigma_\alpha^2)$ with $\mu_\alpha = 0$ and σ_α small. For our example, we let $\sigma_\alpha^2 = (.01)^2$. Since betas vary around the market value $\beta = 1$, we assume that $\beta \sim N(\mu_\beta, \sigma_\beta^2)$ with $\mu_\beta = 1$ and set $\sigma_\beta^2 = (.5)^2$ as a not too unreasonable value. The S+Bayes computations to fit this Bayes model are given in Code 7.15, where the first argument to `bayes.normal` specifies the mean vector $(\mu_\alpha, \mu_\beta) = (0,1)$, and our use of a vector for the `covmat =` argument specifies that the covariance matrix of the joint prior is diagonal. Since a `priorSigma` argument is not used, S+Bayes uses the default of a noninformative prior for σ^2.

```
returns <- largecap.ts@data
prior.mean <- bayes.normal(c(0,1),
  covmat = c((.01)^2,(.5)^2))
my.prior <- blm.prior(priorBeta = prior.mean)
msft.fit <- blm(MSFT-t90 ~ market-t90,
  prior = my.prior,
  data = returns)
msft.fit@model[[1]]@betaMatrix.names <-
  c("ALPHA","BETA")
msft.fit
```

Code 7.15 Bayesian Alpha and Beta with Informative Semi-conjugate Prior

The resulting posterior mean and standard deviation values are:

```
coefficients:
                    mean        stdev        Bayes factor
(Intercept)    0.0037402  0.008497108     0.5191809
     market    1.5398064  0.250437392     0.0000000

scale:
            mean       stdev
sigma   0.1198259  0.01111548
```

The effect of using the informative prior versus a noninformative prior for alpha and beta has been to shift both marginal posterior distributions toward the prior means, with the posterior mean for alpha shifting from .01 to .004 and the posterior mean for beta shifting from 1.69 to 1.54. Marginal and joint posterior plots can be made by adding the line `plot(msft.returns, which = "bivariate")` to the code.

7.3.1.3 Example 7.9: Alpha and Beta with t-Likelihood

In Example 7.5, we saw that use of a *t* distribution resulted in robustness toward outlier returns. Code 7.16 computes alpha and beta posteriors for the stock with ticker EVST shown in Figure 6.1 using normal and *t* distribution likelihoods with the same prior as in Example 7.8.

```
returns <- microcap.ts@data
prior.mean <- bayes.normal(c(0,1),
  covmat = c((.01)^2,(.5)^2))
my.prior <- blm.prior(priorBeta = prior.mean)
evst.fit <- blm(EVST-t90 ~ market-t90,
  prior = my.prior, data = returns)
likelihood <- blm.likelihood("t")
evst.fit.robust <- blm(EVST-t90 ~ market-t90,
  prior = my.prior, likelihood = likelihood,
  data = returns)
plot(evst.fit, which = "box")
plot(evst.fit.robust, which = "box")
evst.fit
evst.fit.robust
```

Code 7.16 EVST Posteriors with Normal and *t* Distribution Likelihoods

Here are the numerical results from Code 7.16, rearranged to facilitate comparisons:

```
Normal likelihood
                    mean         stdev      Bayes factor
ALPHA           0.000801839  0.01011735   1.05542104
BETA            1.117184615  0.49114618   0.01442915

t-likelihood
                    mean         stdev      Bayes factor
ALPHA           -0.01003383  0.00959187   0.15208213
BETA             0.84098359  0.37490413   0.01684809

Normal likelihood
```

```
            mean       stdev
sigma   0.9123989  0.08349846

t-likelihood
            mean       stdev
sigma   0.1647692 0.02185158
```

We see that the posterior means of alpha and beta are both smaller with the t likelihood than with the normal likelihood. The posterior standard deviation of alpha is about the same for both likelihoods, while that of beta is noticeably smaller. These results are as expected because the outlier has less influence on the posterior distribution when using a t likelihood than when using a normal likelihood. The reduction in the values of the posterior mean and standard deviation of σ are much greater than those of alpha and beta because the outlier has a greater influence on the posterior for σ when a normal likelihood is used. We recommend that the reader examine the boxplots produced by Code 7.16 (not reproduced here) to get a clear visual comparison of the posterior distributions associated with the normal and t distribution likelihoods.

7.3.2 Working with the S+Bayes Gibbs Sampler Output

Often we will find it is advantageous to work directly with the Gibbs sampler output from the S+Bayes modeling functions. For example, we may want to produce stylized plots to convey results more clearly than when using a built-in S+Bayes plotting method such as the method `plot.blm` invoked when the argument of `plot` is a Bayes linear model object. Or we may wish to compute the posterior distribution of some nonlinear function of model parameters such as a Sharpe ratio or information ratio. So we explain how the Gibbs sampler sample paths are obtained from a fitted model of class `blm`, and give two examples of applications.

An S+Bayes linear model object of class `blm`, such as `msft.fit`, contains all the Gibbs sampler results in one or more components of an ordinary list object named `model`, which itself is a component of the `blm` object. When using a single MCMC chain, as we have been doing thus far, the model list contains only the single component `model[[1]]`, and this component contains various MCMC objects. We display the names of these MCMC objects by using the extractor function `slotNames`, for example as follows[36]:

```
> slotNames(msft.fit@model[[1]])
[1]"call"                    "betaMatrix"
[3]"betaMatrix.names"        "scaleMatrix"
[5]"scaleMatrix.names"       "tau2ErrorScale"
```

```
[7]"tau2ErrorScale.names"  "gibbs.drawing.stats"
[9]"mixture.drawing.stats"."prior"
[11]"control"
```

The components of `msft.fit@model[[1]]` are themselves S Version 4 objects that are again accessed using the "@" symbol. For example, we display the names of the Gibbs sampler sample paths matrices for alpha and beta (two columns) and sigma (one column) with

```
> msft.fit@model[[1]]@betaMatrix.names
[1] "ALPHA" "BETA"
> msft.fit@model[[1]]@scaleMatrix.names
[1] "SIGMA"
```

We display the first few values of these matrices:

```
> msft.fit@model[[1]]@betaMatrix[1:5,]
             [,1]        [,2]
[1,] 0.019858159    1.617498
[2,] 0.005506613    1.451466
[3,] 0.011686599    2.067712
[4,] 0.015837226    1.439625
[5,] 0.001776106    1.524045

> msft.fit@model[[1]]@scaleMatrix[1:5,]
[1] 0.1311373 0.1306821 0.1347287 0.1192026
     0.1073955
```

In the latter case, we get a vector object because we extracted a single column from a matrix object.

7.3.2.1 Example 7.10: Paired Posteriors Boxplots for Normal and *t* Likelihoods

Revisiting the *t* likelihood (Example 7.9), we provide Code 7.17 to display side-by-side boxplots for alpha, beta, and sigma in pairs corresponding to normal and *t* likelihoods.

```
prior.mean <- bayes.normal(c(0,1),
  covmat = c((.01)^2,(.5)^2))
my.prior <- blm.prior(priorBeta = prior.mean)

evst.fit <- blm(EVST-t90 ~ market-t90,prior =
```

```
  my.prior,data = returns)
evst.fit@model[[1]]@betaMatrix.names <-
  c("ALPHA","BETA")
likelihood <- blm.likelihood("t")
evst.fit.robust <- blm(EVST-t90 ~ market-t90,
  prior = my.prior, likelihood = likelihood,
  data = returns)
evst.fit.robust@model[[1]]@betaMatrix.names <-
  c("ALPHA","BETA")
betas <- evst.fit@model[[1]]@betaMatrix
sigma <- evst.fit@model[[1]]@scaleMatrix
rob.betas <- evst.fit.robust@model[[1]]@betaMatrix
rob.sigma <- evst.fit.robust@model[[1]]@scaleMatrix
names <- c("NORMAL","t")
par(mfrow = c(1,3))
boxplot(betas[,1],rob.betas[,1],names = names,
  main="ALPHA POSTERIOR")
boxplot(betas[,2],rob.betas[,2],names = names,
  main="BETA POSTERIOR")
boxplot(sigma,rob.sigma,names = names,
  main = "SIGMA POSTERIOR")
par(mfrow = c(1,1))
```

Code 7.17 Using Gibbs Sampler Output for Paired Boxplots for Posteriors

The resulting display in Figure 7.26 gives us an immediate grasp of the differences in the marginal posteriors due to using a normal likelihood versus a t likelihood. The reader is encouraged to use Code 7.17 to explore the impact of changing the values of the hyperparameters in the prior for alpha and beta.

7.3.2.2 Example 7.11: Posterior Distribution of an Information Ratio

Consider the beta regression model for portfolio returns $r_{P,t}$ of a portfolio P relative to the returns $r_{B,t}$ of a benchmark B:

$$r_{P,t} = \alpha_P + r_{B,t} \cdot \beta_P + \varepsilon_{P,t}, \quad t = 1,2,\cdots,T \tag{7.48}$$

Here α_P is the expected residual return (alpha) and $\sigma_{\varepsilon,P} = \text{var}^{1/2}(\varepsilon_{P,t})$ is the residual risk. Suppose we we have monthly returns and want to compute the Bayes posterior distribution of the annualized information ratio

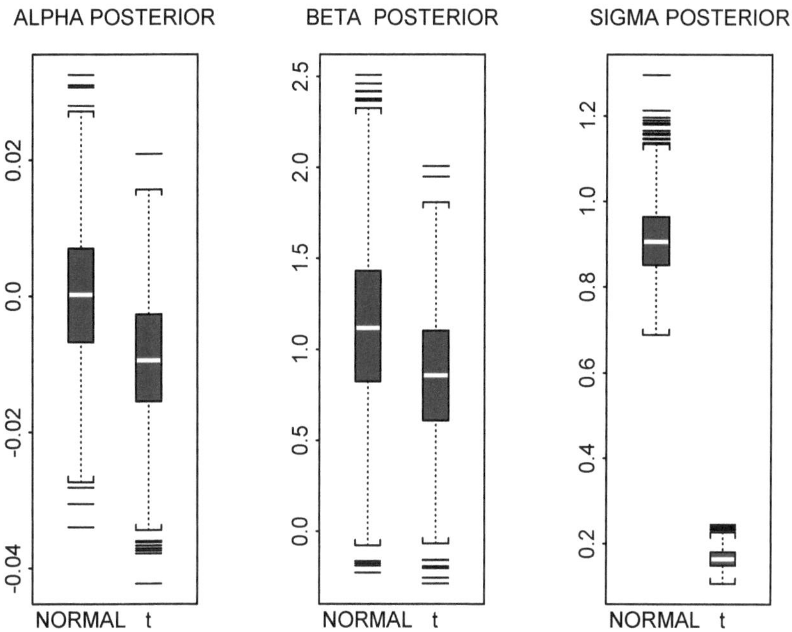

Figure 7.26 EVST Posterior Distributions for Normal and t-Likelihoods

$$\begin{aligned} IR_P &= \frac{\text{Annualized Residual Return}}{\text{Annualized Residual Risk}} \\ &= \sqrt{12}\frac{\alpha_P}{\sigma_{\varepsilon,P}} . \end{aligned} \tag{7.49}$$

We can do this quite easily by fitting a Bayes linear model for $(\alpha_P, \beta_P, \sigma_{\varepsilon,P})$ and using the Gibbs sampler posterior path values for α_P and $\sigma_{\varepsilon,P}$ in the expression for IR_P to estimate the desired posterior distribution.

For the sake of illustration, let's assume that our benchmark portfolio is an equal- weighted portfolio of the twenty stocks in `largecap.ts` and that the portfolio whose information ratio we are interested in is an equal-weighted portfolio of the twenty stocks in `smallcap.ts`. Code 7.18 computes the large cap benchmark and microcap portfolio and plots the returns of these portfolios shown in Figure 7.27. The code also computes and makes boxplots of posterior residual risk, residual return, and information ratios for both normal and *t* likelihoods, as shown in Figure 7.28. In this case, the medians of the information ratio posteriors are rather close to zero, with values of –.07 and –.13 for the normal likelihood and *t* likelihood respectively. The reader is encouraged to use Code 7.18 for other benchmarks and better-performing portfolios.

```
# Make benchmark and portfolio time series object
```

```
positions <- positions(largecap.ts)
BENCHMARK <- apply(largecap.ts@data[,-(1:2)],
  1,mean)
bm.port <- data.frame(BENCHMARK)
bm.port.ts <- timeSeries(bm.port,positions =
  positions)
PORTFOLIO <- apply(microcap.ts@data[,-(1:2)],
  1,mean)
port <- data.frame(PORTFOLIO)
port.ts <- timeSeries(port,positions = positions)
data.ts <- seriesMerge(bm.port.ts,port.ts)
y.name <- colIds(data.ts)
seriesPlot(data.ts,one.plot=F,strip.text=y.name,
  col = 1)
# Bayes models for normal and t-likelihoods
prior.mean <- bayes.normal(c(0,1.5),
  covmat = c((.01)^2,(.5)^2))
my.prior <- blm.prior(priorBeta = prior.mean)
t90 <- largecap.ts@data[,2]
port.fit <- blm(PORTFOLIO-t90 ~ BENCHMARK-t90,
  prior = my.prior, data = data.ts)
likelihood <- blm.likelihood("t")
port.fit.t <- blm(PORTFOLIO-t90~BENCHMARK-t90,
  prior = my.prior,likelihood = likelihood,
  data = data.ts)

# Compute posterior residual returns and risks and
# IR's
res.ret <- 12*port.fit@model[[1]]@betaMatrix[,1]
res.risk <-
     sqrt(12)*port.fit@model[[1]]@scaleMatrix
IR <- res.ret/res.risk
res.ret.t <-
     12*port.fit.t@model[[1]]@betaMatrix[,1]
res.risk.t <-
     sqrt(12)*port.fit.t@model[[1]]@scaleMatrix
IR.t <- res.ret.t/res.risk
names <- c("NORMAL","t")
boxplot(IR,IR.t,names = names,
  main = "NORMAL AND t-LIKELHOOD INFORMATION
     RATIOS")
names <- c("RRET (NL)","RRISK (NL)","RRET (TL)",
  "RRISK (TL)")
```

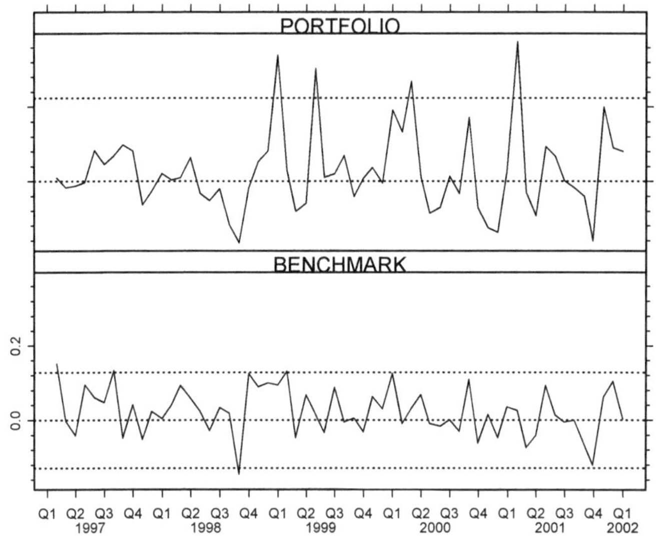

Figure 7.27 Returns of Large cap Benchmark and Microcap Portfolio

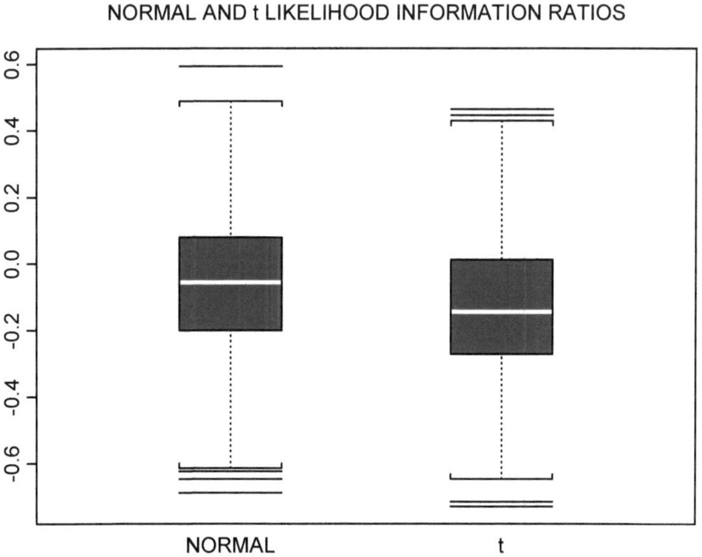

Figure 7.28 Boxplot of Information Ratio Posterior Distributions

```
boxplot(res.ret,res.risk,res.ret.t,res.risk.t,
  names = names)
title(main = "RESIDUAL RISK AND RESIDUAL RETURN\n
  Normal Likelhood (NL) and t-Likelihood (TL)")
apply(cbind(IR,IR.t),2,median)
```

Code 7.18 Information Ratio Posteriors

7.3.3 Cross-Section Regression

We provide a simple example of an S+Bayes cross-section regression for monthly returns for 469 stocks in the STOXX index for January 1993. The data frame containing these data is `stoxx.cs`, and the variable names are:

```
> names(stoxx.cs)
[1] "Returns" "Size"  "Mkt.cap"  "PE"   "PC"
[6] "Sector"  "Country" "Country.two"
```

`Returns` is the response variable, and we choose `Size`, `log(PE)` and `Country.two` as the risk-factor predictor variables. `Country.two` is a factor variable with two levels `"GB"` (Great Britain) and `"OTHER"` (any one of 14 European countries). Our experience leads us to use the mixed prior $p(\boldsymbol{\beta},\sigma^2) \propto MVN(\boldsymbol{\beta} \mid \boldsymbol{\beta}_o, \mathbf{V}_o)$ with hyperparameters $\boldsymbol{\beta}_o = (.1,-.02,0,0)$ and $\mathbf{V}_o = diag((.01)^2,(.01)^2,(.01)^2,(.01)^2)$, where by mixed we mean that the prior is informative for $\boldsymbol{\beta}$ and noninformative for σ^2. Code 7.19 computes a Bayesian model fit via the Gibbs sampler and produces the plots displayed in Figure 7.29 and Figure 7.30. The code also computes least squares coefficient estimates for comparison.

```
form <- formula(Returns ~ Size+log(PE)+Country.two)
prior.mean <- bayes.normal(c(.1,-.02,0,0),
  covmat = (.01)^2)
my.prior <- blm.prior(priorBeta = prior.mean)
stoxx.fit <- blm(form,prior = my.prior,
  data = stoxx.cs, na.action=na.omit)
plot(stoxx.fit, which = c("box","bivariate"))
stoxx.fit
coef(summary(lm(form, data = stoxx.cs,
  na.action = na.omit)))
```

Code 7.19 Bayes Model for STOXX Cross-Section Regression

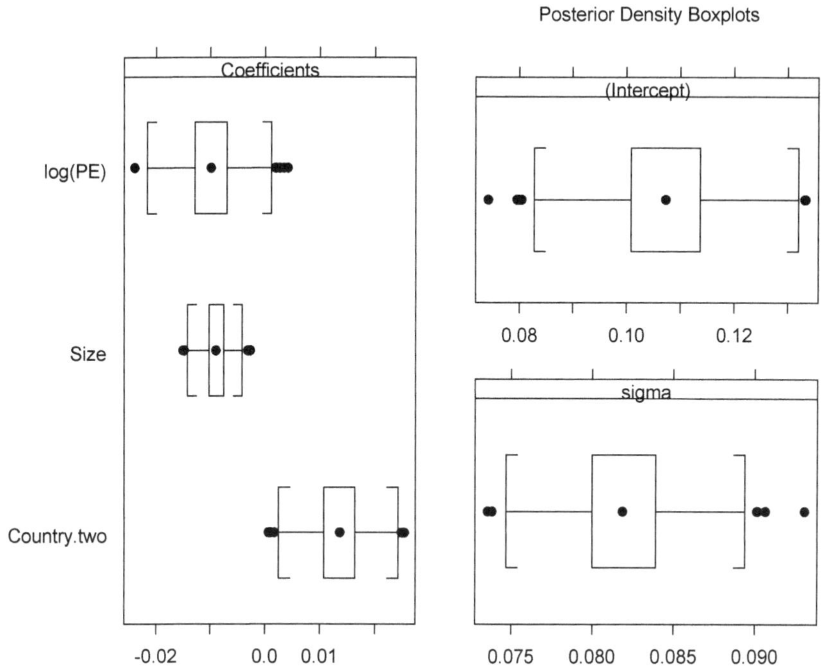

Figure 7.29 Boxplot of Marginal Posteriors for STOXX Cross-Sectional Model

In this example the `Size` and `log(PE)` factors have a substantial negative impact on returns, while `Country.two` has a substantially positive impact and the `Intercept` parameter is quite positive. Several of the posterior correlations are negative.

7.4 Black-Litterman Models

A Black-Litterman (BL) model for portfolio optimization uses an estimate of mean returns that is a weighted combination of (1) a mean vector $\boldsymbol{\mu}_o$ in a prior distribution for mean returns based on an equilibrium model and (2) an "estimate" that reflects an investor's "views" about mean returns.[37] The mean vector $\boldsymbol{\mu}_o$ is usually taken to be the implied returns obtained by reversing the quadratic optimization result used to obtain optimal portfolio weights, instead taking the weight vector to be the market capitalization weights $\mathbf{w}_{MC}$; i.e.,

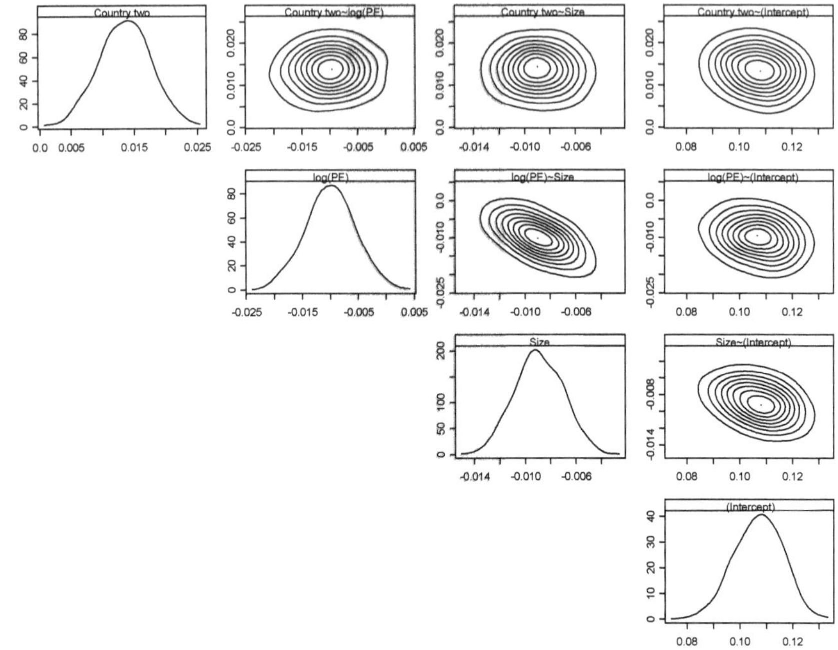

Figure 7.30 Bivariate Marginal Posteriors for STOXX Cross-Section Model

$$\mu_O = \lambda \mathbf{\Omega} \mathbf{w}_{MC}, \tag{7.50}$$

where $\mathbf{\Omega}$ is the covariance matrix of the returns and λ is a risk aversion parameter. Alternatively, μ_o might be obtained from the CAPM

$$\mu_O = \mathbf{1} \cdot r_f + \boldsymbol{\beta} \cdot (\mu_M - r_f), \tag{7.51}$$

where μ_M is the mean return of the market and r_f is the risk-free rate.

The investor's views are uncertain and typically subjective opinions about mean returns, and these views may be absolute or relative (see the discussion in Section 1.2.3). For example, an absolute view might be "stock A will give 1% monthly returns while stock B will give 2% monthly return," and a relative view might be "stock A will outperform stock B by .5%." Investor views can be expressed with a linear regression model

$$\boldsymbol{g} = \mathbf{P}\boldsymbol{\mu} + \boldsymbol{\varepsilon}, \tag{7.52}$$

where $\boldsymbol{\mu}$ is a vector of true mean returns, $\mathbf{P}$ and $\boldsymbol{g}$ are used to define an error-free version of investor views, and $\boldsymbol{\varepsilon}$ represents errors in the investor's views.[38]

It is assumed that $\boldsymbol{\varepsilon}$ has zero mean and has a known covariance matrix that represents the investor's degree of confidence in his or her views.

For example, with four assets A, B, C, and D and the absolute view that stock A will have 1% returns and stock B will have 2% returns, we set $\mathbf{g}' = (1,2)$, $\boldsymbol{\mu}' = (\mu_A, \mu_B, \mu_C, \mu_D)$ and $\mathbf{P} = \begin{pmatrix} 1 & 0 & 0 & 0 \\ 0 & 1 & 0 & 0 \end{pmatrix}$. In the case of the relative view that stock A outperforms stock B by .5%, we set $\boldsymbol{g} = g = .5$ and $\mathbf{P} = \boldsymbol{p}' = (1,-1,0,0)$. In examples such as these, $\mathbf{P}$ is rank-deficient, but as we noted at the beginning of Section 7.3, this is not a problem.

This linear regression model implementation of investor views could also incorporate historical mean returns of some stocks as part of an investor set of views by setting some of the components of $\boldsymbol{g}$ equal to the sample means of the stocks, with appropriate row entries in $\mathbf{P}$ that contain a single entry value of one and remaining entries equal to zero and with an appropriate historical covariance matrix estimate for the corresponding components of $\boldsymbol{\varepsilon}$.

The Black-Litterman estimate is easily derived in one of two ways: (a) as the solution of a constrained least squares problem, as in Lee (2000), or (b) as the posterior mean in a Bayes version of the linear regression model above, as is implicit in Black and Litterman (1992) and He and Litterman (1999).[39] The Bayes derivation of Black-Litterman uses a multivariate normal prior $MVN(\boldsymbol{\mu} \mid \boldsymbol{\mu}_o, \tau \cdot \boldsymbol{\Omega})$ where $\boldsymbol{\mu}_o$ is the equilibrium mean returns, $\boldsymbol{\Omega}$ is assumed to be known, and τ an investor-determined tuning parameter. By applying the posterior mean formula from the beginning of Section 7.3 with the appropriate substitutions, we obtain the posterior mean shrinkage formula

$$\hat{\boldsymbol{\mu}} = \left((\tau \cdot \boldsymbol{\Omega})^{-1} + \mathbf{P}'\boldsymbol{\Sigma}^{-1}\mathbf{P} \right)^{-1} \cdot \left((\tau \cdot \boldsymbol{\Omega})^{-1} \cdot \boldsymbol{\mu}_o + \mathbf{P}'\boldsymbol{\Sigma}^{-1}\boldsymbol{g} \right). \tag{7.53}$$

The size of τ controls the relative weight put on the equilibrium returns $\boldsymbol{\mu}_o$, with smaller values of τ resulting in more weight on $\boldsymbol{\mu}_o$.

7.4.1 Black-Litterman Application Example

We illustrate the use of the Black-Litterman estimate in optimizing a small eight-asset portfolio. Each of the assets is an equally weighted portfolio of ten stock returns from one of the stock returns objects `micrcocap.ts`, `smallcap.ts`, `midcap.ts` and `largecap.ts`. We formed two portfolios from each market cap group by sorting on mean returns and putting the stocks with the ten highest returns in one equally weighted portfolio and those with the ten lowest returns in the other portfolio. Portions of the rather mundane Code 7.20 show how to construct these portfolios.

```
mu1 <- apply(microcap.ts@data[,1:20],2,mean)
sort(mu1)
names1 <- names(sort(mu1)[1:10])
names2 <- names(sort(mu1)[11:20])
...  ...   ...   ...
mu4 <- apply(largecap.ts@data[,1:20],2,mean)

sort(mu1)
names7 <- names(sort(mu4)[1:10])
names8 <- names(sort(mu4)[11:20])
pos <- positions(microcap.ts)
mc1 <- timeSeries(apply(
  microcap.ts[,names1]@data,1,mean),positions =
     pos)
...  ...   ...   ...
mc8 <- timeSeries(apply(
  largecap.ts[,names8]@data,1,mean),positions =
     pos)
mc.slices <- seriesMerge(mc1,mc2,mc3,mc4,mc5,
  mc6,mc7,mc8)
dimnames(mc.slices@data)[[2]] <- paste("MC",1:8,
  sep = "")
mc.slices <- seriesMerge(microcap.ts[,20:21],
  mc.slices)
dimnames(mc.slices@data)[[2]][1:2] <-
  c("MARKET","RF")
```

Code 7.20 Create Eight Equal Weights Portfolios

After running Code 7.20, we get the plots of the returns series of the eight portfolios in Figure 7.32 with the commands:

```
y.name = colIds(mc.slices[,-2])
seriesPlot(mc.slices[,-2],one.plot=F,
  strip.text=y.name,col = 1)
```

The mean returns versus standard deviation plot of the market and portfolios returns in Figure 7.31 is obtained with Code 7.21.

```
mc.ret <- mc.slices@data[,-(1:2)]
mu <- apply(mc.ret,2,mean)
sigma <- apply(mc.ret,2,stdev)
xlim <- c(0,max(sigma))
plot(sigma,mu,pch = 16, xlim = xlim,
  main = "MARKET CAP SLICE PORTFOLIOS")
text(sigma +.003,mu, names(mu),adj = 0, cex = .8)
```

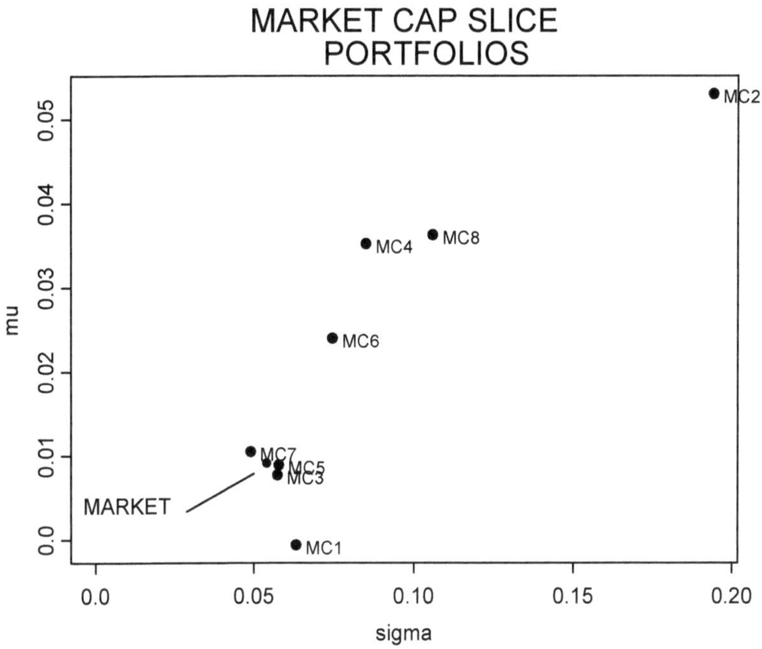

Figure 7.31 Means and Standard Deviations of the Eight Portfolio Returns

```
mkt <- mc.slices@data[,1]
mu.mkt <- mean(mkt)
sigma.mkt <- stdev(mkt)
points(sigma.mkt,mu.mkt,pch = 16)
text(locator(1), "MARKET")
lines(locator())
out <- rbind(mu,sigma,sigma/sqrt(60))
row.names(out) <- c("  MEAN","SIGMA","S.E.(MEAN)")
round(out,3)
```

Code 7.21 Plot Means and Standard Deviations of Portfolios

The last line of Code 7.21 gives the values of the sample means, standard deviations, and standard errors of the sample means of the eight portfolios:

	MC1	MC2	MC3	MC4	MC5	MC6	MC7	MC8
MEAN	-0.001	0.053	0.008	0.035	0.009	0.024	0.009	0.032
SIGMA	0.063	0.194	0.057	0.085	0.058	0.075	0.045	0.085
S.E.(MEAN)	0.008	0.025	0.007	0.011	0.007	0.010	0.006	0.011

Code 7.22 uses NuOPT to compute optimal long-only portfolio weights with a monthly target return of 3% for both the original returns and for the returns with

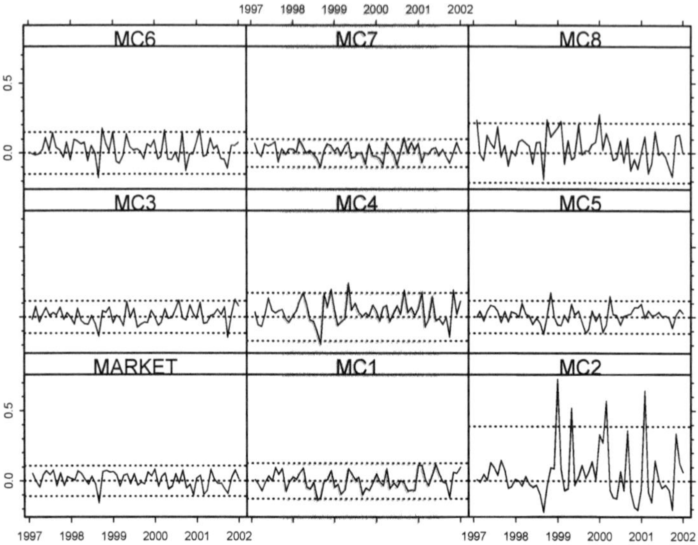

Figure 7.32 Returns of Market and Equally Weighted Portfolios of Ten Stocks

altered mean values. We perturb the mean value of the returns by increasing MC1 and MC3 returns by one standard deviation of their estimted values and decreasing MC4 and MC8 returns by one standard deviation of their estimated values.

```
mc.ret <- mc.slices@data[,-(1:2)]
n <- ncol(mc.ret)
mu.target <- 0.03
mu <- apply(mc.ret,2,mean)
Cov <- var(mc.ret)
A <- rbind(mu,1)
cLO <- c(mu.target,1)
cUP <- c(Inf,1)
bLO <- rep(0, n)
bUP <- rep(Inf, n)
solution <- solveQP(Cov,, A, cLO, cUP, bLO, bUP,,)
wts <- solution$variables$x$current
# Weights with perturbed mu's
mc.ret[,1] <- mc.ret[,1] + .008
mc.ret[,3] <- mc.ret[,3] + .007
mc.ret[,4] <- mc.ret[,4] - .011
mc.ret[,8] <- mc.ret[,8] - .011
mu <- apply(mc.ret,2,mean)
```

```
A <- rbind(mu,1)
solution <- solveQP(Cov,, A, cLO, cUP, bLO, bUP,,)
wts.pert <- solution$variables$x$current
mu.labels <- factor(c(rep("ORIGINAL MU'S",8),
  rep("PERTURBED MU'S",8)))
weights <- c(wts,wts.pert)
wts.data <- data.frame(wt.names =
  rep(names(mc.ret),2), weights,mu.labels)
barchart(wt.names ~ weights|mu.labels,data =
  wts.data)
```

Code 7.22 Markowitz Weights with Original and Perturbed Data

The results in Figure 7.33 display the well-known fact that mean-variance-optimized portfolios are often poorly diversified and the fact that optimal portfolio weights are often highly sensitive to small perturbations of the mean values of returns.

Now suppose a manager has the absolute view that the mean return of MC1 and MC2 will be 4% and the relative view that MC3+MC4 will outperform MC5+MC6 by 1%. The following code bits will create the necessary S-PLUS matrices `p` and `g`, labeled **P** and ***g***, above:

```
> p.abs = c(0.5, 0.5, rep(0, 6))
> p.rel = c(0, 0, 1, 1, -1, -1, 0, 0)

> p = rbind(p.abs, p.rel)
> p
       [,1] [,2] [,3] [,4] [,5] [,6] [,7] [,8]
p.abs   0.5  0.5    0    0    0    0    0    0
p.rel   0.0  0.0    1    1   -1   -1    0    0
> g = rbind(0.04, 0.01)
> g
      [,1]
[1,]  0.04
[2,]  0.01
```

In addition, we assume the manager has the following error covariance matrix (labeled $\mathbf{\Sigma}$ above):

```
> cov.investor = diag(c((0.01)^2, (0.005)^2))
```

Code 7.23 provides a function `bl.means` to compute a Black-Litterman estimate of the posterior shrinkage estimate of mean returns using the CAPM equilibrium model. The first argument, `xret`, is a data frame of returns, with market returns and the risk-free rate in the first two columns and asset returns in

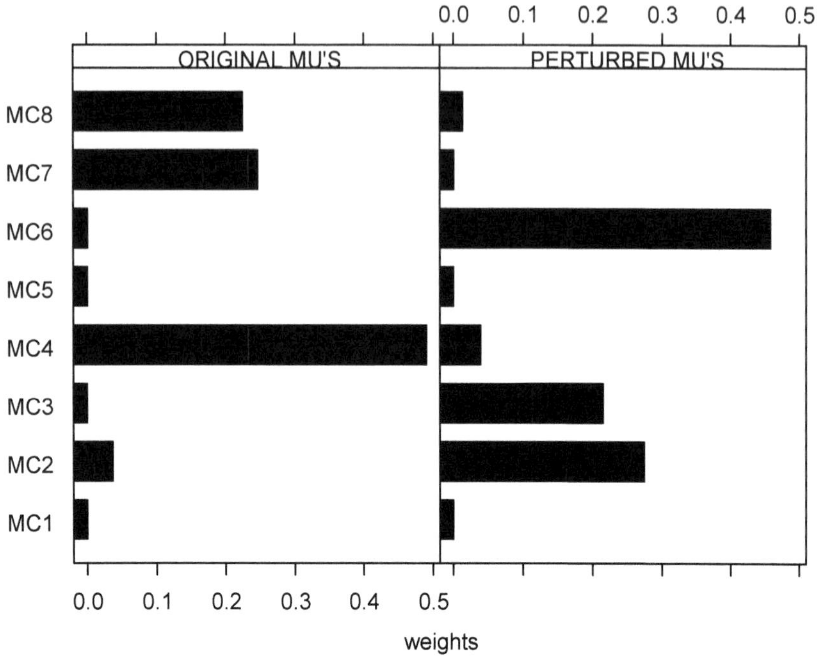

Figure 7.33 Markowitz Weights for Original and Perturbed Data

the remaining columns (the reader can easily change this convention as needed). Additional required arguments are an estimate `cov0` of the covariance $\mathbf{\Omega}$ and the investor's views expressed through the values of `covi`, `p`, `g`, and `tau`.

```
bl.means <- function(xret,cov0,covi,p,g,tau,
  delta.betas=0,print.mus = T) {
  # xret is a data frame of returns with market
  # returns in column 1 and risk-free rate and in
  # column 2
  # Compute betas and equilibrium returns via CAPM
  mkt <- xret[,1]
  rf <- xret[,2]
  betas <- rep(0,8)
  names(betas) <- dimnames(xret[,-(1:2)])[[2]]
  for(i in 1:8){
     betas[i] <- coef(lm(xret[,i+2]-rf~mkt-rf))[2]
  }
  betas <- betas + delta.betas
  mu0 <- rf[60] + betas*(mean(mkt)-rf[60])

  # Compute Black-Litterman mean returns
```

```
  a1 <- solve(cov0)/tau
  a2 <- t(p)%*%solve(covi)
  a3 <- a2%*%p
  muhat <- t(solve(a1 + a3)%*%(a1%*%mu0 + a2%*%g))
  # Output
  mu <- apply(xret[,-(1:2)],2,mean)
  mus <- rbind(mu,betas,mu0,muhat)
  row.names(mus) <- c("HIST. MU'S","BETAS",
     "EQUIL. MU'S","B-L MU'S")
  if(print.mus) print(round(mus,5))
  as.matrix(mus[4,])
}
```

Code 7.23 Black-Litterman Mean Return Estimate

The CAPM equilibrium prior for mean returns has two sources of uncertainty about the value of μ_o: (a) uncertainty in the estimate of market returns and (b) uncertainty in the beta estimates. We now examine the sensitivity of the BL estimate to each of these uncertainties separately in turn.

The S-PLUS commands

```
mean(ret[,1])
```

and

```
stdev(ret[,1])/sqrt(60)
```

compute .009 as the value of the sample mean of the market returns and .007 as the standard error of this sample mean. We evaluate the impact of a perturbation of mean market returns from its estimated value of .009 to the one standard error increase in value to .016. Code 7.24 uses the function `bl.means`, and parts of Code 7.22 that use NUOPT (not all of which is displayed below), to compute optimal portfolio weights for both the orginal sample mean of market returns and the perturbed sample mean of market returns. The results are displayed in Figure 7.34.

```
ret <- mc.slices@data
# Mean returns prior covariance
cov0 <- var(ret[,-(1:2)]) #Prior covariance

tau <- 1
# Define investor's views
p.abs <- c(.5,.5,rep(0,6))
p.rel <- c(0,0,1,1,-1,-1,0,0)
p <- rbind(p.abs,p.rel)
```

```
g <- rbind(.04,.01)
#Investor covariance
covi <- diag(c((.01)^2,(.005)^2))
# Black-Litterman Means
bl.means(ret,cov0,covi,p,g,tau)
# Optimal Black-Litterman Weights
mc.ret <- ret[,1:]
n <- ncol(mc.ret)
mu.target <- 0.03
mu <- bl.means(ret,cov0,covi,p,g,tau)
...  ...   ...   ...
# barplot(wts, names = names(mc.ret))
# Weights with perturbed mu's
ret[,1] <- ret[,1] + .008
mu <- bl.means(ret,cov0,covi,p,g,tau)
...  ...   ...   ...
barchart(wt.names ~ weights|mu.labels,
  data = wts.data)
```

Code 7.24 BL Optimal Weights for Original and Perturbed Data

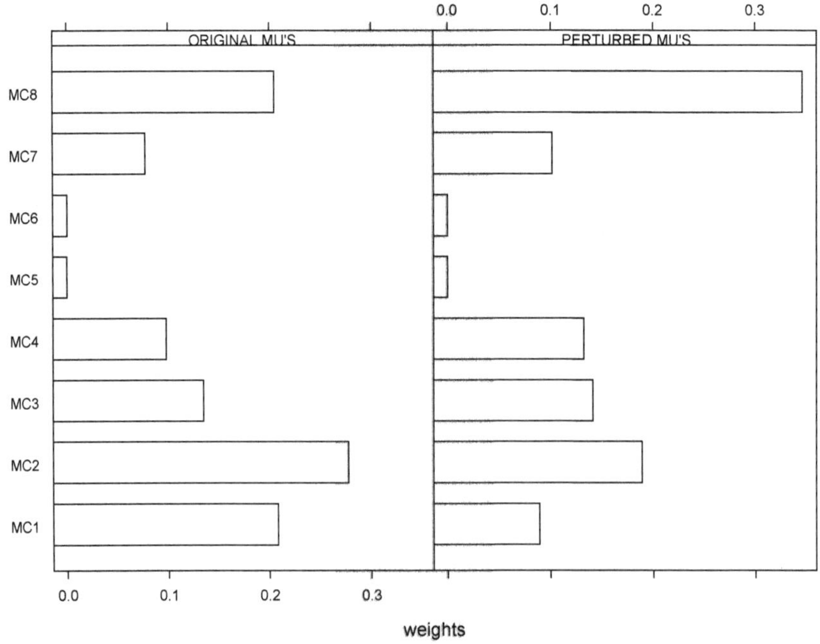

Figure 7.34 BL Weights for Original and Perturbed Market Mean

We see in the left panel of Figure 7.34 that the BL estimate leads to a substantially more diversified portfolio than when using historical sample means, as is clear by comparison with the left panel of Figure 7.33. We also see in the right panel of Figure 7.34 that BL leads to optimal portfolios that are much more stable under perturbations of the data. (Compare it with the right panel of Figure 7.33.)

Table 7.1 displays the historical (sample) means, estimated betas, CAPM equilibrium means, and BL means for the original sample mean of market returns. Table 7.2, produced using the command

```
> round(betas.eqret(mc.slices@data),3)
```

gives the same quantities for the perturbed sample mean of market returns.

Note the relatively large changes in the equilibrium means, and the relatively small changes in the BL means.

Now we perturb the estimated values of the betas. The function `betas.eqret` in Code 7.25 computes betas and their standard errors, along with CAPM equilibrium returns:

```
betas.eqret <- function(xret) {
  # xret is a data frame of returns with market
  # returns in the first column and the risk-free
```

Table 7.1 BL Means for Original and Perturbed Market Mean

	MC1	MC2	MC3	MC4	MC5	MC6	MC7	MC8
HIST. MU'S	-0.001	0.053	0.008	0.035	0.009	0.024	0.009	0.032
BETAS	0.604	1.803	0.509	1.028	0.433	0.994	0.550	1.366
EQUIL. MU'S	0.006	0.015	0.005	0.009	0.005	0.009	0.006	0.012
BL MU'S	0.015	0.064	0.011	0.024	0.005	0.020	0.008	0.022
	MC1	MC2	MC3	MC4	MC5	MC6	MC7	MC8
HIST. MU'S	-0.001	0.053	0.008	0.035	0.009	0.024	0.009	0.032
BETAS	0.604	1.803	0.509	1.028	0.433	0.994	0.550	1.366
EQUIL. MU'S	0.010	0.028	0.009	0.017	0.008	0.016	0.010	0.021
BL MU'S	0.017	0.063	0.013	0.028	0.008	0.023	0.011	0.029

Table 7.2 Betas with Standard Errors and Equilibrium Mean Returns

	MC1	MC2	MC3	MC4	MC5	MC6	MC7	MC8
BETAS	0.604	1.803	0.509	1.028	0.433	0.994	0.550	1.366
STD. ERRORS	0.132	0.407	0.123	0.156	0.129	0.126	0.081	0.102
EQ. RETURNS	0.006	0.015	0.005	0.009	0.005	0.009	0.006	0.012

```
  # rate in the second column
  mkt <- xret[,1]
  n <- nrow(xret)
  rf <- xret[,2]
  mu.mkt <- mean(mkt)
  out <- data.frame(rbind(rep(0,8),
     rep(0,8),rep(0,8)))
  names(out) <- names(xret[,-(1:2)])

  row.names(out) <- c("BETAS","STD. ERRORS",
     "EQ. RETURNS")
  for(i in 1:8){
     out[1:2,i] <- summary(lm(ret[,i+2]-rf ~
       mkt-rf))$coefficients[2,1:2]
     out[3,i] <- rf[n] + out[1,i]*(mu.mkt - rf[n])
  }
  out
}
```

Code 7.25 Estimated Betas and CAPM Equilibrium Returns

We modify Code 7.24 to reflect perturbations in the estimated betas by plus or minus one standard deviation, with random distribution of the signs, by replacing the lines

```
ret[,1] <- ret[,1] + .007
mu.pert <- bl.means(ret,cov0,covi,p,g,tau)
```

with the lines

```
delta.betas <- c(-0.13,0.41,0.12,-0.16,-0.13,0.13,
  0.08,-0.10)
mu.pert <- bl.means(ret,cov0,covi,p,g,tau,
  delta.betas = delta.betas)
```

This produces the result in Figure 7.35, where comparison of the right-hand and left-hand panels shows that the weights are moderately stable with respect to pertubations of one standard deviation in all the betas.

The historical means, estimated betas, equilibrium means, and BL means for the original betas are provided in the first section of Table 7.3, followed by the values for the perturbed betas in the second section of Table 7.3 for comparison.

In the example we have been discussing, the results appear to be rather insensitive to the choice of τ (`tau`), with values of `tau = 5` and `tau = .5` giving almost the same results as `tau = 1`. The results in this example seem to be relatively more sensitive to the choice of $\mathbf{\Sigma}$. The reader is encouraged to experiment with the values of τ and $\mathbf{\Sigma}$ (Exercise 13).

Table 7.3 BL Means for Original and Perturbed Betas

	MC1	MC2	MC3	MC4	MC5	MC6	MC7	MC8
HIST. MU'S	-0.001	0.053	0.008	0.035	0.009	0.024	0.009	0.032
BETAS	0.604	1.803	0.509	1.028	0.433	0.994	0.550	1.366
EQUIL. MU'S	0.006	0.015	0.005	0.009	0.005	0.009	0.006	0.012
B-L MU'S	0.015	0.064	0.011	0.024	0.005	0.020	0.008	0.022
	MC1	MC2	MC3	MC4	MC5	MC6	MC7	MC8
HIST. MU'S	-0.001	0.053	0.008	0.035	0.009	0.024	0.009	0.032
BETAS	0.474	2.213	0.629	0.868	0.303	1.124	0.630	1.266
EQUIL. MU'S	0.005	0.018	0.006	0.008	0.004	0.010	0.006	0.011
B-L MU'S	0.014	0.065	0.012	0.023	0.004	0.020	0.008	0.021

7.4.2 Extending Black-Litterman with S+Bayes

The BL method is based on the use of a normal likelihood and a conjugate normal prior with known error covariance and produces only a posterior mean and posterior covariance matrix based on this model assumption. S+Bayes allows us to extend the Black-Litterman method in several ways. On the one hand, the Gibbs sampler output allows one to compute posterior distributions of optimal portfolio quantities such as the posterior distribution of the maximum Sharpe ratio, an attractive possibility not available with the standard BL, even with its simple normal distribution conjugate prior. S+Bayes allows us to use a more realistic semi-conjugate prior in the linear regression model, which may already be a significant extension of the basic BL method. Furthermore, S+Bayes supports use of a multivariate t prior for equilibrium returns, a potentially desirable choice to reflect the possibility of an unusually large change in the equilibrium mean (e.g., due to a significant short-term market or market segment dislocation). Finally, S+Bayes also supports use of a t likelihood for the linear regression model to account for the small probability of one or more large errors in the investor's views.

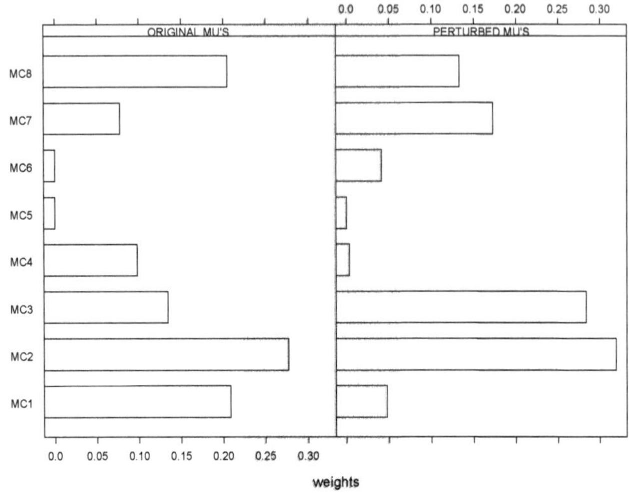

Figure 7.35 BL Weights for Original and Perturbed Betas

First we use S+Bayes to compute the posterior distribution of mean returns for the eight portfolios assets in Figure 7.32 using the same prior and model parameters as before. Code 7.26 provides the computations needed.

```
ret <- mc.slices@data
# Compute equilibrium returns empirical
# hyperparameters
mu0 <- as.matrix(betas.eqret(ret)[3,])
cov0 <- var(ret[,-(1:2)])
# Define investors views
p.abs <- c(.5,.5,rep(0,6))
p.rel <- c(0,0,1,1,-1,-1,0,0)
p <- rbind(p.abs,p.rel)
g <- rbind(.04,.01)
gp <- data.frame(g,p)
names(gp) <- c("g",paste("MU",1:8,sep=""))
#Investor covariance
covi <- diag(c((.01)^2,(.005)^2))
# S+Bayes computation
prior.mean <- bayes.normal(mu0,cov0)
prior.var <- bayes.invChisq(500,1)
my.prior <- blm.prior(priorBeta = prior.mean,
  priorSigma = prior.var)
like <- blm.likelihood(errorCov = covi)
my.samp <- blm.sampler(init.point = "user's
     choice",
```

```
  beta.init = t(mu0), sigma.init = 1)
form = g ~ MU1+MU2+MU3+MU4+MU5+MU6+MU7+MU8 - 1
mu.fit = blm(form,prior = my.prior,likelihood =
     like,
  sampler = my.samp,data = gp)
plot(mu.fit, which = c("box"))
round(summary.blm(
  mu.fit)@results[[1]]@basicStats[[1]][,1],3)
```

Code 7.26 S+Bayes Semi-conjugate Prior Generalization of Black-Litterman

The main ingredients of the code are: (a) use of `betas.eqret` and `var` to compute the hyper-parameter `mu0` ($\boldsymbol{\mu}_0$) and `cov0` ($\boldsymbol{\Omega}$), (b) creation of the data frame `gp` to specify the investor views, (c) creation of a normal prior for mean returns with the function `bayes.normal(mu0,cov0)`, (d) use of `bayes.invChisq(500,1)` to force a a very nearly fixed value of one for the variance σ^2 that is the multiplier of the user-specified covariance matrix for the errors[40], (e) use of `blm.likelihood(errorCov = covi)` to fix the covariance matrix for the errors, and (f) use of `blm.sampler` to specify a starting point for the Gibbs sampler. The resulting marginal posterior distributions for the means and for σ are shown in Figure 7.36.

The corresponding posterior mean values are:

```
  MU1   MU2   MU3   MU4   MU5   MU6   MU7   MU8
0.016 0.064 0.011 0.022 0.005 0.018 0.007 0.024
```

Note that these posterior mean values are relatively quite close to the BL posterior mean values obtained earlier, which is what we expect for a properly functioning S+Bayes computation.

Now we make simple modifications of Code 7.26 to get posterior distributions using (a) a t likelihood with a normal prior, (b) a t prior with a normal likelihood, and (c) a t prior and a t likelihood. For case (a) we replace the code line

```
like = blm.likelihood(errorCov = covi)
```

with

```
like = blm.likelihood(type = "t", errorCov = covi)
```

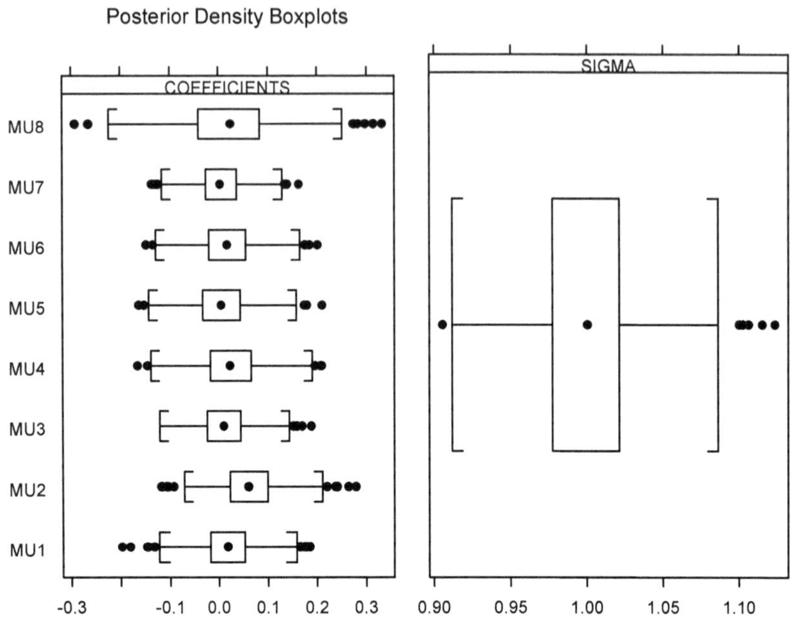

Figure 7.36 Marginal Posterior Distributions via S+Bayes Generalization of Black-Litterman (BL)

The latter uses the degrees of freedom default `df = 3`. For case (b), we replace the code line `prior.mean = bayes.normal(mu0,cov0)` with `prior.mean = bayes.t (mu0,cov0,df=3)`. And, for case (c), we make both replacements. Here are the resulting posterior means:

```
t-likelihood and normal prior
   MU1  MU2   MU3   MU4   MU5   MU6   MU7   MU8
 0.019 0.06 0.013 0.027 0.007 0.022 0.008 0.024

normal likelihood and t-prior
  MU1   MU2  MU3   MU4   MU5   MU6   MU7   MU8
 0.014 0.065 0.01 0.024 0.003 0.021 0.006 0.018

t-likelihood + t-prior
   MU1   MU2   MU3   MU4   MU5   MU6   MU7   MU8
 0.017 0.062 0.012 0.025 0.006 0.021 0.008 0.022
```

None of the resulting posterior means differ greatly from those obtained above for a normal prior and likelihood relative to the posterior variability of the latter (compare with Figure 7.36).

7.5 Bayes-Stein Estimators of Mean Returns

We have seen in Section 7.3 that for regression models with conjugate priors the posterior mean for the vector of regression coefficients is a shrinkage estimator. Therefore, it should not be very surprising to find that if we use a Bayesian model for the multivariate mean returns without recourse to use of a regression model, we once again obtain a shrinkage estimator. The most famous class of shrinkage estimators for a multivariate mean were discovered in a non-Bayesian decision theory framework. This line of research was initiated by James and Stein (1961) and motivated by the earlier work of Stein (1955), who showed that the usual sample mean estimate of a multivariate normal mean is inadmissible in dimensions three and higher. Subsequent decision-theory-oriented research papers on the topic by L. D. Brown and by B. Efron and C. Morris, are documented in Jorion (1986), who used an empirical Bayesian framework to derive a shrinkage estimator of the mean returns vector. See also Frost and Savarino (1986), who derived an empirical Bayes shrinkage estimator of the mean returns vector. Here we present the version derived by Jorion and provide an S-PLUS implementation.

As before, let $\mathbf{r}_t = (r_{t1}, r_{t2}, \cdots, r_{tK})$ be the row vector of returns on K assets, and let $\mathbf{R}$ be the $T \times K$ matrix of such row vectors. It is assumed that these row vectors of returns are independent of one another and that each has a multivariate normal distribution with unknown mean and covariance $\boldsymbol{\mu}$ and $\boldsymbol{\Omega}$. Thus the likelihood for $(\boldsymbol{\mu}, \boldsymbol{\Omega})$ is

$$\begin{aligned} L(\boldsymbol{\mu}, \boldsymbol{\Omega}) &= p(\mathbf{R} \mid \boldsymbol{\mu}, \boldsymbol{\Omega}) \\ &= \prod_{t=1}^{T} N(\boldsymbol{r}_t \mid \boldsymbol{\mu}, \boldsymbol{\Omega}) . \end{aligned} \tag{7.54}$$

It is assumed that the prior for $\boldsymbol{\mu}$ has the conjugate form

$$p(\boldsymbol{\mu} \mid \eta, \lambda, \boldsymbol{\Omega}) = N(\boldsymbol{\mu} \mid \eta \mathbf{1}, \lambda^{-1} \boldsymbol{\Omega}), \tag{7.55}$$

where the hyperparameter η has a noninformative hyperprior

$$p(\eta \mid \boldsymbol{\mu}, \boldsymbol{\Omega}) \propto constant .$$

Let $\tilde{\mathbf{r}} = (\tilde{r}_1, \tilde{r}_2, \cdots, \tilde{r}_K)$ be a new returns vector, and consider the posterior predictive density

$$
\begin{aligned}
p(\tilde{\boldsymbol{r}}|\mathbf{R},\boldsymbol{\Omega},\lambda) &= \int p(\tilde{\boldsymbol{r}} \mid \boldsymbol{\mu},\eta,\mathbf{R},\boldsymbol{\Omega},\lambda)p(\boldsymbol{\mu},\eta \mid \mathbf{R},\boldsymbol{\Omega},\lambda)\,d\boldsymbol{\mu}\,d\eta \\
&= \int p(\tilde{\boldsymbol{r}} \mid \boldsymbol{\mu},\boldsymbol{\Omega})p(\boldsymbol{\mu},\eta \mid \mathbf{R},\boldsymbol{\Omega},\lambda)\,d\boldsymbol{\mu}\,d\eta \\
&= \int p(\tilde{\boldsymbol{r}} \mid \boldsymbol{\mu},\boldsymbol{\Omega})\Big[\int p(\boldsymbol{\mu},\eta \mid \mathbf{R},\boldsymbol{\Omega},\lambda)\,d\eta\Big]d\boldsymbol{\mu} \\
&= \int p(\tilde{\boldsymbol{r}} \mid \boldsymbol{\mu},\boldsymbol{\Omega})p(\boldsymbol{\mu} \mid \mathbf{R},\boldsymbol{\Omega},\lambda)\,d\boldsymbol{\mu}\,.
\end{aligned}
\tag{7.56}
$$

In Appendix 7C, it is shown that $p(\boldsymbol{\mu} \mid \mathbf{R},\boldsymbol{\Omega},\lambda) = N(\boldsymbol{\mu} \mid \boldsymbol{\mu}_{post},\boldsymbol{\Omega}_{post})$, where the posterior mean is

$$\hat{\boldsymbol{\mu}}_{post} = \frac{T}{T+\lambda}\overline{\boldsymbol{r}} + \frac{\lambda}{T+\lambda}\mathbf{1}\hat{\mu}_{mv}, \tag{7.57}$$

with $\hat{\mu}_{mv}$ the estimated mean return of the global minimum variance portfolio

$$\hat{\mu}_{mv} = \frac{\mathbf{1}'\boldsymbol{\Omega}^{-1}\overline{\boldsymbol{r}}}{\mathbf{1}'\boldsymbol{\Omega}^{-1}\mathbf{1}}. \tag{7.58}$$

The corresponding posterior covariance is

$$\boldsymbol{\Omega}_{post} = \frac{1}{T+\lambda}\boldsymbol{\Omega} + \frac{\lambda}{T(T+\lambda)}\frac{\mathbf{1}\mathbf{1}'}{\mathbf{1}'\boldsymbol{\Omega}^{-1}\mathbf{1}}. \tag{7.59}$$

It then follows that $p(\tilde{\boldsymbol{r}}|\mathbf{R},\boldsymbol{\Omega},\lambda) = N(\tilde{\boldsymbol{r}} \mid \boldsymbol{\mu}_{pred},\boldsymbol{\Omega}_{pred})$ with

$$\hat{\boldsymbol{\mu}}_{pred} = \hat{\boldsymbol{\mu}}_{post} = \frac{T}{T+\lambda}\overline{\boldsymbol{r}} + \frac{\lambda}{T+\lambda}\mathbf{1}\hat{\mu}_{mv} \tag{7.60}$$

$$\boldsymbol{\Omega}_{pred} = \boldsymbol{\Omega} + \boldsymbol{\Omega}_{post} = \frac{T+\lambda+1}{T+\lambda}\boldsymbol{\Omega} + \frac{\lambda}{T(T+\lambda)}\frac{\mathbf{1}\mathbf{1}'}{\mathbf{1}'\boldsymbol{\Omega}^{-1}\mathbf{1}}. \tag{7.61}$$

When $\lambda = 0$, the prior for $\boldsymbol{\mu}$ is uninformative, giving $\hat{\boldsymbol{\mu}}_{pred} = \overline{\boldsymbol{r}}$ and $\boldsymbol{\Omega}_{post} = (1+1/T)\boldsymbol{\Omega}$. On the other hand, when λ is large relative to T the prior is highly informative, and in the limit $\hat{\boldsymbol{\mu}}_{pred} = \mathbf{1}\hat{\mu}_{mv}$ and $\boldsymbol{\Omega}_{post} = \boldsymbol{\Omega}$.

In order to implement this predictive mean and predictive covariance Jorion (1986) proposed to use empirical sample-based estimates of λ, $\boldsymbol{\Omega}$, and $\boldsymbol{\Omega}^{-1}$. Jorion showed that the conditional density $p(\lambda \mid \boldsymbol{\mu},\eta,\boldsymbol{\Omega})$ is Gamma with shape parameter $\alpha = (K+2)/2$ and scale parameter $\sigma = 2/d$, where $d = (\boldsymbol{\mu} - \eta\mathbf{1})'\boldsymbol{\Omega}^{-1}(\boldsymbol{\mu} - \eta\mathbf{1})$. Jorion's proposal was to replace the unkown λ by its expected value $\alpha \cdot \sigma = (K+2)/d$ in the conditional distribution $p(\lambda \mid \boldsymbol{\mu}, \eta, \boldsymbol{\Omega})$ and use sample-based estimates of the unknown parameters $\boldsymbol{\mu}, \eta, \boldsymbol{\Omega}^{-1}$ in d. Jorion used $\hat{\boldsymbol{\mu}} = \bar{\boldsymbol{r}}$, $\hat{\eta} = \hat{\mu}_{mv}$ and the unbiased estimate of the inverse of the covariance matrix:

$$\hat{\boldsymbol{\Omega}}^{-1} = \frac{T-K-2}{T-1}\mathbf{S}^{-1}, \tag{7.62}$$

where

$$\mathbf{S} = \frac{1}{T-1}\sum_{t=1}^{T}(\boldsymbol{r}_t - \bar{\boldsymbol{r}})'(\boldsymbol{r}_t - \bar{\boldsymbol{r}}) \tag{7.63}$$

is the unbiased sample covariance estimate of $\boldsymbol{\Omega}$. Note that $\hat{\boldsymbol{\Omega}}^{-1}$ denotes an estimate of the inverse of the covariance matrix, while $\mathbf{S}^{-1}$ is the inverse of an estimate of the covariance matrix.[41]

Plugging the results into the expression for expected value of λ gives

$$\hat{\lambda} = \frac{K+2}{(\bar{\boldsymbol{r}} - \mathbf{1}\hat{\mu}_{mv})'\hat{\boldsymbol{\Omega}}^{-1}(\bar{\boldsymbol{r}} - \mathbf{1}\hat{\mu}_{mv})}, \tag{7.64}$$

and substituting $\hat{\lambda}$ into the expression for the posterior predictive mean gives

$$\hat{\boldsymbol{\mu}}_{pred} = \hat{\boldsymbol{\mu}}_{post} = (1-\hat{w})\cdot\bar{\boldsymbol{r}} + \hat{w}\cdot\mathbf{1}\hat{\mu}_{mv}, \tag{7.65}$$

where

$$\hat{w} = \frac{K+2}{K+2+T\cdot(\bar{\boldsymbol{r}} - \mathbf{1}\hat{\mu}_{mv})'\hat{\boldsymbol{\Omega}}^{-1}(\bar{\boldsymbol{r}} - \mathbf{1}\hat{\mu}_{mv})}. \tag{7.66}$$

The estimated covariance matrix for $\hat{\boldsymbol{\mu}}_{pred}$ is

$$\hat{\boldsymbol{\Omega}}_{pred} = \hat{\boldsymbol{\Omega}} + \hat{\boldsymbol{\Omega}}_{post} = \mathbf{S} + \frac{1}{T+\hat{\lambda}}\mathbf{S} + \frac{\hat{\lambda}\cdot(T-1)}{T(T+\hat{\lambda})(T-K-2)}\frac{\mathbf{1}\mathbf{1}'}{\mathbf{1}'\mathbf{S}^{-1}\mathbf{1}}. \tag{7.67}$$

Now we apply the estimator to the eight equal-weighted portfolio assets of Section 7.4 (see Figure 7.32). First we use Code 7.27 to compute $\hat{\boldsymbol{\mu}}_{pred} = \hat{\boldsymbol{\mu}}_{post}$ and the corresponding posterior and posterior predictive standard deviations.

```
xret <- mc.slices@data[,3:10]
print.corr <- F
```

```
n <- nrow(xret)  # n = T
k <- ncol(xret)
# Compute posterior predictive mean
mu <- as.matrix(apply(xret,2,mean))
one <- as.matrix(rep(1,k))
S <- var(xret)   # V = OMEGA
a <- solve(S,one)
mu.prior <- one*as.numeric(t(mu)%*%a/t(one)%*%a)
S.inv <- solve(S)
d <- t(mu - mu.prior)%*%S.inv%*%(mu - mu.prior)
d <- as.numeric(d)
lambda <- (k+2)/d
w <- lambda/(n+lambda)
mu.pred <- (1-w)*mu + w*mu.prior
# Compute s.e.'s, correlations of post. pred. mean
wc1 <- 1/(n+lambda)
wc2 <- lambda*(n-1)/(n*(n+lambda)*(n-k-2))
wc2 <- wc2/as.numeric(t(one)%*%a)
V.post <- wc1*S + wc2*one%*%t(one)
V.pred <- S+V.post
sigma.post <- sqrt(diag(V.post))
sigma.pred <- sqrt(diag(V.pred))
out <- rbind(t(mu.pred),sigma.post,sigma.pred)
row.names(out) = c("POST./PRED. MEAN",
  "POST. STD.ERR","PRED. STD.ERR")
round(out,3)
round(w,3)
if(print.corr){
  rho.pred = diag(1/sigma)%*%V.pred%*%diag(1/sigma)
  round(rho.pred,3)
}
```

Code 7.27 Empirical Bayes Version of Bayes-Stein Estimator

The code computes $w = .32$ and gives the values in Table 7.4.

Table 7.4 Bayes Posterior Predictive Mean and Standard Errors

	MC1	MC2	MC3	MC4	MC5	MC6	MC7	MC8
POST PRED MEAN	0.001	0.037	0.006	0.025	0.007	0.018	0.007	0.023
POST. STD.ERR	0.007	0.021	0.007	0.010	0.007	0.009	0.006	0.010
PRED. STD.ERR	0.064	0.195	0.058	0.086	0.058	0.075	0.045	0.086

These posterior predictive means are substantially different from the BL estimates based on the same data in the top portions of Table 7.1 and Table 7.3.

Upon careful inspection, we see that the values MC1 and MC2 in those tables strongly reflect the investor view that the average of the returns of these assets is 4%. On the other hand, the empirical Bayes estimate above does not impose such views, and it is therefore not surprising to find that the posterior mean returns of MC1 and MC2 differ, having an average value of only 1.9%. Similarly, the values of MC3, MC4, MC5, and MC6 in Tables 7.1 and 7.3 almost exactly reflect the investor view that the sum of the returns for the first two are 1% greater than the sum of the returns for the second two, while this difference is only .6% for the empirical Bayes estimate. On the other hand, the posterior mean returns for MC7 and MC8 are almost the same for both estimates, which is consistent with the fact that they are not influenced by investor views in the BL estimate.

One might wonder whether Jorion's assumption that the prior is centered at the vector of identical returns equal to that of the minimum variance portfolio is reasonable. Perhaps the use of a BL equilibrium mean centering of the prior would make more sense. As a quick check, we altered Code 7.27, using `betas.eqret` to compute the vector of equilibrium returns for centering the prior. The resulting posterior/predictive mean values shown in Table 7.5 indicate that the estimator is not very sensitive to this type of change in centering the prior, at least for this example.

Table 7.5 Posterior Predictive Means with BL Equilibrium Centering of Prior

	MC1	MC2	MC3	MC4	MC5	MC6	MC7	MC8
POST. PRED. MEAN	0.002	0.039	0.007	0.025	0.007	0.018	0.007	0.025
POST. STD.ERR	0.007	0.020	0.007	0.009	0.007	0.008	0.006	0.009
PRED. STD.ERR	0.064	0.195	0.058	0.086	0.058	0.075	0.045	0.086

It should be noted that the posterior predictive standard errors in Table 7.4 have values quite close to the standard deviations obtained from the diagonal of the estimated covariance matrix (and are always at least this large). As is often the case, these standard errors are much larger than the posterior standard errors. In any event, one is stuck with a posterior predictive covariance matrix estimate that is at least as large as the returns sample covariance matrix estimate.

In order to reduce the errors in sample covariance estimates, one may want to consider using an improved estimate based on an appropriate Bayes method. Ledoit and Wolfe (2003) proposed an estimate that shrinks the sample covariance matrix toward a prior covariance matrix based on the single-factor market model, with the shrinkage determined by some large-sample frequentist arguments. It remains to be seen whether there is a Bayes rationale for such an approach.

7.6 Appendix 7A: Inverse Chi-Squared Distributions

A standard inverse chi-squared distribution with ν_o degrees of freedom (dof) is the distribution of $W = 1/Y$, where Y is a chi-squared random variable with ν_o dof. The density of the latter is

$$\chi^2(y;\nu_o) = \frac{1}{\Gamma(\nu_o/2)\cdot 2^{\nu_o/2}} \cdot y^{\nu_o/2-1}e^{-y/2}, \tag{7.68}$$

and a standard change of variable calculation to obtain the density of W gives

$$Inv\chi^2(w;\nu_o) = \frac{1}{\Gamma(\nu_o/2)\cdot 2^{\nu_o/2}} \cdot \frac{1}{w^{\nu_o/2+1}} e^{-1/2w}. \tag{7.69}$$

To get a feeling for the shape of the densities for standard inverse chi-squared random variables, use Code 7.28, which uses the function `density.invchisq` defined in Code 7.4 to get the plots of Figure 7.37.

```
x <- seq(.01,.4,.005)
#Set scaled = F for std. inv-chi-squared density
scaled <- F
plot(x,density.invchisq(x,30,sigmasq,scaled),
  type="l",ylab="DENSITY")
title(main="STANDARD INVERSE CHI-SQUARED
  DENSITIES")
lines(x,density.invchisq(x,20,sigmasq,scaled),
  lty=2, lwd=3)
lines(x,density.invchisq(x,10,sigmasq,scaled),
  lty=3, lwd=3)
lines(x,density.invchisq(x,5,sigmasq,scaled),
  lty=4, lwd=3)
legend(locator(1),
  legend=c("dof=50","dof=15","dof=5","dof=1"),
  lty=1:4)
```

Code 7.28 Plot Standard Inverse Chi-Squared Densities

A *scaled* inverse chi-squared random variable with *scale parameter* $\nu_o\sigma_o^2$ is obtained by multiplying a standard inverse chi-squared random variable by

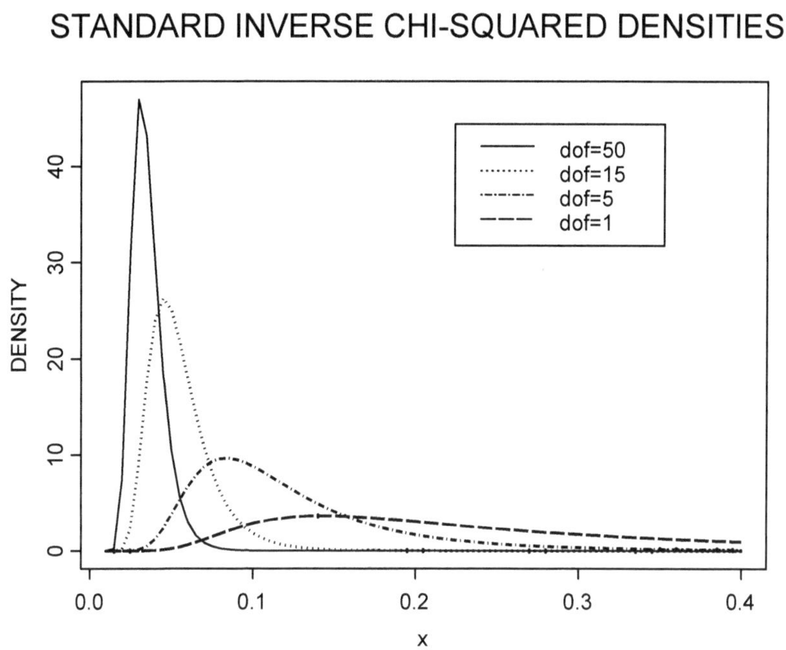

Figure 7.37 Standard Inverse Chi-Squared Densities

$\nu_o\sigma_o^2$ (e.g., $W = \nu_o\sigma_o^2 / Y$ is a scaled inverse chi-squared random variable). The probability density of a scaled inverse chi-squared random variable is easily obtained from the standard inverse chi-squared density by the usual transformation of a scaled random variable:

$$Inv\chi^2(w;\nu_o,\sigma_o^2) = \frac{1}{\Gamma(\nu_o/2)\cdot 2^{\nu_o/2}}\cdot\frac{(\nu_o\sigma_o^2)^{\nu_o/2}}{w^{\nu_o/2+1}}e^{-\nu_o\sigma_o^2/2w}. \qquad (7.70)$$

A few scaled inverse chi-squared densities, shown in Figure 7.38, are produced with Code 7.29, obtained by slightly modifying Code 7.28 (note the change in σ_o^2 and in dof values).

```
x <- seq(.01,.4,.005)
sigmasq <- .1
plot(x,density.invchisq(x,50,sigmasq),type="l",
  ylab="DENSITY")
title(main="SCALED INVERSE CHI-SQUARED DENSITIES")
lines(x,density.invchisq(x,15,sigmasq),lty=2,lwd=3)
```

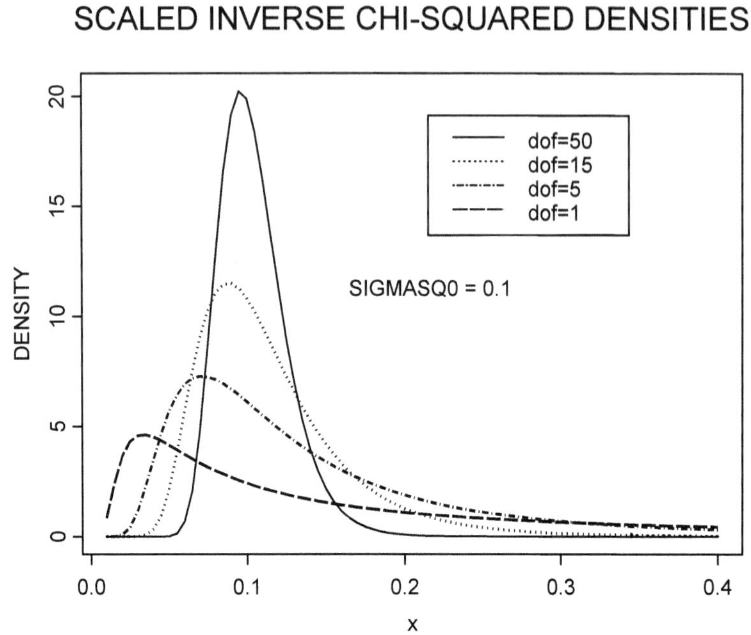

Figure 7.38 Scaled Inverse Chi-Squared Densities

```
lines(x,density.invchisq(x,5,sigmasq),lty=3,lwd=3)
lines(x,density.invchisq(x,1,sigmasq),lty=4,lwd=3)
legend(locator(1),legend=c("dof=50","dof=15",
  "dof=5","dof=1"),lty=1:4)
text(locator(1), paste("SIGMASQ0 =",sigmasq))
```

Code 7.29 Scaled Inverse Chi-Squared Densities

The mean μ_{σ^2} and variance V_{σ^2} of a scaled inverse chi-squared random variable are given by

$$\mu_{\sigma^2} = \frac{\nu_o \sigma_o^2}{\nu_o - 2}, \quad \nu_o > 2 \tag{7.71}$$

and

$$V_{\sigma^2} = \frac{2\nu_o^2 \sigma_o^4}{(\nu_o - 2)^2 (\nu_o - 4)}, \quad \nu_o > 4. \tag{7.72}$$

These equations may be inverted to give ν_o and σ_o^2 in terms of the mean μ_{σ^2} and standard deviation $SD_{\sigma^2} = \sqrt{V_{\sigma^2}}$ of σ^2:

$$\nu_o = 2R^2 + 4, \tag{7.73}$$

$$\sigma_o^2 = \mu_{\sigma^2} \frac{R^2 + 1}{R^2 + 2}, \tag{7.74}$$

where $R = \mu_{\sigma^2} / SD_{\sigma^2}$.

7.6.1.1.1 Scaled Inverse Gamma Density Equivalent to an Inverse Chi-Squared Density

Many textbooks and research papers work with inverse Gamma priors rather than inverse chi-squared priors. The relationship between the two is as follows. The formula for a standard Gamma density is

$$p(x;\alpha) = \frac{1}{\Gamma(\alpha)} x^{\alpha-1} e^{-x}, \quad x > 0, \tag{7.75}$$

and the formula for a scaled inverse Gamma density is

$$p(x;\alpha,s) = \frac{s^{\alpha}}{\Gamma(\alpha) \cdot x^{\alpha+1}} e^{-s/x}, \quad x > 0. \tag{7.76}$$

From this it is clear that the scaled inverse chi-squared density with parameters ν_o and σ_o^2 is the same as a Gamma density with parameters $\alpha = \frac{\nu_o}{2}$ and $s = \frac{\nu_o \sigma_o^2}{2}$. Conversely, a scaled inverse Gamma density with parameters α and s is the same as a scaled inverse chi-squared density with parameters $\nu_o = 2\alpha$ and $\sigma_o^2 = \frac{s}{\alpha}$.

7.7 Appendix 7B: Posterior Distributions for Normal Likelihood Conjugate Priors

For the case of the normal likelihood with a conjugate normal inverse chi-squared prior, it may be shown that the joint posterior is a normal inverse chi-squared distribution, with Student's t- and inverse chi-squared distributions for the marginal posteriors for the mean and variance.[42] Specifically

$$\begin{aligned}
&\mu,\sigma^2 \mid \mathbf{r} \sim NInv\chi^2\left(\mu_T,\frac{\sigma_T^2}{k_T},\nu_T,\sigma_T^2\right)\\
&\mu \mid \mathbf{r} \sim ST_{\nu_T}\left(\mu_T,\sigma_T^2\right)\\
&\sigma^2 \mid \mathbf{r} \sim Inv\chi^2(\nu_T,\sigma_T^2),
\end{aligned} \tag{7.77}$$

where

$$\begin{aligned}
k_T &= k_o + T\\
\nu_T &= \nu_o + T\\
\mu_T &= \frac{k_o}{k_o+T}\mu_o + \frac{T}{k_o+T}\overline{r}\\
\sigma_T^2 &= \frac{1}{\nu_T}\left(\nu_o\sigma_o^2 + (T-1)s^2 + \frac{k_o T}{k_o+T}(\overline{r}-\mu_o)^2\right).
\end{aligned} \tag{7.78}$$

Using (7.77) and (7.78), it is quite easy to write an S-PLUS function to compute the joint and marginal posterior densities, an exercise that we leave to the reader.

7.8 Appendix 7C: Derivation of the Posterior for Jorion's Empirical Bayes Estimate

Given the above assumptions about the likelihood and prior, we have:

$$\begin{aligned} p(\boldsymbol{\mu},\eta|\mathbf{R},\boldsymbol{\Omega},\lambda) &\propto p(\mathbf{R}\,|\,\boldsymbol{\mu},\eta,\boldsymbol{\Omega},\lambda)p(\boldsymbol{\mu},\eta\,|\,\boldsymbol{\Omega},\lambda) \\ &= p(\mathbf{R}\,|\,\boldsymbol{\mu},\boldsymbol{\Omega})p(\boldsymbol{\mu}\,|\,\eta,\boldsymbol{\Omega},\lambda)p(\eta\,|\,\boldsymbol{\Omega},\lambda) \\ &\propto \prod_{t=1}^{T} N(\boldsymbol{r}_t\,|\,\boldsymbol{\mu},\boldsymbol{\Omega})N(\boldsymbol{\mu}\,|\,\eta\mathbf{1},\lambda^{-1}\boldsymbol{\Omega}) \\ &\propto \exp\left(-\frac{1}{2}Q\right), \end{aligned} \tag{7.79}$$

where

$$Q = T\cdot(\boldsymbol{\mu}-\bar{\boldsymbol{r}})'\boldsymbol{\Omega}^{-1}(\boldsymbol{\mu}-\bar{\boldsymbol{r}}) + \lambda(\boldsymbol{\mu}-\eta\mathbf{1})'\boldsymbol{\Omega}^{-1}(\boldsymbol{\mu}-\eta\mathbf{1}). \tag{7.80}$$

Straightforward rearrangement shows that

$$(\boldsymbol{\mu}-\eta\mathbf{1})'\boldsymbol{\Omega}^{-1}(\boldsymbol{\mu}-\eta\mathbf{1}) = \mathbf{1}'\boldsymbol{\Omega}^{-1}\mathbf{1}(\eta-\mu_{mv})^2 + \boldsymbol{\mu}'\boldsymbol{\Omega}^{-1}\boldsymbol{\mu} - \frac{(\mathbf{1}'\boldsymbol{\Omega}^{-1}\boldsymbol{\mu})^2}{\mathbf{1}'\boldsymbol{\Omega}^{-1}\mathbf{1}},$$

where

$$\mu_{mv} = \frac{\mathbf{1}'\boldsymbol{\Omega}^{-1}\boldsymbol{\mu}}{\mathbf{1}'\boldsymbol{\Omega}^{-1}\mathbf{1}} \tag{7.81}$$

is the mean return of the global minimum variance portfolio.

We can now integrate out η in $p(\boldsymbol{\mu}\,|\,\mathbf{R},\boldsymbol{\Omega},\lambda) = \int p(\boldsymbol{\mu},\eta\,|\,\mathbf{R},\boldsymbol{\Omega},\lambda)d\eta$, which gives

$$p(\tilde{\boldsymbol{r}}|\mathbf{R},\boldsymbol{\Omega},\lambda) = \int p(\tilde{\boldsymbol{r}}\,|\,\boldsymbol{\mu},\boldsymbol{\Omega})p(\boldsymbol{\mu}\,|\,\mathbf{R},\boldsymbol{\Omega},\lambda)d\boldsymbol{\mu}, \tag{7.82}$$

where

$$p(\boldsymbol{\mu}\,|\,\mathbf{R},\boldsymbol{\Omega},\lambda) \propto \exp\left(-\frac{1}{2}\tilde{Q}\right) \tag{7.83}$$

with

$$\tilde{Q} = T\cdot(\boldsymbol{\mu}-\bar{\boldsymbol{r}})'\boldsymbol{\Omega}^{-1}(\boldsymbol{\mu}-\bar{\boldsymbol{r}}) + \lambda\boldsymbol{\mu}'\boldsymbol{\Omega}^{-1}\boldsymbol{\mu} - \lambda\frac{(\mathbf{1}'\boldsymbol{\Omega}^{-1}\boldsymbol{\mu})^2}{\mathbf{1}'\boldsymbol{\Omega}^{-1}\mathbf{1}}. \tag{7.84}$$

Now some tedious rearrangement, which includes use of a matrix inversion lemma, allows one to complete the square in the quadratic form $\tilde{Q}$ to get

$$\tilde{Q} = (\boldsymbol{\mu} - \boldsymbol{\mu}_{post})'\boldsymbol{\Omega}_{post}^{-1}(\boldsymbol{\mu} - \boldsymbol{\mu}_{post}) + constant\,. \tag{7.85}$$

So we have $p(\boldsymbol{\mu} \mid \mathbf{R}, \boldsymbol{\Omega}, \lambda) = N(\boldsymbol{\mu} \mid \boldsymbol{\mu}_{post}, \boldsymbol{\Omega}_{post})$, where

$$\hat{\boldsymbol{\mu}}_{post} = \frac{T}{T+\lambda}\overline{\boldsymbol{r}} + \frac{\lambda}{T+\lambda}\mathbf{1}\hat{\mu}_{mv} \tag{7.86}$$

with $\hat{\mu}_{mv}$ the estimated mean return of the global minimum variance portfolio

$$\hat{\mu}_{mv} = \frac{\mathbf{1}'\boldsymbol{\Omega}^{-1}\overline{\boldsymbol{r}}}{\mathbf{1}'\boldsymbol{\Omega}^{-1}\mathbf{1}} \tag{7.87}$$

and

$$\boldsymbol{\Omega}_{post} = \frac{1}{T+\lambda}\boldsymbol{\Omega} + \frac{\lambda}{T(T+\lambda)}\frac{\mathbf{1}\mathbf{1}'}{\mathbf{1}'\boldsymbol{\Omega}^{-1}\mathbf{1}}\,. \tag{7.88}$$

We are now in the standard situation of a normal likelihood and a multivariate normal posterior for $\boldsymbol{\mu}$. For this situation, it is well-known that the posterior predictive density $p(\tilde{\boldsymbol{r}} | \mathbf{R}, \boldsymbol{\Omega}, \lambda) = \int p(\tilde{\boldsymbol{r}} \mid \boldsymbol{\mu}, \boldsymbol{\Omega}) p(\boldsymbol{\mu} \mid \mathbf{R}, \boldsymbol{\Omega}, \lambda) d\boldsymbol{\mu}$ is multivariate normal $N(\tilde{\boldsymbol{r}} \mid \boldsymbol{\mu}_{pred}, \boldsymbol{\Omega}_{pred})$ with $\hat{\boldsymbol{\mu}}_{pred} = \hat{\boldsymbol{\mu}}_{post}$ and $\boldsymbol{\Omega}_{pred} = \boldsymbol{\Omega} + \boldsymbol{\Omega}_{pred}$, so we have[43]:

$$\hat{\boldsymbol{\mu}}_{pred} = \frac{T}{T+\lambda}\overline{\boldsymbol{r}} + \frac{\lambda}{T+\lambda}\mathbf{1}\hat{\mu}_{mv} \tag{7.89}$$

$$\boldsymbol{\Omega}_{pred} = \frac{T+\lambda+1}{T+\lambda}\boldsymbol{\Omega} + \frac{\lambda}{T(T+\lambda)}\frac{\mathbf{1}\mathbf{1}'}{\mathbf{1}'\boldsymbol{\Omega}^{-1}\mathbf{1}}\,. \tag{7.90}$$

Exercises

1. Show that $\sum_{t=1}^{T}(r_t - \mu)^2 = \sum_{t=1}^{T}(r_t - \overline{r})^2 + n(\overline{r} - \mu)^2$.

2. Complete that square as suggested to obtain the expression for $p(\mu \mid \mathbf{r}) = N(\mu \mid \mu_T, \sigma_T^2)$ in Section 7.2.1.

3. For the normal model of mean returns with known variance in Section 7.2.1, show that the posterior predictive mean is $\mu_P = \mu_T$ and the posterior predictive variance is $\sigma_P^2 = \sigma^2 + \sigma_T^2$. Hint: Use the iterated conditional expectation formula $E(Y) = E(E(Y \mid X))$ and the variance decomposition $\text{var}(X) = E(\text{var}(X \mid Y)) + \text{var}(E(X \mid Y))$.

4. Use Code 7.2 on the returns data sets provided with this book to check on whether or not it is safe to assume that $\mu_o = 0$.

5. Modify Code 7.4 so that it plots the likelihood as well as the prior and posterior.

6. Modify Code 7.4 so that it displays priors and posteriors for volatility rather than variance.

7. Implement the Gibbs sampler for the semi-conjugate mean variance model, and use it to estimate the marginal and joint posterior distributions of the mean and variance for a stock returns set or portfolio returns set of your choice. Plot your marginal posterior densities using an S-PLUS kernel density estimate, and plot a visualizaton of the joint posterior using the S-PLUS `contour` function.

8. Use the Gibbs sampler of Exercise 7 to estimate the posterior distribution of the Sharpe ratio for the same set of stock returns or portfolio returns you used in that exercise.

9. Use Code 7.9 to compute the posterior distribution of mean returns for a stock of your choice, using what you regard as reasonable values for the hyperparameters. Experiment with the sensitivity of your result to your choice of hyperparameters by varying the latter. Try the various plotting options available for plotting the Bayes posterior results.

10. Use Code 7.10 with twelve months of returns for a stock of your choice, with several "reasonable" choices of the mixture probability, the two mean parameters, and two variance parameters. Explore possible unimodal and bimodal posterior densities that are achievable by using a normal mixture prior.

11. For the generalized one-way random effects model for estimating the hyper-parameter τ_0^2, show that the estimate $\hat{\tau}_0^2$ is unbiased.

12. Use Code 7.13 on twenty stock returns in the data set `smallcap.ts`, experimenting with the trimming capability as needed to get a good chi-squared density fit to the bulk of the data. How often is trimming needed?

13. Experiment with the parameter τ in the Black-Litterman weight calculation to determine how sensitive the resulting weights are to the choice of this parameter.

Endnotes

[1] See, for example, the early works by Zellner (1996) and Leamer (1978) or the recent book by Gelman et al. (2004).

[2] See, for example, Gelman et al. (2004) and Carlin and Louis (2000) for basic introductions to Bayesian models, including basic MCMC methods, and Gilks et al. (1996) for an extensive treatment of MCMC methods. See also Tanner (1996).

[3] See, for example, Carlin and Louis (2000, p. 142) for a convergence theorem.

[4] The use of the first four years of monthly returns to estimate values for μ_o and σ represents a very crude form of the *empirical Bayes* method. The choice of σ_o on the other hand represents a *subjective* Bayes method, so the overall choice here is a combined empirical-subjective method.

[5] An *noninformative* prior for μ is formally defined by $p(\mu) \propto \text{constant}$. Such a prior is improper in that $\int p(\mu)d\mu = \infty$. Nonetheless, such formal improper priors have often been used in Bayesian analysis for the sake of simplicity and to give probabalistic inference statements as an alternative to the frequentist inference statements, along with the argument that improper noninformative priors can be approximated by proper priors with large standard deviations such that the prior is relatively flat over the region where the likelihood is significantly nonzero. We do not view uninformative priors as being of much use in finance, the point being that investors often have informative prior information of one form or another with which they can improve their investment decision payoffs.

[6] We show shortly that it usually suffices to assume that $\mu_0 = 0$ when estimating volatility.

[7] We are following the terminology and notation convention of Gelman et al. (2004), who represent the inverse Gamma form of conjugate prior in the form of an inverse chi-squared density.

[8] This form of noninformative prior for the variance can be justified in several ways, one of which is a data-translated likelihood rationale. See, for example, Box and Tiao (1973).

[9] The marginal noninformative prior for σ^2 is a limiting form of inverse chi-squared distribution obtained with $\nu_o = 0$ degrees of freedom. The marginal noninformative prior for μ can be thought of as the limiting form of a normal prior for μ with $\sigma_o^2 = \infty$.

[10] These have the same distributional form as the sampling distributions for the Student's t-statistics and sample variance. However, the interpretation is different in that these posteriors provide probability statements (e.g., highest posterior density intervals) and do not require the repeated experiments interpretation of the frequentist sampling distribution approach. This is perhaps the strongest justification for the use of a joint noninformative prior for the mean and variance, which is otherwise rather uninteresting.

[11] The normal inverse chi-squared terminology and notation are due to Gelman et al. (2004).

[12] Another reason that people seem to feel comfortable with the conjugate prior is that the prior has the interpretation of having k_o additional observations from a process with the same variance as the likelihood. While this explanation no doubt brings comfort to many users of Bayesian methods, it is usually a quite unrealistic rationale.

[13] The fact that higher mean returns are typically associated with higher volatility begs the question of whether one can construct a useful dependent joint prior for the mean and variance of asset returns by allowing the hyperparameter μ_o to depend upon the variance or standard deviation of returns in a monotonic manner.

[14] See Gelman et al. (2004). In the semi-conjugate case, the marginal density $p(\sigma^2 \mid \mathbf{r})$ also has an analytic form known up to a normalizing constant.

[15] The noninformative prior is also relatively uninteresting because it results in marginal posterior distributions that coincide with what one obtains using a sampling theory approach. Of course, there is still the added value of having posterior probability densities.

[16] It may also be that empirical evidence does not support the use of the inverse chi-squared distribution as the prior for the variance.

[17] To get results more like those in Figure 7.6 you will need to obtain large numbers of stocks in each market cap group and apply Code 7.6.

[18] You will find similar results upon examining the universe of U.S. stocks in these market cap groups.

[19] Probabilities for the inverse chi-squared posterior are calculated in an obvious manner using the S-PLUS function `pchisq` for computing chi-squared probabilities.

[20] Sampling from a scaled inverse chi-squared distribution is easily accomplished by sampling from a chi-squared distribution using the S-PLUS random number generator function `rchisq`, taking the reciprocals of the random numbers obtained and scaling by multiplying by $\nu_o \sigma_o^2$.

[21] See, for example, Gelman et al. (2004).

[22] For further details on composition sampling, see Tanner (1996, p. 52).

[23] For other forms of MCMC such as the Metropolis and Metropolis-Hastings algorithms, see Gilks et al. (1996) and Gelman et al. (2004).

[24] See, for example, Carlin and Louis (2000, p. 142), and Gilks et al. (1996).

[25] For further details, see the S+Bayes help files for these functions.

[26] The posterior densities for the Gibbs sampler are kernel density estimates based on the sampler output after burn-in for the mean and (square root of) variance. Posteriors for standard deviations rather than variances are computed in S+Bayes because standard deviations are easier to interpret.

[27] One can in principle estimate k_o using the marginal maximum likelihood approach as in Frost and Savarino (1986). However, the method is complex.

[28] In the classical analysis of variance for the one-way random effects model with $\sigma_k^2 \equiv \sigma^2$, the well-known results are that $E(\hat{\sigma}_{\mu_k}^2) = \tau_o^2 + \sigma^2 / T$ and $E(\bar{\sigma}^2) = \sigma^2$, and in this case the result follows. For the case of unequal asset variances, we replace

σ^2 by the mean of the true variances (the "pooled" variances) $\sigma_P^2 = \frac{1}{K}\sum_{k=1}^{K} \sigma_k^2$, and one can then show that $\hat{\tau}_0^2$ is unbiased (Exercise 11).

[29] The sample variance of the sample variances may need a correction factor.

[30] Of course, this does not solve the problem of getting a good estimate of the density in the tails of the density.

[31] In the case of covariance matrix estimates for multidimensional parameters, it often happens that the estimates fail to be positive definite, and ad hoc methods are often used to force positive definiteness. See, for example, Appendix C of Fama and French (1997).

[32] This can be seen as part of the derivation in Exercise 11, where the unrealistic assumption of independence of returns across assets is used.

[33] The assumption is not justified in the case of illiquid assets where serial correlation arises because of asynchronous trading.

[34] In practice, one needs to use an exponentially-weighted moving average (EWMA) or Generalized Autoregressive Conditional Heteroskedastic (GARCH) volatility clustering model to account for time-varying volatility of returns.

[35] Vasicek (1973) reparameterized the model by centering the excess market returns so that they had zero mean, in which case the intercept no longer represents only the excess returns deviation from the CAPM.

[36] To see the names of an S Version 3 list object we use the function names. Here we must use `slotNames` because a `blm` object such as `msft.fit` is an S Version 4 object with so-called "slots" for components. Note that components of an S Version 4 objects are accessed with the "`@`" symbol rather than the "`$`" or "`[[j]]`" symbol as in the case of an S Version 3 object (such as a data frame or model object).

[37] See, for example, Black and Litterman (1992) and He and Litterman (1999).

[38] We use **P** rather than **X** to conform to the notation of Section 1.2.3 and that of some other authors such as Lee (2000).

[39] Somewhat surprisingly, the word "Bayes" seldom if ever appears in these works.

[40] For the specification of an S+Bayes linear model, it is assumed that the error covariance matrix is of the form $\mathbf{\Sigma} = \sigma^2 \tilde{\mathbf{\Sigma}}$ where σ^2 is a random variable with an inverse chi-squared prior (or noninformative prior limiting case), and $\tilde{\mathbf{\Sigma}}$ is a known error covariance matrix supplied by the user. In this example, we are assuming that $\sigma^2 = 1$, which is obtained by setting ν_o at a very large value and $\sigma_o^2 = 1$ in the inverse chi-squared prior.

[41] It is a somewhat overlooked fact that while $\mathbf{S}$ is an unbiased estimate of $\mathbf{\Omega}$, $\mathbf{S}^{-1}$ is a biased estimate of $\mathbf{\Omega}^{-1}$: $(T-1)\cdot \mathbf{S}^{-1}$ has an inverse Wishart distribution, with $E\left((T-1)\cdot \mathbf{S}^{-1}\right) = \frac{1}{T-K-2}\mathbf{\Omega}^{-1}$. See, for example, Lemma 7.7.1 of Anderson (1984).

[42] See Gelman et al. (2004).

[43] See Gelman et al. (2004)

Bibliography

Acerbi, C., C. Nordio and C. Sirtori (2001). "Expected Shortfall as a Tool for Financial Risk Management." http://arxiv.org/PS_cache/cond-mat/pdf/0102/0102304.pdf.

Acerbi, C. and D. Tasche (2001). "Expected Shortfall: a Natural Coherent Alternative to Value at Risk." http://arxiv.org/PS_cache/cond-mat/pdf/0105/0105191.pdf.

Anderson, T. W. (1984). *An Introduction to Multivariate Statistical Analysis*, John Wiley & Sons, New York.

Artzner P, F. Delbaen, J. Eber, and D. Heath (1997). "Thinking Coherently," Risk, 10 (11), 68–71.

Artzner, P, F. Delbaen, J. Eber and D. Heath (1999). "Coherent Measures of Risk," Mathematical Finance 9 (3), 203–228.

Atkinson, A. C. (1985). *Plots, Transformations and Regression: an Introduction to Graphical Methods of Diagnostic Regrssion Analysis*, Oxford University Press, Oxford.

Birge, J. R. and F. Louveaux (1997). *Introduction to Stochastic Programming*, Springer, New York.

Black, F. and R. Litterman (1992). "Global Portfolio Optimization," Financial Analysts Journal, Sept.–Oct., 28–43.

Blume, M. E. (1971). "On the assessment of risk," Journal of Finance, 26, 1–10.

Box, G. E. P. and G. Tiao (1973). *Bayesian Inference in Statistical Analysis,* Addison–Wesley, Reading, MA.

Bradley, B. O. and M. S. Taqqu (2003). "Financial Risk and Heavy Tails," in S. T. Rachev, ed., *Handbook of Heavy Tailed Distributions in Finance*, Elsevier/North–Holland, Amsterdam.

Brandimarte, P. (2002). *Numerical Methods in Finance: a MATLAB-based Introduction*, John Wiley & Sons, New York.

Campbell, J. and L. Viceira (2002). *Strategic Asset Allocation: Portfolio Choice for Long Term Investors*, Oxford University Press, Oxford.

Carlin, B. P. and T. A. Louis (2000). *Bayes and Empirical Bayes Methods for Data Analysis*, Chapman & Hall/CRC, London.

Chow, G., E. Jacquier, M. Krizmann, and K. Lowry (1999). "Optimal Portfolios in Good Times and Bad," Financial Analysts Journal, 55 (3), 65–73.

Chow, G. and M. Kritzman (2001). "Risk Budgets," Journal of Portfolio Management, 27 (2), 56–60.

Davison, A. C. and D. V. Hinkley (1999). *Bootstrap Methods and Their Application*, Cambridge University Press, Cambridge.

De Bever, L., W. Kozun, and B. Zwan (2000). "Risk Budgeting in a Pension Fund," in L. Rahl, ed., *Risk Budgeting: A New Approach to Investing*, Risk Books, London.

Dert, C. L. (1995). "Asset Liability Management for Pension Funds: a Multistage Chance Constrained Programming Approach," Ph.D. dissertation, Erasmus University, Rotterdam.

Efron, B. and R. J. Tibshirani (1998). *An Introduction to the Bootstrap*, Chapman and Hall/CRC, Boca Raton, FL.

El Karoui, N., A. Frachot, and H. Geman (1998). "On the Behavior of Long Zero Coupon Rates in a No Arbitrage Framework," Review of Derivatives Research 1, 351–369.

Embrechts P., A. McNeil, and D. Straumann (2002). "Correlation and Dependency in Risk Management: Properties and Pitfalls," in M. A. H. Dempster, ed., *Risk Management: Value at Risk and Beyond*, Cambridge University Press, Cambridge, pp. 176–223.

Fama, E. F. and K. R. French (1997). "Industry Costs of Equity," Journal of Financial Economics, 43 (2), 153–193.

Feinstein C. and M. Thapa (1993). "A Reformulation of a Mean-Absolute Deviation Portfolio Optimization Model," Management Science, 39 (12), 1552–1553.

Frost, P.A. and J. E. Savarino (1986). "An Empirical Bayes Approach to Efficient Portfolio Selection," Journal of Financial and Quantitative Analysis, 21 (3), 293–305.

Gelman, A., J. B. Carlin, H. S. Stern, and D. B. Rubin (2004). *Bayesian Data Analysis*, 2nd edition, Chapman & Hall/CRC, Boca Raton, FL.

Gilks, W. R., S. Richardson and D. J. Spiegelhalter, eds. (1996). *Markov Chain Monte Carlo in Practice*, Chapman & Hall/CRC, Boca Raton, FL.

Grinold, R. (1999). "Mean-Variance and Scenario-Based Approaches to Portfolio Selection," Journal of Portfolio Management, 25 (2), 10–22.

Grinold, R. and K. Easton (1998). "Attribution of Performance and Holdings," in W. Ziemba and J. Mulvey, eds., *Worldwide Asset and Liability Modeling*, Cambridge University Press, Cambridge.

Grinold, R. and R. Kahn (2000). *Active Portfolio Management: A Quantitative Approach for Providing Superior Returns and Controlling Risk*, 2nd edition, McGraw-Hill, New York.

Hampel, F. R., E. M. Ronchetti, P. J. Rousseeuw, and W. A. Stahel (1986). *Robust Statistics: the Approach Based on Influence Functions*, John Wiley & Sons, New York.

Hamza, F. and J. Janssen (1995). "Linear Approach for Solving Large-Scale Portfolio Optimization Problems in a Lognormal Market." Paper presented at AFIR Colloquium, Nürnberg, Germany, October 1-3, 1996. http://www.actuaries.org/members/en/AFIR/colloquia/Nuernberg/Hamza_Janssen.pdf.

He, G. and R. Litterman (1999). "The Intuition Behind Black–Litterman Model Portfolios," Goldman Sachs Quantitative Resources Group, Goldman Sachs, New York.

Hillier, F. and G. Lieberman (1995). *Introduction to Mathematical Programming*, 2nd edition, McGraw-Hill, New York.

Huang, C. and R. Litzenberger (1998). *Foundations of Financial Economics,* North-Holland, New York.

Huber, P. J. (1964). "Robust Estimation of a Location Parameter," Annals of Math. Statistics, 35 (1), 73–101.

Huber, P. J. (1973). "Robust regression: Asymptotics, Conjectures and Monte Carlo," Annals of Statistics, 1 (5), 799–821.

Huber, P. J. (2004). *Robust Statistics*, John Wiley & Sons, New York.

Ingersoll Jr., J. E. (1987). *Theory of Financial Decision Making,* Rowman & Littlefield Publishing Inc., Totowa, NJ.

Insightful Corporation (2002). *S–PLUS 6 Robust Library User's Guide*, Insightful Corporation, Seattle, WA. Included with S-PLUS.

Insightful Corporation (2004). *S+Bayes Module and Preliminary Documentation,* Insightful Corporation, Seattle, WA. Available for free from http://www.insightful.com/downloads/libraries/default.asp.

James, W. and C. Stein (1961). "Estimation with Quadratic Loss," In Neyman, J., ed., *Proceedings of the Fourth Berkeley Symposium on Mathematical Statistics and Probability Volume 1*, University of California Press, Berkeley, pp. 361-379.

Jorion, P. (1986). "Bayes–Stein Estimation for Portfolio Analysis," Journal of Financial and Quantitative Analysis, 21 (3), 279–292.

Jorion, P. (1992). "Portfolio Optimization in Practice," Financial Analysts Journal, 48 (1), 68–74.

Judd, K. L. (1998). *Numerical Methods in Economics*, MIT Press, Cambridge.

Kass, R. E. and A. E. Raftery (1995). "Bayes Factors," Journal of the American Statistical Association, 90 (430), 773–795.

Kim, J. and C. Finger (2000). "A Stress Test To Incorporate Correlation Breakdown," RiskMetrics Journal, 1 (May), 61-75.

Klein, R. and V. S. Bawa (1976). "The Effect of Estimation Risk on Portfolio Choice," Journal of Financial Economics, 3, 215–231.

Konno, H. and H Yamazaki (1991). "Mean-Absolute Deviation Portfolio Optimization Model and Its Applications to the Tokyo Stock Market," Management Science 37 (5), 519–531.

Larsen, N., H. Mausser, and S. Uryasev (2002). "Algorithms for Optimization of Value at Risk," in P. Pardalos and V. K. Tsitsiringos, eds., *Financial Engineering, e-Commerce and Supply Chains*, Kluwer Academic, Dordrecht, pp. 129-157.

Leamer, E. E. (1978). *Specification Searches: Ad Hoc Inference with Nonexperimental Data*, John Wiley & Sons, New York.

Ledoit, O. and M. Wolf (2003). "Improved Estimation of the Covariance Matrix of Stock Returns with an Application to Portfolio Selection," Journal of Empirical Finance, 10 (5), 603–621.

Lee, W. (2000). *Theory and Methodology of Tactical Asset Allocation*, Frank J. Fabozzi Associates, New Hope, PA.

Leibowitz, M., L. Bader, and S. Kogelman (1996). *Return Targets and Shortfall Risks: Studies in Strategic Asset Allocation*, Irwin Professional, Chicago.

Lo, A.W. (2002). "The Statistics of Sharpe Ratios," Financial Analysts Journal, 58 (4), 36–52.

Lobo, M., M. Fazel, and S. Boyd (2002). "Portfolio Optimization with Linear and Fixed Transaction Costs." http://www.stanford.edu/~boyd/reports/portfolio.pdf.

Mallows, C. L. (1975). "On some topics in robustness," Technical memorandum, Bell Laboratories, Murray Hill, NJ.

Markowitz, H. (2000). *Mean-Variance Analysis in Portfolio Choice and Capital Markets*, Frank J. Fabozzi Associates, New Hope, PA.

Martin, R. D. and T. T. Simin (2003). "Outlier Resistant Estimates of Beta," Financial Analysts Journal, 59 (5), 56–69.

Martin, R. D., V. J. Yohai, and R. H. Zamar (1989). "Min–max Bias Robust Regression," Annals of Statistics, 17 (4), 1608–1630.

Martin, R. D. and R. H. Zamar (1993). "Efficiency constrained bias robust estimation," Annals of Statistics, 21 (1), 338–354.

Martin, R. D. and S. Zhang (2004). "Influence functions for portfolios," manuscript, Department of Statistics, University of Washingon.

McCarthy, M. (2000). "Risk Budgeting for Pension Funds and Investment Managers Using VaR," in L. Rahl, ed., *Risk Budgeting: A New Approach to Investing*, Risk Books, London.

Memmel, C. (2003). "Performance Hypothesis Testing with the Sharpe Ratio," Finance Letters, 1 (1), 21–23.

Menn, C. (2003). Unpublished note (christian.menn@statistik.uni–karlsruhe.de).

Michaud, R. (1998). *Efficient Asset Management: A Practical Guide to Stock Portfolio Optimization and Asset Allocation*, Harvard Business School Press, Boston.

Mitchell, J. and S. Braun (2002). "Rebalancing an Investment Portfolio in the Presence of Transaction Costs." http://www.rpi.edu/~mitchj/papers/exact.pdf.

Mitra, G., T. Kyriakis, C. Lucas, and M. Pirbhai (2003). "A Review of Portfolio Planning: Models and Systems," in S. Satchell and A. Scowcroft, eds., *Advances in Portfolio Construction and Implementation,* Butterworth-Heinemann, Amsterdam.

Nankervis, J. (2002). "Stopping Rules for Double Bootstrap Confidence Intervals." http://www.bus.qut.edu.au/esam02/program/papers/Nankervis_John.pdf

Pliska, S. (1997). *Introduction to Mathematical Finance: Discrete Time Models*, Blackwell, Malden, MA.

Rachev, S. and S. Mittnik (2000). *Stable Paretian Models in Finance*, John Wiley & Sons, New York.

Rockafellar, R. T. and S. Uryasev (2000). "Optimization of Conditional Value–at–Risk," Journal of Risk, 2 (3), 21–41.

Rocke, D. M. and D. L. Woodruff, (1996). "Identification of Outliers in Multivariate Data," Journal of the American Statistical Association, 91(435), 1047–1061.

Rousseeuw, P. J., and K. Van Driessen, (1999). "A Fast Algorithm for the Minimum Covariance Determinant Estimator," Technometrics, 41 (3), 212–223.

Rustem, B. and R. Settergren (2002). "Robust Portfolio Analysis," in E. Kontoghiorghes, B. Rustem, and S. Siokos, eds., *Computational Methods in Decision-Making, Economics and Finance*, Kluwer, Dordrecht.

Satchell, S. and A. Scowcroft (2000). "A Demystification of the Black-Litterman Model," Journal of Asset Management, 1 (2), 138–150.

Satchell, S. and F. Sortino (2001). *Managing Downside Risk in Financial Markets: Theory, Practice and Implementation*, Butterworth-Heinemann, Oxford.

Scherer, B. (2004). *Portfolio Construction and Risk Budgeting*, 2nd edition, Risk Books, London.

Sharpe, W. F. (1994). "The Sharpe Ratio," Journal of Portfolio Management, Fall, 49–58.

Shectman, P. (2001). "Multiple Benchmarks and Multiple Sources of Risk." Paper presented at the Northfield Research Seminar, London England, March 19, 2001. Available from http://www.northinfo.com/Papers/.

Sklar, A. (1996). "Random Variables, Distribution Functions and Copulas—A Personal Look Backward and Forward," in L. Rüschendorff et al., eds., *Distributions with Fixed Marginals and Related Topics*, Institute of Mathematical Statistics, Hayward, CA.

Sortino, F. A., and L. N. Price (1994). "Performance Measurement in a Downside Risk Framework," Journal of Investing, Fall, 59–65.

Stambaugh, R. (1997). "Analysing Investments Whose Histories Differ in Length," Journal of Financial Economics, 45, 285–311.

Stein, C. (1956). "Inadmissability of the Usual Estimator for the Mean of a Multivariate Normal Distribution," *Proceedings of the 3rd Berkeley Symposium on Mathematical Statistics and Probability Volume I*, University of California Press, Berkeley, pp. 197–206.

Tanner, M. A. (1996). *Tools for Statistical Inference: Methods for the Exploration of Posterior Distributions and Likelihood Functions*, 3rd edition, Springer, New York.

Tanner, M. A. and W. H. Wong (1987). "The Calculation of Posterior Distributions by Data Augmentation," Journal of the American Statistical Association, 82 (398), 528–540

Vasicek, O. A. (1973). "A note on using cross–sectional information in Bayesian estimation of security betas," Journal of Finance, 28 (5), 1233–1239.

Wang, M. (1999). "Multiple Benchmark and Multiple Portfolio Optimization," Financial Analysts Journal, 55 (1), 63–72.

Yohai, V. J. and R. H. Zamar (1997). "Optimal Locally Robust M–Estimates of Regression," Journal of Statistical Planning and Inference, 64, 309–323.

Yohai, V. J., W. Stahel, and R. H. Zamar (1991). "A procedure for robust estimation and inference in regression," in W. Stahel and S. Weisberg, eds., *Directions in Robust Statistics and Diagnostics, Vol. II*, IMA Volumes in Mathematics and its Applications 34, Springer-Verlag, New York, pp. 365–374.

Young, M. (1998). "A Minimax Portfolio Selection Rule with Linear Programming Solution," Management Science, 44, 673–683.

Zellner, A. (1996). *An Introduction to Bayesian Inference in Econometrics*, 3rd edition, John Wiley & Sons, New York.

Ziemba, W., ed. (2003). *The Stochastic Programming Approach to Asset, Liability and Wealth Management*, AIMR/Blackwell, Malden, MA. http://www.nccr-finrisk.unizh.ch/media/pdf/wtz_alm_app.pdf.

Ziemba, W. and J. Mulvey (1998). *Worldwide Asset and Liability Modeling*, Cambridge University Press, Cambridge.

Zimmermann, H. (1998). *State-Preference Theorie und Asset Pricing—Eine Einführung*, Physica Verlag, Heidelberg.

Index

MIX
Papier aus verantwortungsvollen Quellen
Paper from responsible sources
FSC® C105338

If you have any concerns about our products,
you can contact us on
ProductSafety@springernature.com

In case Publisher is established outside the EU,
the EU authorized representative is:
Springer Nature Customer Service Center GmbH
Europaplatz 3, 69115 Heidelberg, Germany

Printed by Libri Plureos GmbH
in Hamburg, Germany